THE COMPLETE HANDBOOK OF LOCKS & LOCKSMITHING

No. 920
$10.95

THE COMPLETE HANDBOOK OF LOCKS & LOCKSMITHING

BY C. A. ROPER

TAB BOOKS
Blue Ridge Summit, Pa. 17214

Preface

This handbook on locksmithing is written for the student, the professional locksmith, and for everyone who is interested in the fascinating world of locks.

There is enough material here to get you started in the business of selling, servicing, and installing locks. Besides the challenge of the work, locksmithing is a very lucrative profession. According to the *U.S. Occupation Outlook Handbook* and the *Locksmith Ledger*, a self-employed locksmith can look forward to earning twice the hourly wage of an auto mechanic and three times as much as a watchmaker. And locksmithing is rewarding in other ways: no other craft offers such prestige and trust.

The amateur—the student of lock technology and the collector—will find much that is rewarding here. One chapter traces the evolution of the lock from simple Egyptian bolts, through Greek and Roman designs, locks of the Middle Ages, and the Renaissance, the latchstrings of the American frontier, and concludes with the great Nineteenth-Century Revolution in lockmaking. Other chapters carry the story to modern times, including ultra-high security locks used in military installations.

I have tried to include everything that you need to know. Each of the various lock mechanisms—combination, warded, lever, disc, wafer, and pin-tumbler—is discussed in detail with an eye to troubleshooting and servicing. Chapters are devoted to interior and exterior locksets, with special emphasis on the most popular brands. While locksmithing requires surprisingly few tools, each tool that you will need is described and, in most cases, illustrated. Instructions are given on how to make your own tools and how to choose and buy factory-made tools.

Professionals will welcome the chapter on masterkey systems, the in-depth information on Schlage and Corbin locksets, and the discussion of Medeco and other speciality locks. Vending-machine, filing-cabinet, post office box, and automobile locks are illustrated and discussed in detail.

Carl A. Roper

Contents

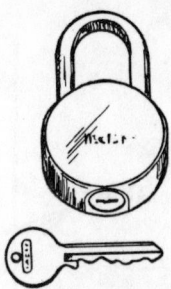

Chapter 1

A Short History of Locks and Keys

Locks and keys have been around for centuries. They have evolved with man as a natural concomitant of his existence, signifying his continual need for protection of life and property. Every civilization in history understood the need for security, and their understanding spawned a persistent search for more reliable locks, more sophisticated keys.

EGYPT

Long before the glories of Greece, Rome, and the Greco-Roman empire, the Egyptians developed a reliable lock that operated in essentially the same way our locks do today. They invented the pin-tumbler lock and key. Old paintings inside pyramids give us a general idea of what the lock was like (Fig. 1-1). The internal form of the lock was similar to that reinvented in the mid-1800s by Linius Yale, Sr., in the United States.

The lock body was made of wood and mounted securely to the door (Fig. 1-1). The bolt passed through the lock body and into a bracket mounted on the wall. In the locked position,

13

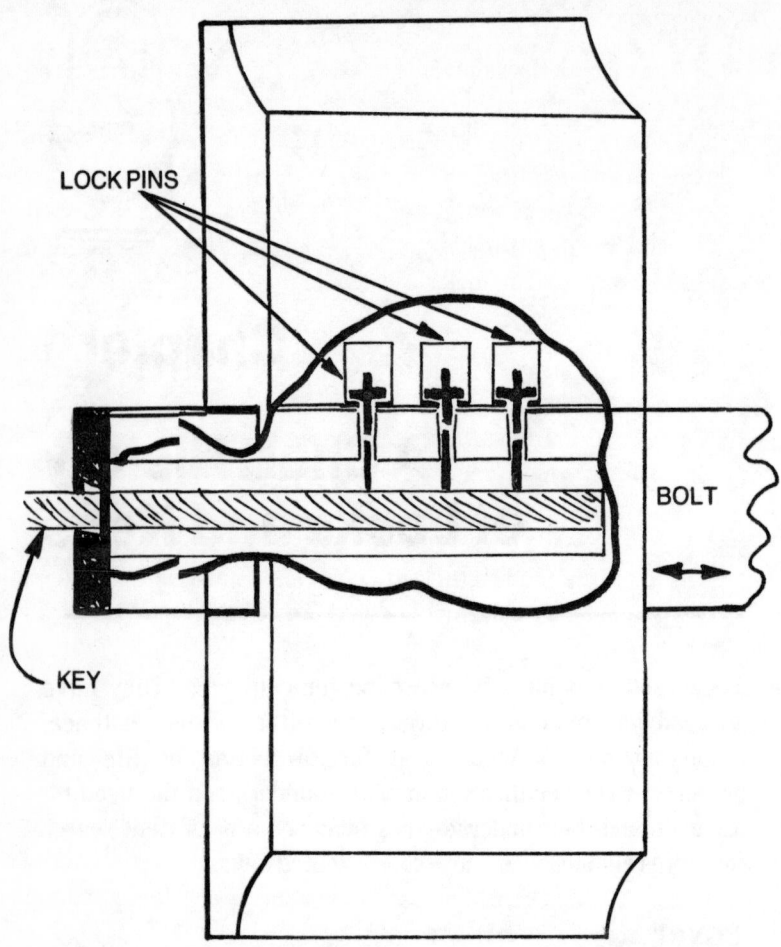

Fig. 1-1. An ancient Egyptian lock. Raising the key brought the pins to the shear line and released the bolt. The same principle was used some 400 years later on the Yale lock.

hardwood or iron pins dropped into holes in the bolt, fixing it securely in the lock body. The key was no more than a length of wood fitted with pins that matched those in the lock body. It was inserted through the hollow end of the bolt (the bolt hole, the forerunner of the keyhole) and raised to disengage the pins from the bolt. The bolt was then free to slide. The size, pattern, number, and length of the key pins varied, even as they do today.

GREECE

Most Greek doors pivoted at the center and were secured by bolts from the inside. In the few cases where locks were used, they were very primitive. The key was a large crescent-shaped affair, resembling a modern sickle. To work the bolt, the owner inserted the key into the keyhole and gave it a twist. The tip of the key aligned with a hole in the bolt and moved it left or right, locking or unlocking the door.

ROME

Unfortunately, most Roman locks have been destroyed by time, but the few keys and locks that remain give testimony to the Roman genius. The basic lock was the warded type. (A ward is a projecting ridge in a lock or on a key designed to permit only the correct key to be inserted in a lock.) Many locks are still made on this principle. In addition, the Romans made ingenious attempts to disguise the keyway as ornamentation or to hide it altogether. One famous Roman lock is in the shape of a fish. The keyway is revealed when one of the fins is turned.

Roman innovations included the use of wards in their own iron case, the metal key, the spring-loaded bolt, the spring-operated padlock, and the first true (i.e., removable) padlock. One curious aspect of Roman lock lore is that the keys were designed to be worn as rings because togas did not have pockets (Fig. 1-2).

After the fall of Rome, Europe was swept by barbarian hordes moving south and east from the Rhineland and Scandinavia. Brute force was the order of the day, and locks meant little. Men put their faith in arms and fortresses.

This period passed as the barbarians learned from their victims and settled down into the routines of farming and fishing. The Vikings who settled in Brittany evolved a new kind of civilization with the king at the apex and power radiating downward from him into various levels of nobility. This social pattern, known as feudalism, gave the nobility access to wealth and property. Thus the art of locksmithing again flourished.

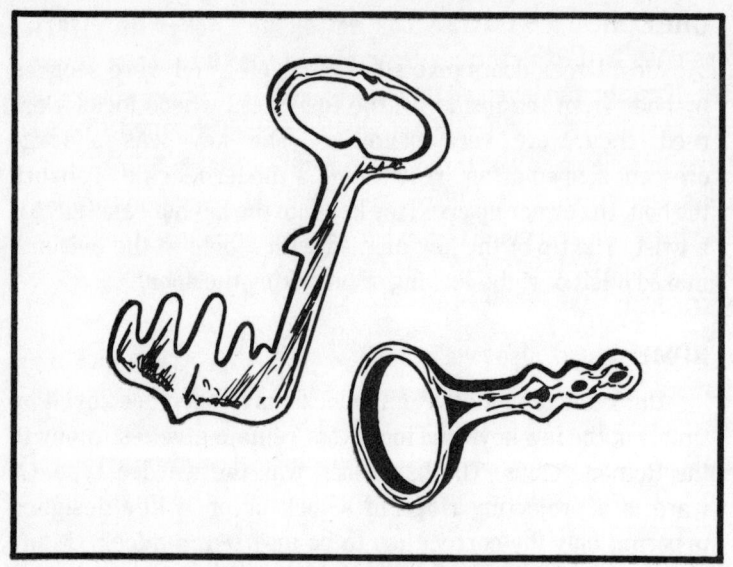

Fig. 1-2. Roman keys were worn as rings.

EUROPE

In Europe, keys were made that could move about a post and shift the position of a movable bar (the locking bolt). In its various forms it worked quite well. The first obstacles to unauthorized people using the lock were the various internal wards. Medieval and Renaissance craftsmen made warded locks and improved upon them. Some were very complex, using many interlocking wards and complicated keys. Since many of these wards could be easily bypassed by almost anyone, newer methods had to be devised.

Later, chests were made entirely of metal and the locksmith reasoned that if one bolt was secure, then a chest with 8, 10, 12, or even 15 locking bolts would be more secure. Thus, the craftsmen created elaborate internal mechanisms to allow the many bolts to be shifted by one or two separate keys. Levers, springs, ratchets, and pinions were employed to do this job. The locksmiths also installed separate locks for separate bolts; the locks had false keyholes and safety devices, such as spring-loaded knives that would injure or kill the thief. Hidden keyholes were extremely popular among the merchant and upper classes.

In France, the treatise *The Art of the Locksmith*[x](1767) was published. It describes examples of the tumbler or lever lock. Exactly who invented this lock is unknown, but gratitude is long overdue. As locksmithing advanced, multiple-lever locks came into being. Two, three, six, or more levers were used. Each lever had to be lifted and when all were in proper alignment, the bolt could be moved, opening or closing the lock.

ENGLAND

The English also worked on newer and better locking devices. Incentive was given in the form of cash awards and honors to people who could successfully open these newer and more complex locks. In the forefront of lock designing were three Englishmen: Robert Barron, Joseph Bramah, and Jeremiah Chubb.

Robert Barron patented the double-action tumbler lock in 1788. Like others before it, this lock employed a series of lever tumblers pegged at one end (Fig. 1-3). One side of these tumblers engaged a notch in the bolt. The key bore against the opposing surface. Notches on the key corresponded to the individual tumblers. The width of the key-bearing surface varied with each tumbler: a wide tumbler required a correspondingly deep notch in the key; a narrow tumbler required a shallow notch. When the key and tumbler stack

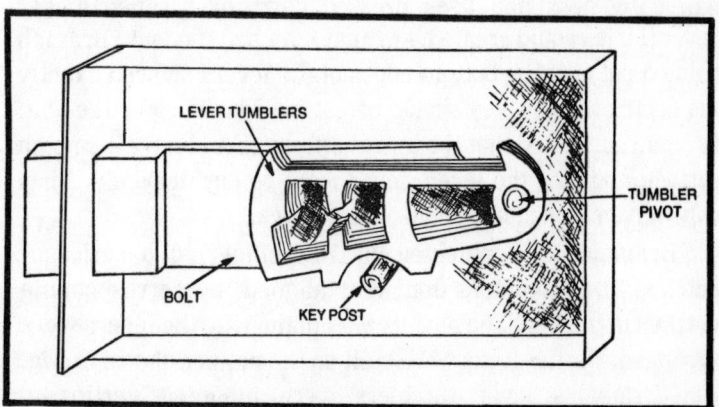

Fig. 1-3. An 18th-century lever-tumbler mechanism. The width of the tumblers corresponded to slots in the key.

matched, all tumblers would move in unison and release the bolt.

Barron's patent involved the key-bearing surface on the tumbler. Earlier locksmiths were content with a simple notch on the lower edge of the tumbler stack. When the tumblers were raised high enough, the bolt would release. Barron pierced the tumbler stack so that the key controlled both up and down movement of the individual tumblers. This refinement worked in conjunction with a more complex gating; the bolt would remain latched until all tumblers were at the same height. Because the tumblers move in two directions, this lock is described as a double-action type. Barron added up to six of these double-lever actions to his lock and thought it virtually impossible to open except with the proper key. He soon found out differently.

Another Englishman, Joseph Bramah, wrote an essay, *A Dissertation on the Construction of Locks*, which exposed the many weaknesses of existing so-called thiefproof locks and pointed out that any of them could be picked by a good lock specialist or criminal with some training in locks and keys. Bramah admitted that Barron's lock had many good points, but he also revealed its major fault: the levers, when in the locked position, gave away the lock's secret. The levers had uneven edges at the bottom; thus, a key coated with wax could be inserted into the lock and a new key could be made by filing where the wax had been pressed down or scraped away. Several tries could create a key that matched the lock. Bramah pointed out that the bottom edges of the levers showed exactly the depths the new key should be cut in order to clear the bolt. Bramah suggested that the lever bottoms should have a smooth surface and that the lever slots should be cut unevenly. Then only master locksmiths could open the lock.

Bramah's lock, patented in 1784, employed a series of notched lever tumblers that were aligned by corresponding notches in the key. The novelty was in the way the levers were arranged. Earlier locks were built on the pattern shown in Fig. 1-3; Bramah's lever tumblers were mounted vertically. Bramah used radial tumblers and a barrel key (Fig. 1-4). The pin on the side of the key rotated the lock.

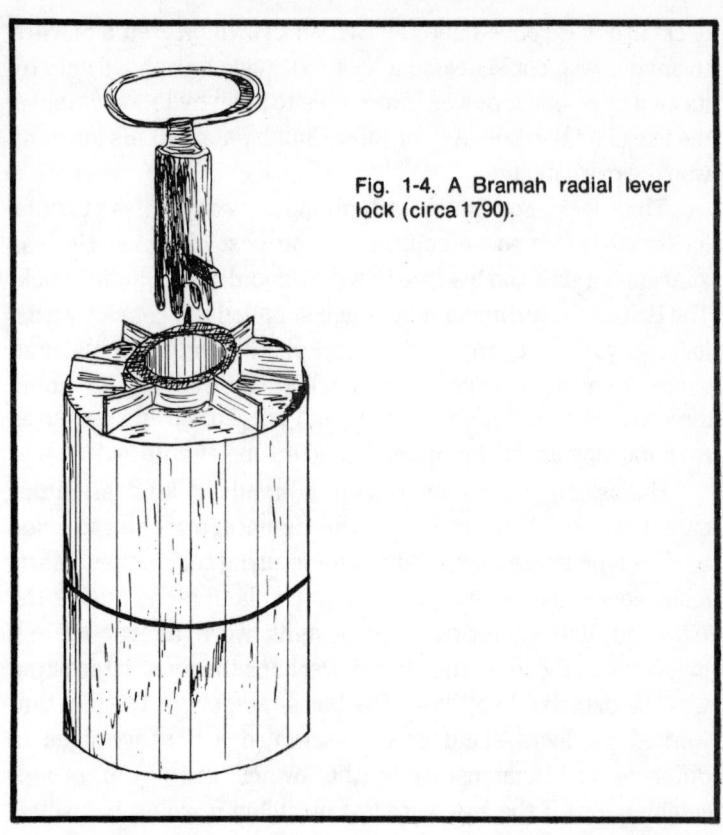

Fig. 1-4. A Bramah radial lever lock (circa 1790).

Though this lock could be picked, the job was beyond the average thief. Each tumbler had to be aligned and the control piece had to be turned in the right direction.

Jeremiah Chubb added refinements to the Barron lock in an attempt to make it more secure. One of Chubb's improvements was a metal "curtain" which fell across the keyhole when the mechanism began to turn, making the lock difficult to pick. He also added a detector lever that indicated whether the lock had been tampered with. A pick or an improperly cut key would raise one of the levers too high for the bolt gate. This movement engaged a pin that locked the detector lever. The detector could be cleared by turning the correct key backwards and then forwards.

During this period, many robberies were committed against property and person. In 1817, the Portsmouth, England dockyard was robbed and the British Crown offered a reward to anyone who could devise a lock that could be opened *only* by its own key—one that was impossible to open by lockpicking or the use of a false key. A year later Chubb patented his lock and won the prize money.

This lock got much attention. A convict, a former locksmith, claimed he could open the best of locks. He was guaranteed $250 and his freedom if he could open Chubb's lock. The British Government and Chubb supplied the convict with a lock, key blanks, appropriate lockpicking implements, and some dismantled locks to work on. After working for some three months, he finally gave up and served out his sentence. England announced that it finally had an unpickable lock.

Hobbs, an American locksmith, made a kind of minor career picking English locks. The manufacturers responded by developing new locks and manufacturing techniques. Many locks—over 3,000—were patented in England alone during the 18th and 19th centuries. Key designs were as varied and ingenious as the locks they fitted. Perhaps the most interesting was the detachable bit key. The bits—the part of the key that worked the lock—could be disassembled and rearranged in different combinatons. Only the owner knew the proper combination; if the key were lost or stolen it would be useless unless bitted correctly. Other keys had projecting pins like the Bramah key, and were intended for rotating locks. Others were flat, with a wide tip or else socketed for detachable bits. Another type had a detachable tip. Elaborately notched keys were not uncommon, and were used with extremely complex locking wards. Each ward had a distinctive shape location in the lock. A false key might turn the lock a few degrees only to be frustrated by additional wards.

The swivel key was another example of 19th century ingenuity. The lever, or the bitting portion of the key, and the barrel, or shank, were moved independently of each other. Each half was turned in opposite directions to open the lock.

AMERICAN LOCKS

Early American lock bolts were mounted on the inside of the door, and could be opened from the outside by means of a latchstring. Hence the phrase, "the latchstring's always out." At night the string would be pulled inside, "locking" the door. Of course, someone had to be inside to release the bolt. An empty house was left unlocked. As the country was settled, theft increased; local merchants and blacksmiths soon got into the lock trade. Small cabinet and cupboard locks were very popular. The lady of the house or, in large establishments, the butler was in charge of the household keys. The doors were unlocked in the morning and the cabinets and cupboards were opened to dispense the day's supplies and immediately relocked. At night the doors were secured.

In the 1850s two inventors, Andrews and Newell were granted patents on an important new feature—removable tumblers. The tumblers could be disassembled and scrambled to make, in effect, a new lock. The keys had interchangeable bits that matched the various tumbler arrangements. After locking up for the night, a prudent owner would scramble the key bits. Even if a thief got possession of the key, it would take him hours to stumble on the right combination. In addition to removable tumblers, Newell's lock featured a double set of internal levers. He was so proud of this lock, that he offered a reward of five hundred dollars to anyone who could open it. A master mechanic took him up on the offer and collected the money.

This experience convinced Newell that the only secure lock would have its internals sealed off from view. Ultimately, the sealed locks appeared on bank safes in the form of combination locks.

As we have seen, Hobbs picked the famed English locks with ease. Until Hobbs' time, locks were opened by making a series of false keys. If the series was complete, one of the false keys would match the original. Of course, this procedure took time. Thousands of hours might pass before the right combination was found. Hobbs depended upon manual dexterity. He applied pressure on the bolt, while manipulating one lever at a time with a small pick inserted through the keyhole. As each lever tumbler unlatched, the bolt moved a hundredth of an inch or so.

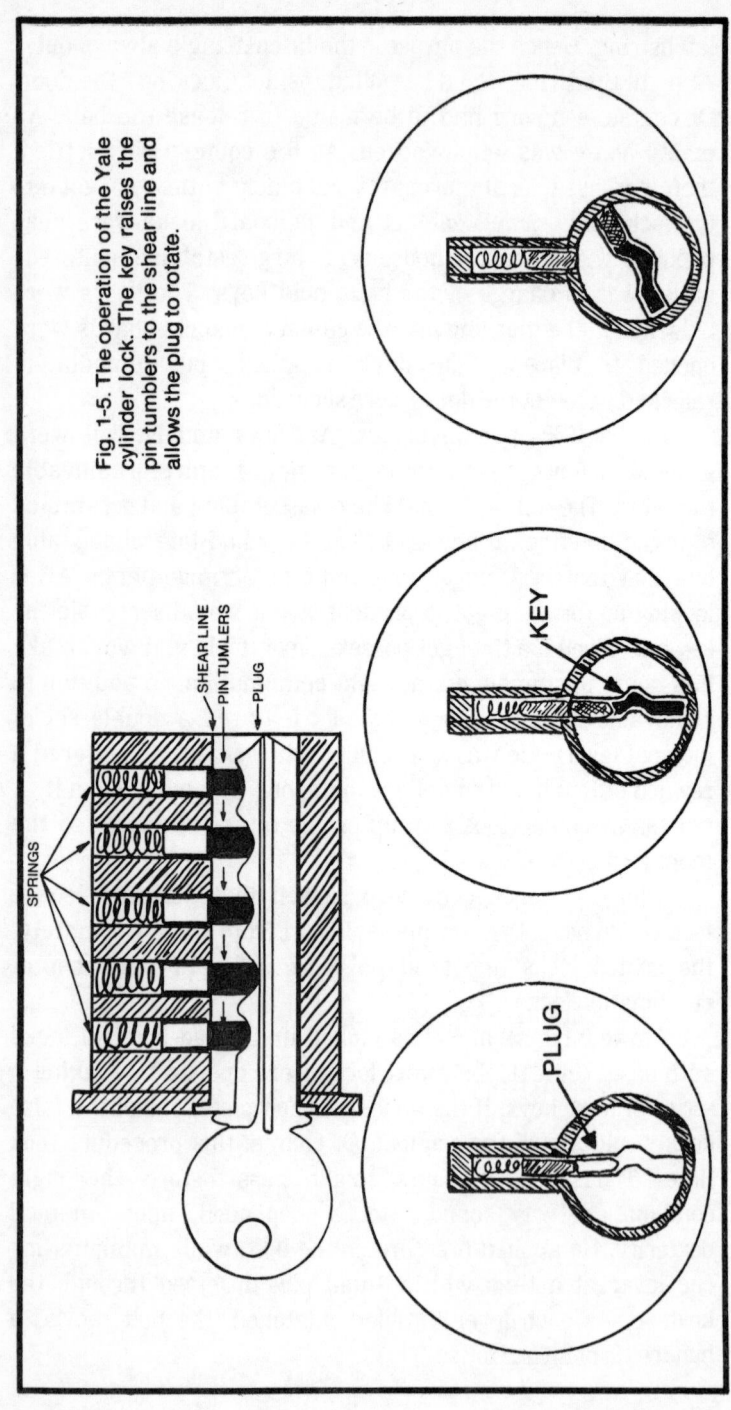

Fig. 1-5. The operation of the Yale cylinder lock. The key raises the pin tumblers to the shear line and allows the plug to rotate.

SPRINGS

SHEAR LINE
PIN TUMLERS
PLUG

KEY

PLUG

Until the early 19th century locks were made by hand. Each locksmith had his own ideas about the type of mechanism, the number of lever tumblers, wards, and internal cams to put into a given lock. Keys showed the same individuality. A lock could have 20 levers and weigh as much as 5 pounds. Linius Yale changed all this. An inventor of milling machinery, Yale devised a simple, safe, and compact key-operated lock that could be made on automatic machinery. This lock was based upon the principle first developed by the Egyptians. Yale improved upon this principle and developed a keying system that is still used today. His son continued in the trade, making further refinements upon his father's work. He developed a simplified cylinder with a rotating internal plug core that advanced lock design to new heights. Other refinements included the solid case and the small keyway that made life difficult for pick artists. The son also developed a series of dies and cutters to mass produce all parts of the lock.

The major structural difference between the Yale lock and the Egyptian prototype was the revolving plug that replaced the sliding bolt. A tailpiece on the back of the plug worked the bolt mechanism. Using machine tools, the Yale family was able to produce these locks with the same dimensions and tolerances. The keyway was ridged to accept a grooved key blank. Only the correctly cut key would turn the plug to open the lock. This lock meant that thieves had to start over and devise new methods and tools. Figure 1-5 illustrates the basic Yale mechanism.

Competitors were quick to copy the Yale cylinder. The reliability of these locks revolutionized the industry. Some of the early lock companies are still in business.

And so it continues, this endless war between lock makers and thieves.

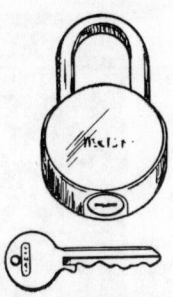

Chapter 2
Tools of the Trade

Having the proper tools and equipment is critical to the locksmithing trade. How much money you spend on tools and equipment depends upon the amount of effort you are willing to put forth to build your own and the level of quality that you will accept in factory-made goods.

Consider some basic requirements. You will need tools and supplies to put the information you learn from this book into practice. The only way you can learn the trade is through practice. Be realistic in the assessment of what you will make; purchase locally, or purchase through locksmith supply houses. (A listing of various suppliers is provided in Appendix III.)

Space is another requirement. You should have a place free from distractions where you can study this book, store your supplies and equipment, develop projects based on the materials presented here, and still have room to be comfortable.

WORKBENCH

The workbench should be:

- Long enough to insure adequate work space.
- Strong enough to support the key machine at one end, out of the way.

Fig. 2-1. Arco bins are made of heavy polypropylene for long, maintenance-free service.

- Solid enough to keep the machine in alignment.
- High enough so you can work without stooping.
- Wide enough to temporarily store parts—30 in. is the comfortable maximum.
- Lit from overhead and behind or from the sides. Never have the light too close to the bench.

The location of the workbench must also be considered. If possible, it should be located in a well ventilated area, away from general traffic, near stored equipment and supplies, and with easy access from several directions. Bench storage facilities should be arranged to give instant access to frequently used tools. In this regard, a tool clipboard against the back of the bench is an excellent feature. Space for storage boxes, bins, and a shelf for reference books and manuals must also be considered (Fig. 2-1). It does no good to have the bench at one end of a room and all supplies and manuals at the other end.

Examples of workbench designs are shown in Figs. 2-2 and 2-3. No one example is ideal for everybody; individual tastes and needs differ.

Fig. 2-2. This simple bench is adequate for a student.

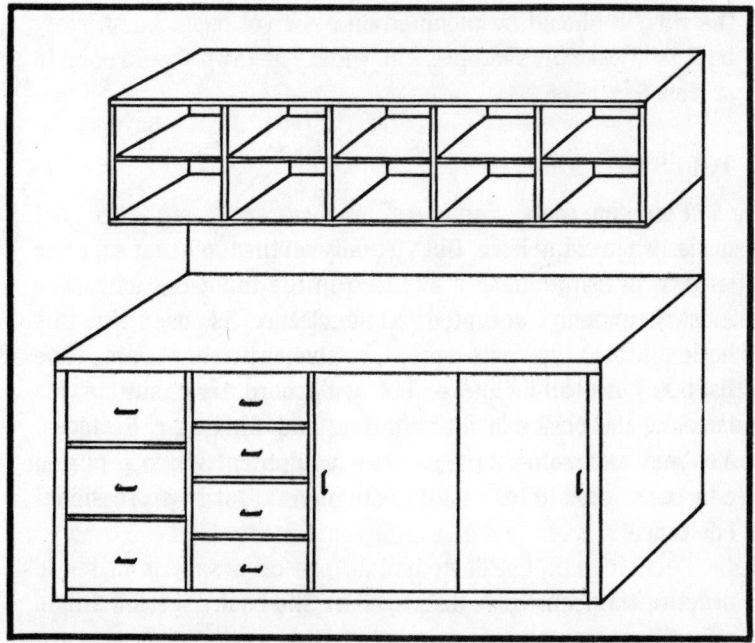

Fig. 2-3. This bench with its full complement of drawers and partitioned overhead bin is ideal for the professional.

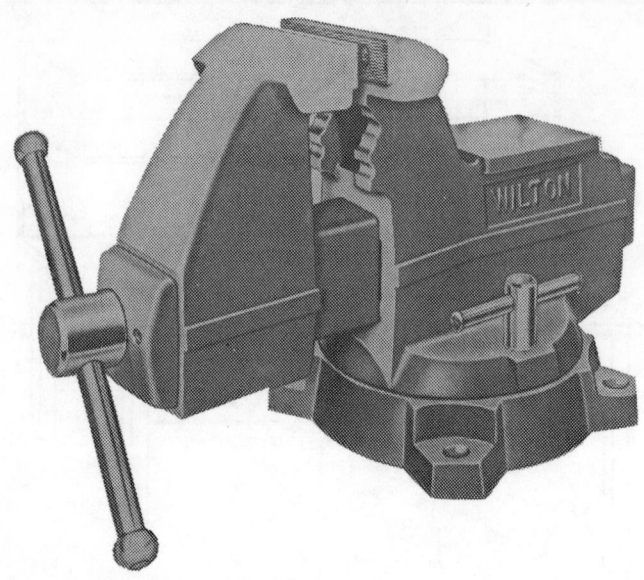

Fig. 2-4. A Wilton professional-quality vise.

A machinist's bench should be installed at one end, out of the way. It should be mounted on a swivel base, be sturdily built, and have jaws at least 3 in. wide. The jaws should open to at least 5 in. (Fig. 2-4).

TOOLS AND SUPPLIES

Common tools—hammers, screwdrivers, etc.—are not discussed in detail here. But virtually all the tools and supplies used in locksmithing are included in the following lists. The most commonly accepted nomenclature is used in this book, although alternate names are given when applicable. The lists are not all-inclusive, but will more than suffice for learning the basics of locksmithing and starting a business. You may have some of the pieces of equipment at home, or you can use substitutes until you can afford professional equipment.

Section 1 is a list of items that are necessary in order to practice the techniques discussed in this book; section 2 lists miscellaneous items that are not essential but are nevertheless useful.

Fig. 2-5. Stanley C-clamps are light and durable.

Section 1

- Ace lock tools

- Assorted springs

- Automotive entrance tools

- Brace an auger or expansion bits (Fig. 2-9)

- Brazing equipment

- Broken key extractors

- C-clamps (Fig. 2-5)

- Cam assortment

- Cam screw assortment

- Circular hole cutter

- Code books

- Coping saw and blades

- Cut-all knives and blades

- Cylinder pin assortment

- Depth gage

- Depth key assortment

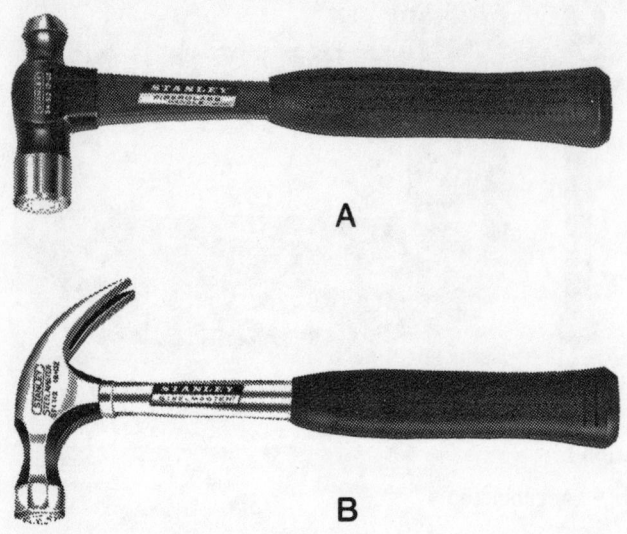

Fig. 2-6. Since a locksmith does both metalwork and simple carpentry, he needs both ball-peen (A) and claw (B) hammers. (Courtesy Stanley Works.)

- Drills (hand and electric) and drill bits (Fig. 2-10)

- Files (warding, round, pippin, and triangular)

- Flashlight

- Flat spring steel stock

- Grinder with wire wheel

- Hacksaw and blades (Fig. 2-13)

- Hammers (ball-peen and claw hammers are shown in Fig. 2-6)

- Handcleaner

- Hollow mill set rivet drills

- Key blanks (assorted)

- Key caliper (with cutter assortment)

- Key clamp

- Key layout board

- Key machine

- Lever tumbler assortment

- Lock-aid tool

- Lock pick set

- Lock-reading tool

- Lubricants (graphite. WD 40. and cup grease)

- Micrometer (Fig. 2-11)

- Piano wire

- Pin tray

- Pliers (adjustable. cutting. and locking lever)

- Plug holder

- Plug followers (Fig. 2-7)

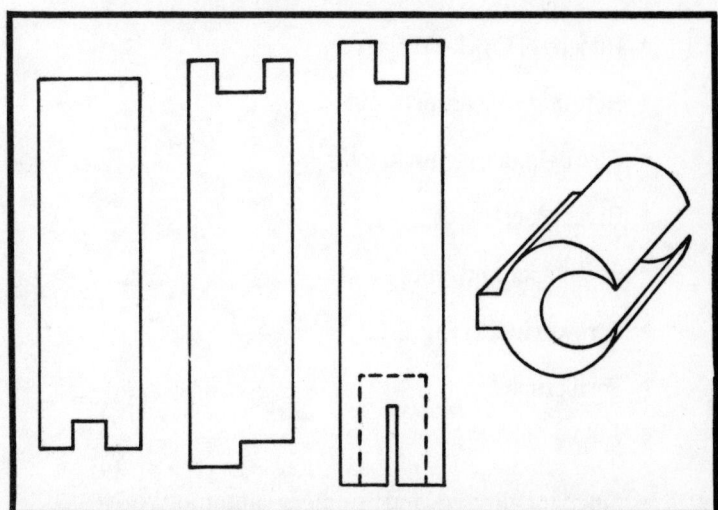

Fig. 2-7. Plug tools. Insert a plug follower into the cylinder chamber as the plug is withdrawn (A). The follower prevents the pins and springs from falling into the chamber. Use the appropriate diameter plug holder as a vise (B).

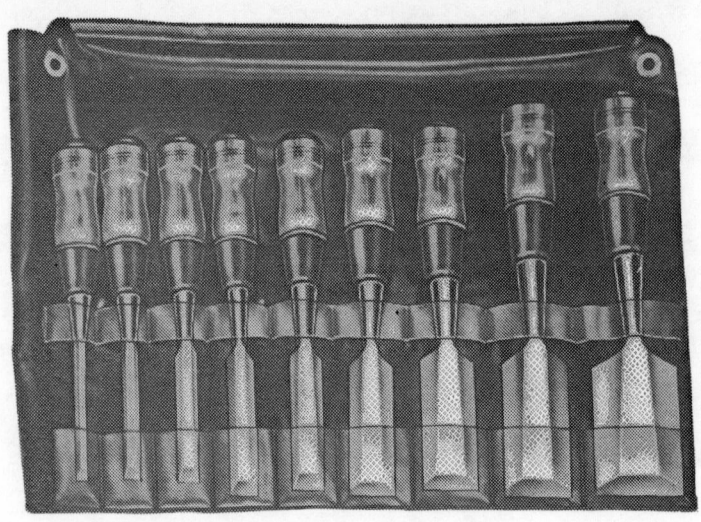

Fig. 2-8. A Stanley wood chisel set.

- Portable tool kit

- Punch and die set in single and double *D* configurations (Fig. 2-14)

- Punches (Fig. 2-15)

- Retainer ring assortment

- Rim cylinder removal tools

- Rivet assortment

- Sandpaper and emery cloth

- Screwdrivers (Fig. 2-12)

- Scribers

- Shims

- Sidebar cylinders and tumblers (automotive)

- Spindle assortment

- Storage trays (Fig. 2-16)

- Tap set

- Turning wrench

- Tweezers

- Vise

- Wafer tumblers (assorted)

- Wood chisel set (Fig. 2-8)

- Wrenches (adjustable and pipe)

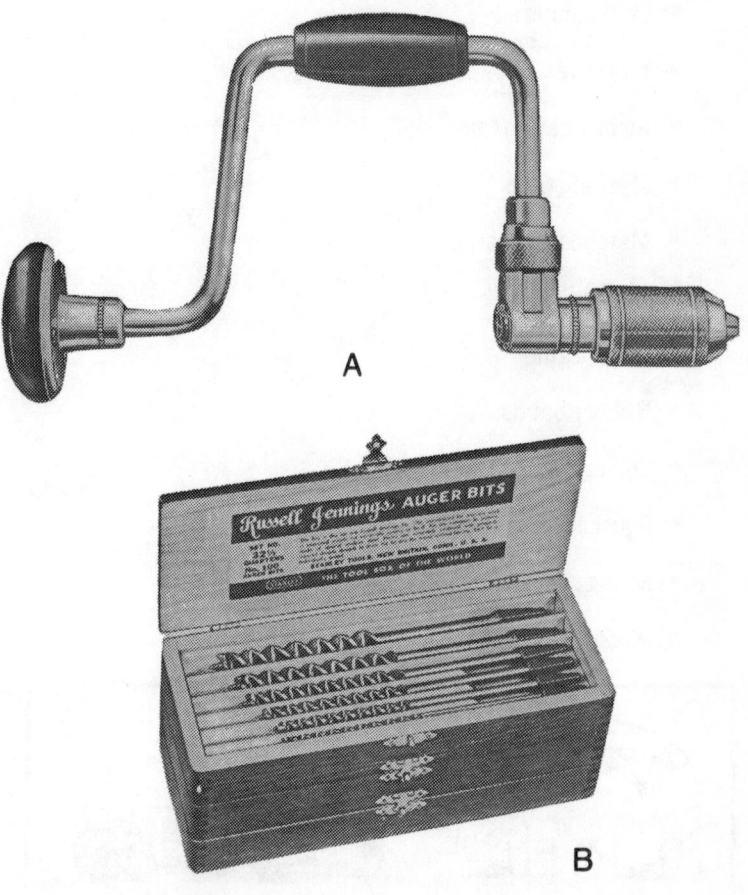

Fig. 2-9. You will need a good brace (A) and collection of auger bits (B). (Courtesy Stanley Works.)

Fig. 2-10. Spade-type bits are used with power tools. (Courtesy Stanley Works.)

Section 2

- Assorted nails and screws
- Baggies
- Cigar boxes
- Coat hangers
- Ice trays
- Identification tags
- Masking tape
- Matches and candles
- Paper clips
- Plastic glue
- Rubber bands
- Soldering iron
- Squeeze tube
- String
- Wood wedges

Fig. 2-11. Key micrometers are available from locksmith supply houses.

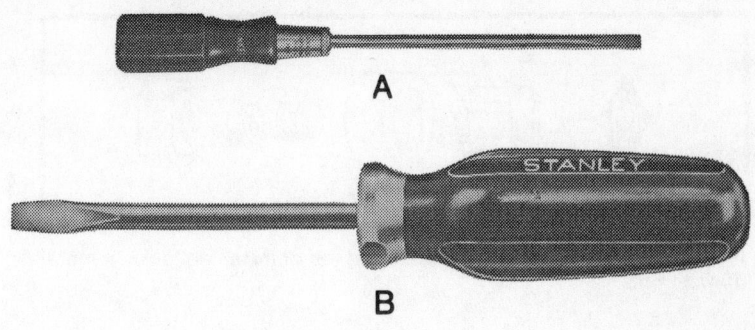

A

B

Fig. 2-12. Screwdrivers. Electrician's screwdrivers are designed for precision work (A); standard screwdrivers are used on the heavier jobs (B). (Courtesy Stanley Works.)

TOOL SPECIFICS

The use of most tools is obvious and needs no discussion. However, a few are worthy of additional comment.

broken key extractor—Used to extract broken keys from cylinder plugs.

cup grease—Used only to lubricate the boltwork on exterior door locks, mortise cylinder locks, etc.

Double D punch and die set—Used for cutting the lock mounting hole in a metal desk drawer.

files—Various types are needed, including an 8 and 6 in. mill, round (Swiss, fine tooth), 6 in. triangular, 4 in. flat warding, 6 in. flat warding, and 6 in. pippin. These files are used to cut keys by hand and are handy for miscellaneous repair jobs.

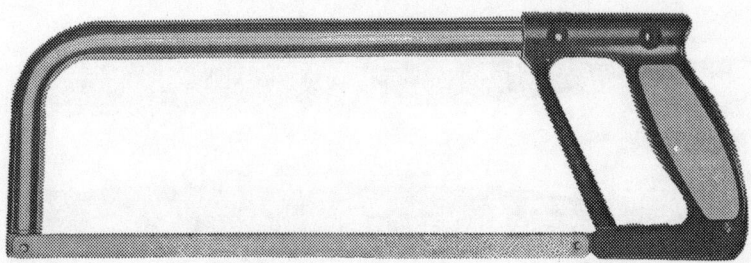

Fig. 2-13. The Stanley hacksaw features a tubular frame for maximum rigidity.

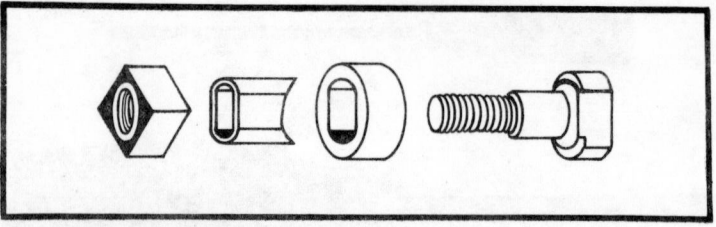

Fig. 2-14. A Double-D punch makes a neat job of installing locks in metal drawer fronts.

graphite gun—Used to force powdered graphite into the tumbler mechanisms of pin and disc locks. (Oil and grease must not be used since these lubricants attract dirt.)

hollow mill set rivet drill—Used for removing rivets from padlocks for access to the internal mechanism.

key caliper—Used for measuring depths of key cuts; a caliper is especially useful when duplicating bit keys by hand. The caliper is set to the depth of the original key cuts and serves as a guide when filing the new key.

layout board—Used to keep track of the pins and springs as they are removed from a lock plug and cylinder. It is important to reassemble these parts in their proper positions; otherwise the lock will have to be rekeyed. A layout board is also used to position masterkey plug setups prior to assembly.

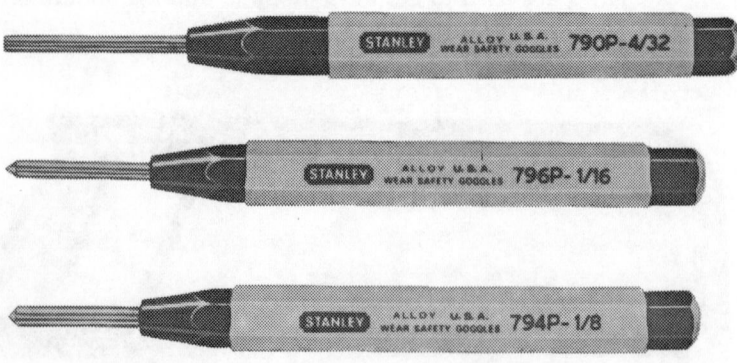

Fig. 2-15. Purchase a good set of pin and center punches. (Courtesy Stanley Works.)

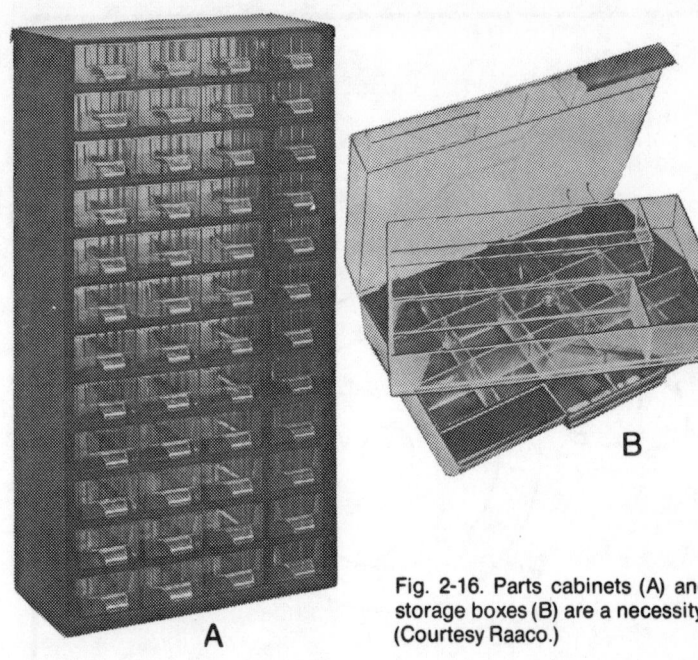

Fig. 2-16. Parts cabinets (A) and storage boxes (B) are a necessity. (Courtesy Raaco.)

machine oil—Used to coat the tumblers of lever locks. It is used as a lubricant and rust preventative on outside locks.

pin tray—Used for holding various pins not necessarily ready for use in a specific lock. Many pin trays are arranged to carry an entire series of pins.

plug followers—Used to secure the upper pins and springs of cylinder locks. The follower is inserted into the plug chamber as the plug is withdrawn. Popular chamber dimensions are given in Appendix IV.

thimble—Used to hold the plug when repairing it. This tool is also called a plug holder.

tweezers—Used to handle small lock parts, especially the pins, during or in conjunction with disassembly or "setting up" of a pin-tumbler lock.

KEY MACHINE

A good key-duplicating machine is an expensive tool but one that you will need if you open your own shop. In addition to

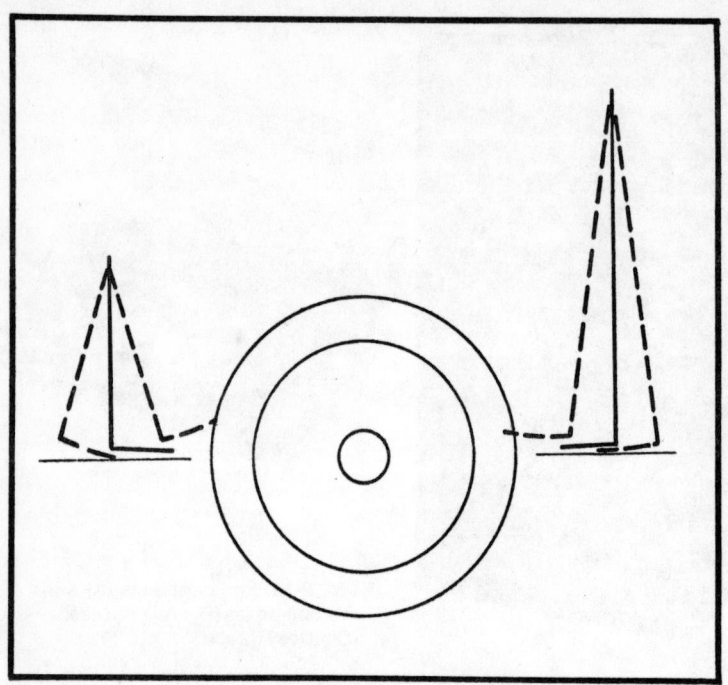

Fig. 2-17. A key machine should be constructed so that the key is almost parallel to the cutter. A long pivot (shown on the right) reduces the angle.

duplicating cylinder and flat keys, the more sophisticated machines will reproduce warded and end-cut (Ace) keys of the type used on vending machines.

A key machine consists of three basic parts:

- A pair of vises coupled together and moving in unison. One vise holds the original key and the other holds the key blank.
- A key guide that reproduces the profile of the original key on the blank.
- A cutter wheel that notches the key blank on orders from the key guide.

Other features are useful but by no means are absolutely essential. Some machines will duplicate a wide variety of keys with only a change in cutter wheel widths. Less sophisticated models require that the key holding vises be changed as well.

Critical Design Features

When evaluating a key machine, look carefully at the pivot mechanism. The key must meet the cutter wheel squarely on a dead parallel with the axis of the wheel. A slight angle is enough to upset the dimensions of the duplicated key (Fig. 2-17).

Some machines don't have pivots and arrange matters so that the vises move laterally into the cutter. As long as the bearings are true this ensures that the duplicated key is a faithful copy of the original.

The jaws of the vises must be carefully engineered to ensure that:

- Both the original and blank key do not shift during the duplicating process.
- The keys are held square against the cutter. Some keys have groove dimensions that make them difficult to grip. A piece of piano wire or a straightened paper clip can be used as a shim.

Cutters

There are three basic types of cutters. Filing cutters and milling cutters are the most popular. A slotter cutter is used on square-ended keys and to make side cuts on keys for lever and warded locks. Filing and most milling cutters are flat on one side and beveled on the other (Fig. 2-18). The flat side makes the right angle cut that is usually required next to the shoulder

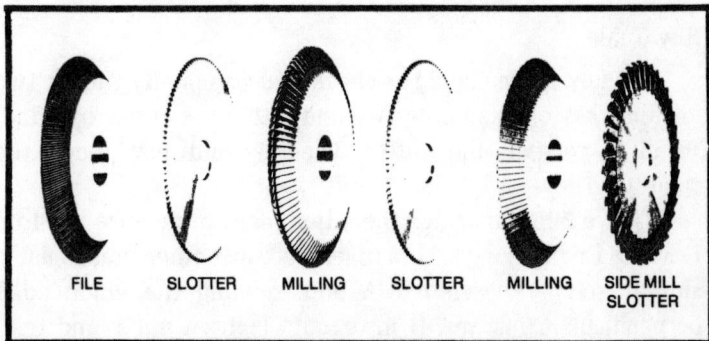

FILE   SLOTTER   MILLING   SLOTTER   MILLING   SIDE MILL SLOTTER

Fig. 2-18. Various cutter wheels.

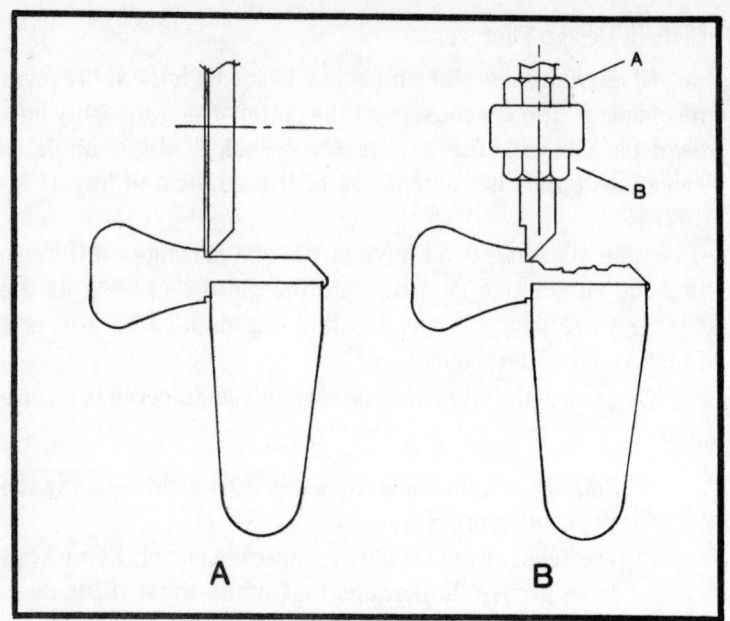

Fig. 2-19. The cutter wheel should barely touch the key blank (A); if adjustment is required, loosen nut A and move nut B in or out. (Courtesy Keynote Engineering.)

of the key. These cutters are always mounted with the flat side towards the original key.

The diameter of the cutter is also important. Large diameter cutters leave a very slight, almost imperceptible, concave on the key blank. Smaller cutters make a deep concave in the key. The greater the concavity, the more rapidly the key wears.

Key Guide

The key guide should be checked occasionally. Mount two identical key blanks in the vise and rotate the cutter by hand. If all is correct, the cutter wheel should just touch the right-hand key.

Figure 2-19 illustrates the adjustment procedure used on Keynote Engineering's M-76 machine. Most other makes use a similar setup. Loosen nut A and, holding the guide edge perpendicular, turn nut B in or out. Tighten nut A and test. Several tries are usually required to get the adjustment right.

Locks For Study

A beginning locksmith needs all kinds of locks to examine, take apart, repair, and cannibalize for parts. Second-hand locks are available from pawn shops, auto wrecking yards, and abandoned buildings. Of course, you must have permission to enter buildings that are to come under the wrecker's ball. Check with the superintendent or, if demolition has already begun, with the job foreman. Office buildings are a particularly rich source of locks. A search of the storage areas may turn up lock parts and spare keys.

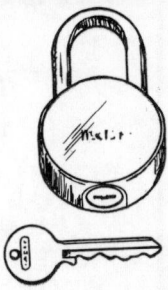

Chapter 3
Types of
Locks and Keys

The locking devices commonly used today can be grouped into a number of general types. These types are:

- Warded locks

- Mortise locks

- Automobile locks

- Key-in-the-knob locks

- Surface-mounted auxiliary locks

- Lever locks

- Luggage locks

- Padlocks

- Combination locks

An example of each type is shown in Fig. 3-1. Within these few major types all design variations fall. Some locks, such as a combination door lock, represent a melding of two different types.

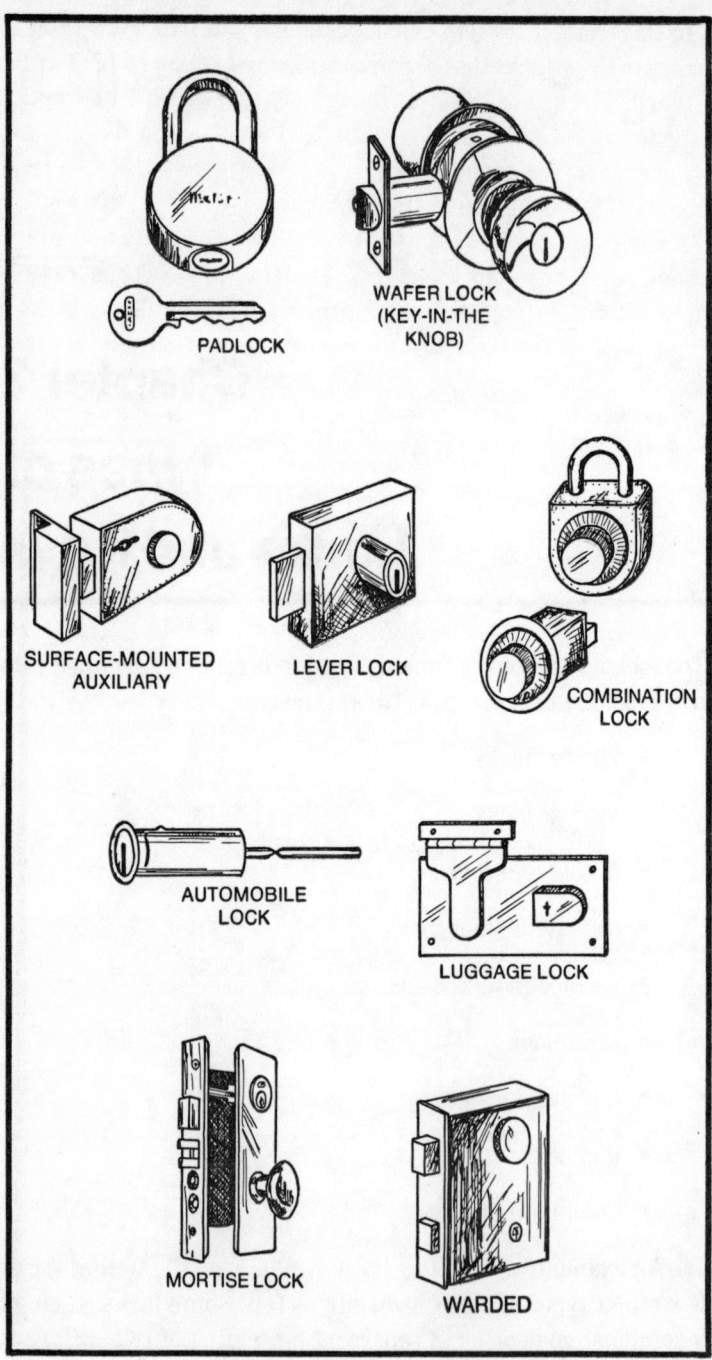

Fig. 3-1. The nine basic locking devices.

In this chapter we will consider the various lock types and the keys to these locks that are in common use today (Fig. 3-2).

Warded locks are either of the rim or mortise type and are among the oldest locks still being made (Figs. 3-3 and 3-4).

Lever and wafer locks are related, although they were introduced to the public at different times.

A variety of padlocks exist—warded, wafer, disc, and pin tumbler. Smaller padlocks have a single ward and take very simple keys. The common railroad lock is a padlock. It is opened with an old-style barrel key and uses a simple lever mechanism. Some railroad locks work dependably even after

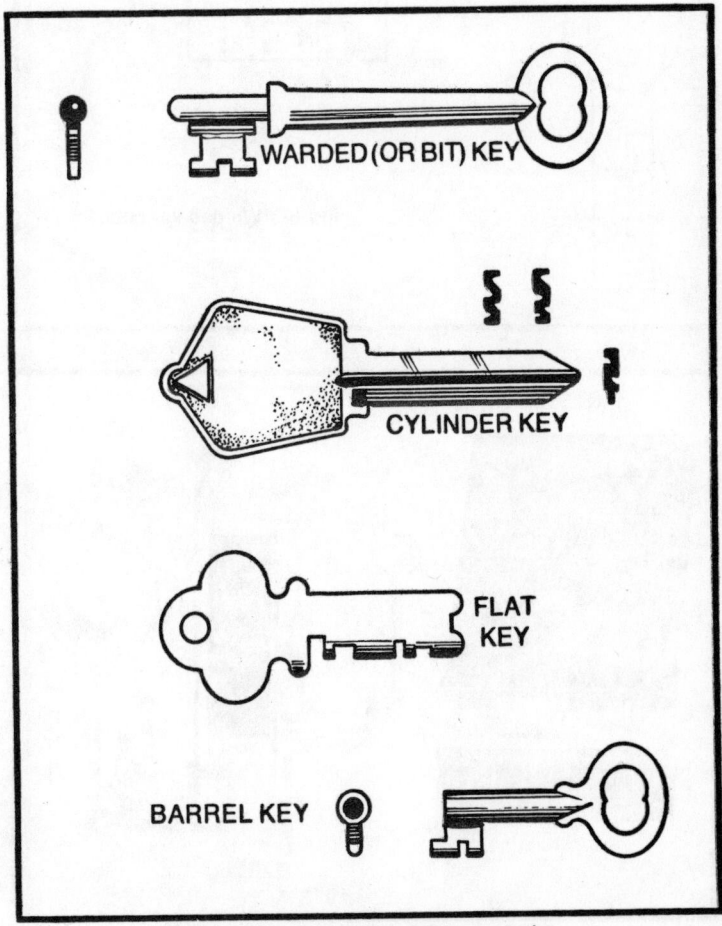

Fig. 3-2. Four basic key types in use today.

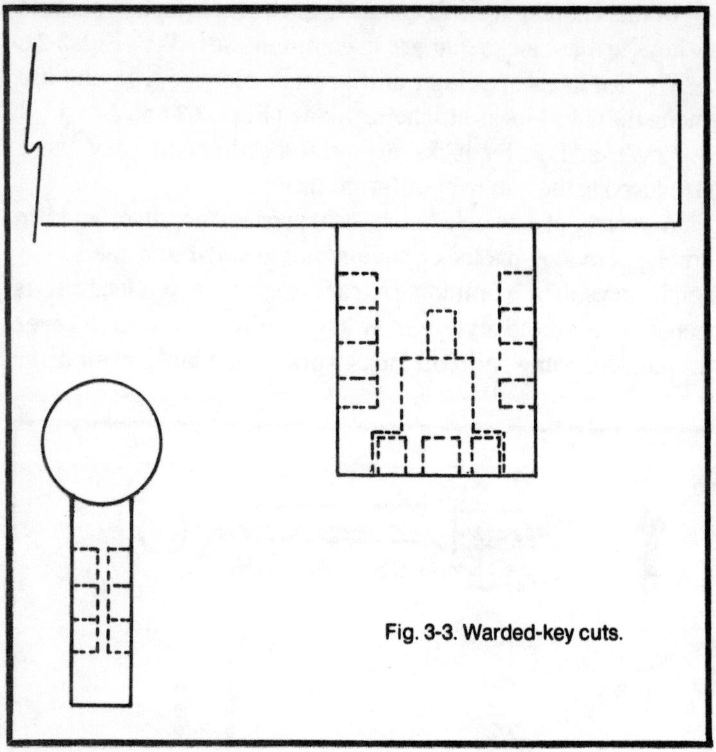

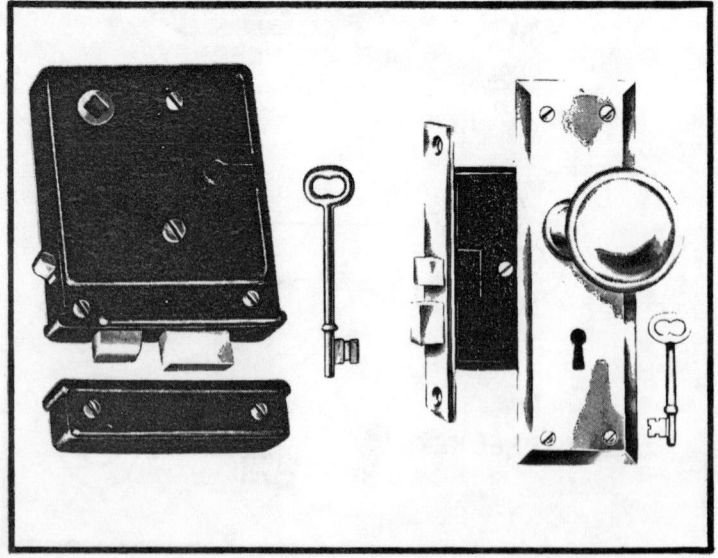

Fig. 3-3. Warded-key cuts.

Fig. 3-4. Typical warded locks. (Courtesy Taylor Lock Company.)

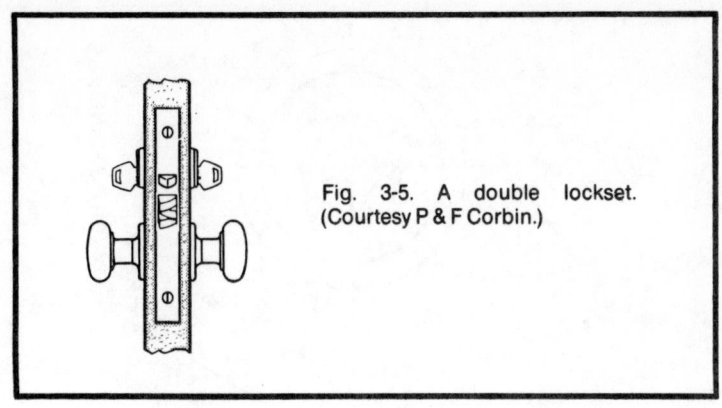

Fig. 3-5. A double lockset. (Courtesy P & F Corbin.)

being exposed to the elements for fifty years. Padlocks can also be combination locks.

The simplicity of the cylinder lock spawned its mass production. The cylinder is a single rotating unit firmly encased within a solid metal housing. The creation of the Yale cylinder lock encouraged several variations: the key-in-the-knob lock, the double lockset (Fig. 3-5), and the desk lock. A variation of the cylinder lock used in coil-operated machines is the Ace circular lock (Fig. 3-6).

A combination lock consists of a series of interconnecting wheels rotating about a central core that is controlled in its revolutions by an outside combination knob. The number of revolutions of the knob necessary to open the lock is pretty much standard: a 3-2-1 sequence is most common; some high-security locks may require a 4-3-2-1 rotation. Combination locks are of two basic types: the hand and the key-change variety. A kind of hand combination lock, not dependent upon internal wheels, is the pushbutton lock, shown in Fig. 3-7.

Surface-mounted auxiliary locks include deadlatches, surface-mounted cylinders (when used separately from another lock unit), chain door guards, rim latches, surface bolts, and chain bolts (Fig. 3-8).

The automotive lock employs the sidebar principle: When the key engages the tumblers, it aligns them so that a sidebar built into the lock drops into place, allowing the key to turn to open the lock. The sidebar principle is also used on certain pin-tumbler cylinders.

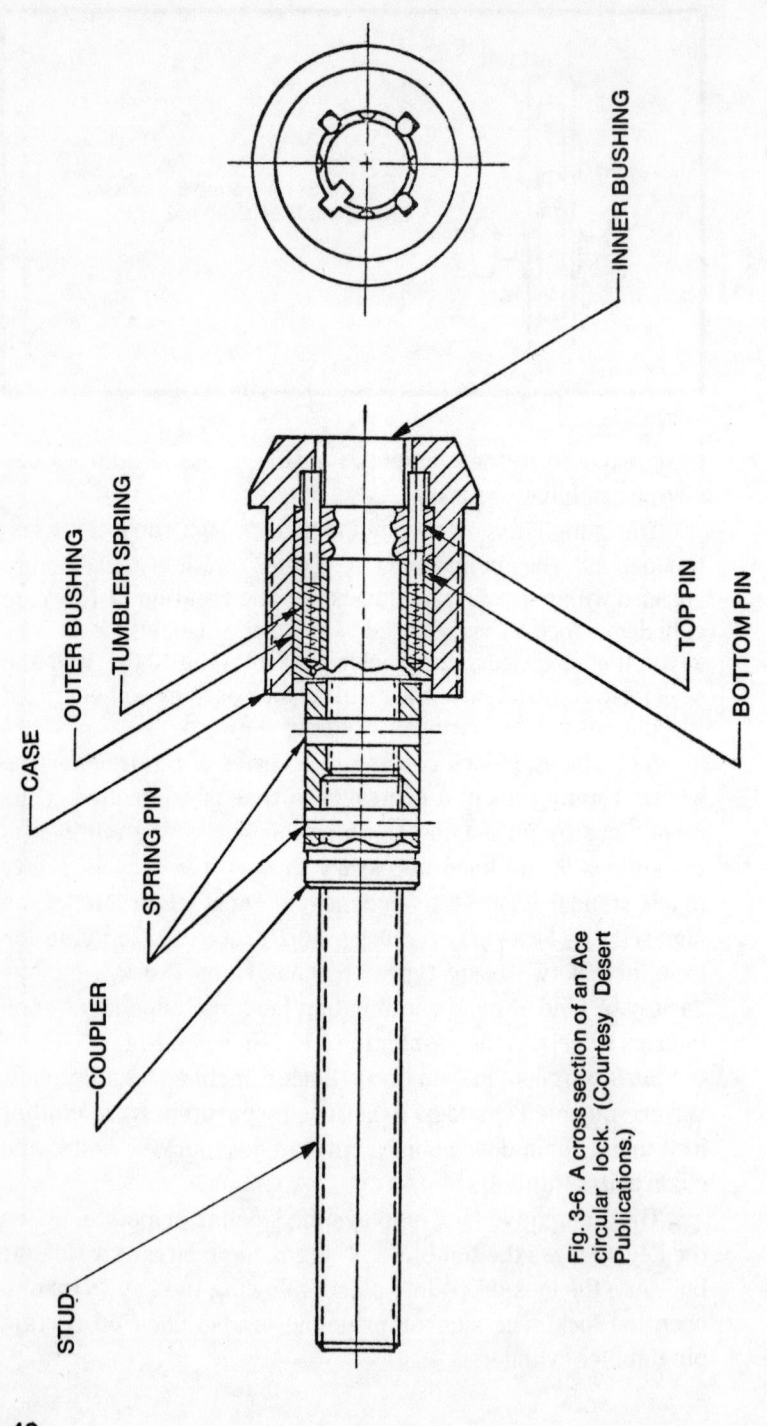

Fig. 3-6. A cross section of an Ace circular lock. (Courtesy Desert Publications.)

INNER BUSHING

OUTER BUSHING

TUMBLER SPRING

CASE

SPRING PIN

COUPLER

STUD

TOP PIN

BOTTOM PIN

48

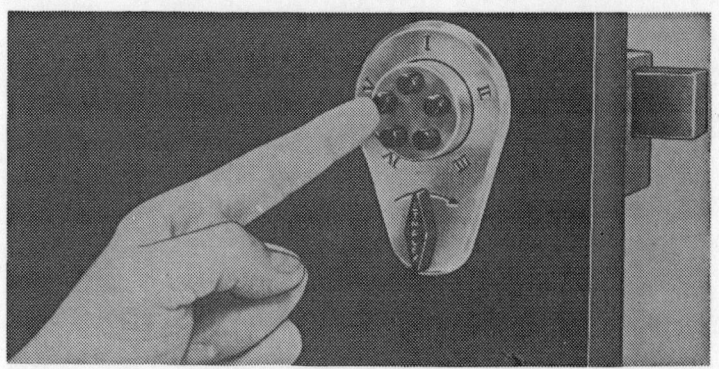

Fig. 3-7. A Simplex pushbutton combination lock.

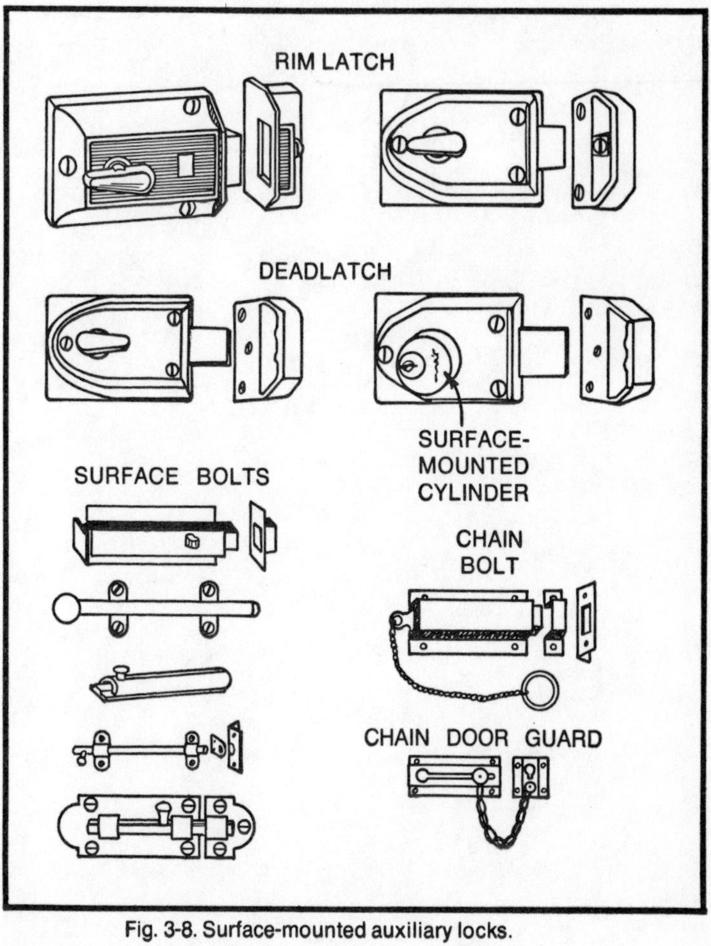

RIM LATCH

DEADLATCH

SURFACE-
MOUNTED
CYLINDER

SURFACE BOLTS

CHAIN
BOLT

CHAIN DOOR GUARD

Fig. 3-8. Surface-mounted auxiliary locks.

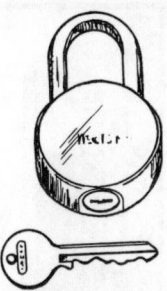

Chapter 4
The Warded
Lock and its Construction

The warded lock is the oldest lock still in use and is found in all corners of the world. It employs a single or multiple warding system. Because of its simple design, its straightforward internal structure, and its easily duplicated key, this lock is an excellent training aid for locksmiths. This same simplicity means that warded locks give very little security. Use these locks in low-risk applications such as storage sheds, and rooms where absolute security is not essential.

At one time warded locks were used on most doors. These locks are still found in abundance in older buildings still standing in decayed metropolitan neighborhoods, such as the center city of Philadelphia, Market Street in San Francisco, the Old Town section of Chicago, and the East Side of New York City.

The oldest of these buildings have the cast iron locks on the doors; some locks date back to the last century. Later types were made of medium gage sheet metal. The casing consists of two stampings: the cover plate and the backplate. The latter mounts the internal mechanism and forms the sides.

The warded lock derives its name from the word "ward," meaning *to guard*. The interior of the lock case has protruding ridges or wards that help to protect against the use of an

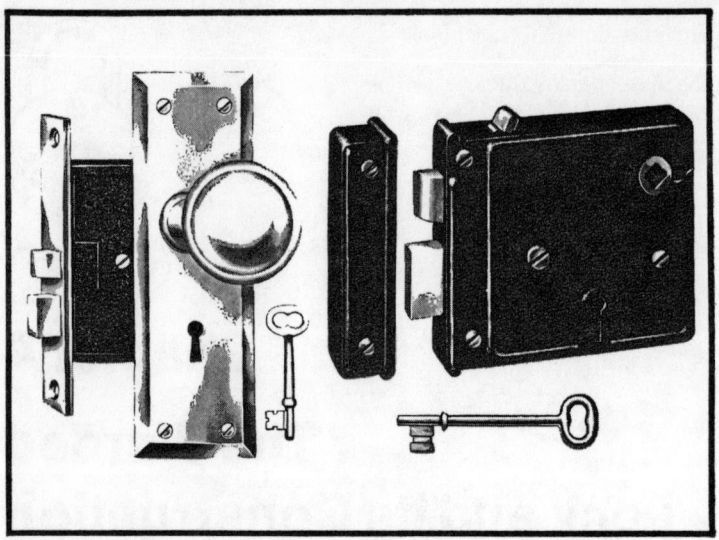

Fig. 4-1. Warded locks are available in mortised (A) and surface-mounted (B) varieties. (Courtesy Taylor Lock Co.)

unauthorized or improperly cut key. Normally, there are two interior wards positioned on the inside of the cover and backing plate, and directly across from each other.

This lock is sometimes mistakenly referred to as a skeleton-key lock. The proper and full name is the warded-bit key lock.

TYPES

Two types of warded locks are currently in use: the surface-mounted, or rim, lock and the mortised lock (Fig. 4-1). While both types are similar in structure and size, the degree of security they give varies. The internal mechanisms of both operate on the same principle, but the mortise lock may have several additional parts. Differences between these locks are:

Surface-mounted (rim) lock	Mortised ward lock
Mounted on door surface	Mounted inside of door
Secured by screws in the doorface	Secured by screws in the side of the door at the lock face-plate

Surface-mounted (rim) lock	Mortised ward lock
Door can be any thickness	Door must be thick enough to accommodate
Thin case	Fairly thick case
Short latchbolt throw	Up to 1 in. latchbolt throw
Locked from either side	Locked from either side
Strike can be removed with door closed	Strike cannot be removed with door closed
Very restricted range of key	Restricted range of key changes
Very weak security	Weak security

CONSTRUCTION

The basic interior mechanism is drawn in Fig. 4-2. The basic parts are:

A Latchbolt
B Latchbolt spring

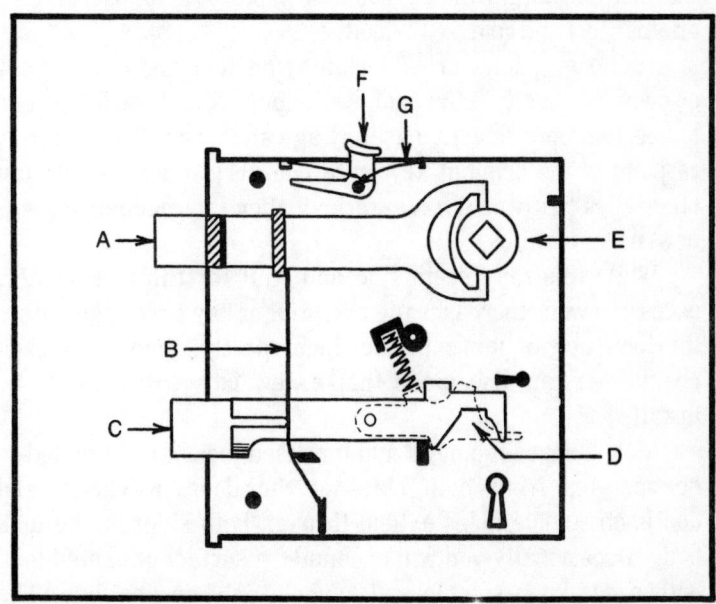

Fig. 4-2. The internals of a typical warded lock. See text for nomenclature.

C Dead bolt (thrown by the key and independent of the knob)

D Lever tumbler (raised by the key to release the bolt)

E Knobspindle hub (turned by the spindle)

F Trip switch (locks and unlocks the latchbolt)

G Trip switch spring (holds the trip switch in either the locked or unlocked position)

Since the relative security of any lock lies in the type of key used, the number of key variations possible, and the amount of access to the locking mechanism afforded by the keyhole, the warded lock is the least secure.

If a lock were designed to have no more than 10 different key patterns (changes), and 1000 locks were made, 10 different keys would open all 1000 locks. By the same token, one key would open the lock it was sold with and 99 others. Furthermore, it is often possible to cut away parts of a key to pass negotiate the wards of all the 1000 locks. From this you can see that lock security is related to the kind and number of key changes built into the system when it is initially designed.

The keyhole is an access route to the interior mechanism of the lock. The larger the keyhole, the easier it is to insert a pick or other tool and release the bolt.

In theory, each warded lock may be designed to accept 50 or even 100 slightly different keys. In practice, these locks tend to become more selective as they age and wear. The lock may respond to the original key or to one very much like it; but other keys that would have worked when the mechanism was new, no longer fit.

While this may seem fine and well for the lock owner, excessive wear increases the potential of key breakage within the lock and of jambs in the open, partly open, or closed position. It can also mean that a new lock will have to be installed.

Most surface-mounted and mortised locks are intended to be operated from both sides of the door. Keyholes and doorknob spindle holes extend through both sides of the lock body. Occasionally you will encounter a surface-mounted lock with a doorknob spindle and keyhole only on one side. The other side is blanked off.

A lock of this type can be modified to accept a key from the other side. This modification entails cutting a keyhole through the door and lock body and may require some filing on the key. Figure 4-3 illustrates the differences in keys. Note the additional cut on the left-hand key.

OPERATION

The key must be cut to correspond to the single or multiple side and end wards that have been designed into the lock. After the key passes these wards, it comes into contact with the locking mechanism. The cuts on the key lift the lever to the correct height and throw the dead bolt into the locked or unlocked position.

Turning the doorknob activates the spindle and, so long as the dead bolt is retracted, releases the door.

Figure 4-4 depicts various keyhole control features that allow only certain types of cut keys to enter the keyhole. Figure 4-5 shows a key entering a keyhole. Notice that the key

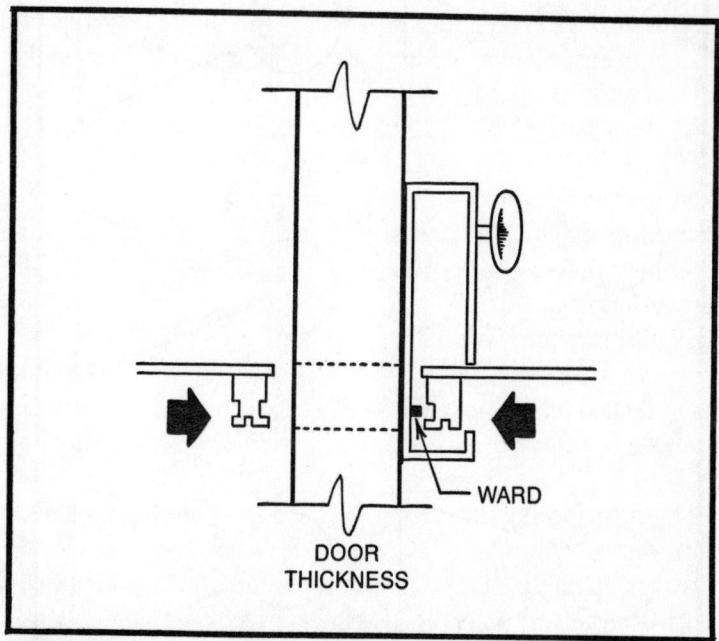

Fig. 4-3. The key on the right is for a lock activated from one side of a door; the key on the left can pass the wards from either side.

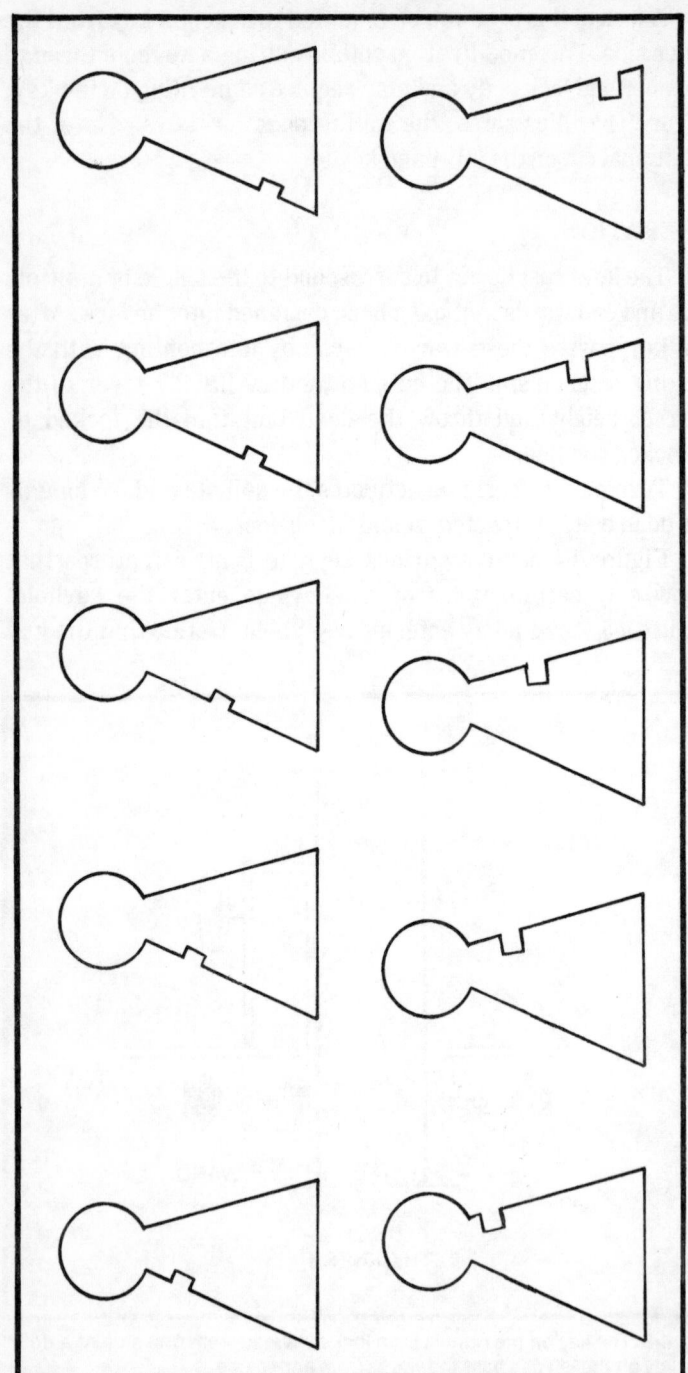

Fig. 4-4 Case ward variations.

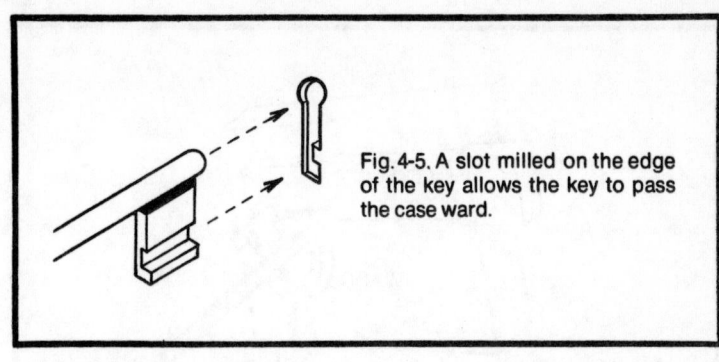

Fig. 4-5. A slot milled on the edge of the key allows the key to pass the case ward.

has the appropriate side groove to allow it to pass through the keyhole and into the lock. If you were to file off this obstruction (called a case ward) any key thin enough to pass could enter. (Some ward bit keys are quite thick.) By the same token, a very thin key is able to pass whether or not the side ward is present (Fig. 4-6). The common skeleton key is a prime

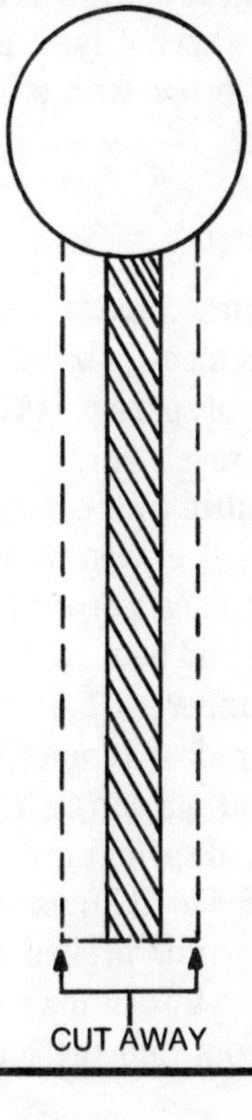

Fig. 4-6. A skeleton key is one that has been ground down to defeat the case ward.

CUT AWAY

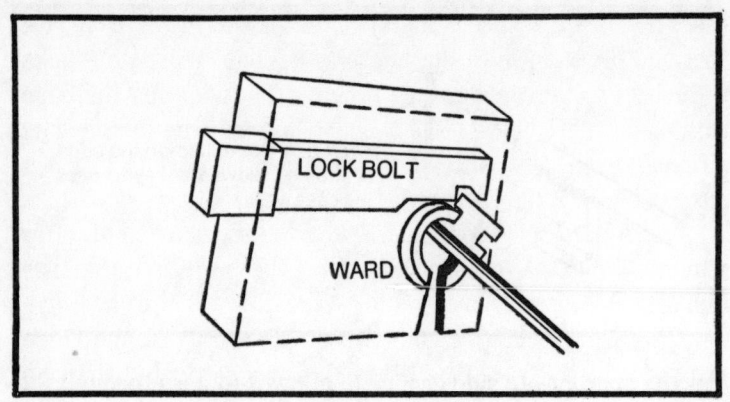

Fig. 4-7. A single ward is better than none.

example of this; it is thin enough to pass most case wards, but it will not necessarily open the lock!

Figure 4-7 shows a key engaging the bolt. While there is only one set of wards in this particular lock, because this ward is set where it is, it gives more security than a lock with no ward, or one with a ward that has been worn down to almost nothing.

REPAIR

Warded locks are not repaired to any great extent since replacement is usually cheaper. You should, however, have a supply of spare parts for these locks. The most frequent failure is a broken spring. Over a period of time, the spring may crystallize where it mates with the bolt. In addition, the wards can break or wear down into uselessness.

Replace a broken spring with a piece of spring stock, cut to length and bent to the correct angle. Some spring stock must be tempered before use; other springs come already tempered. If tempering is necessary, heat the spring cherry red and quench it in oil. You can save time by purchasing standard springs already bent into a variety of shapes that are designed to fit almost all locks.

Worn or broken wards on locks with cast cases can be repaired by drilling a small hole in the case and forcing a short brass pin into the lock case. The best technique for brittle cases is to braze a piece of metal on the case at the appropriate spot and file it down to the appropriate size. Wards on locks with sheet metal cases can be renewed by indenting the case with a punch ground to a fairly sharp point. If the factory has already punched out the wards, it would be best to braze a piece of metal at the proper spot.

Since most of these locks are inexpensive and offer minimal security, you should remind the customer that the cost of repair may far exceed the cost of the lock. Purchasing a new and more secure lock has definite advantages, and since you are already there, the homeowner's cost would then be less than ordering one and having you make a second trip to install it.

Should the homeowner decide to take your advice and purchase a new lock, request that you keep the broken one; it is of no use to him and parts are always nice to have. Sooner or later you will have to repair another lock of the same type; having the correct parts at hand will save you time. Furthermore, you have made a sale, and by being allowed to keep the old lock, have obtained parts at no cost.

Removing paint from these locks is also a form of repair. A major cause of lock failure is the home painter. He does not take the time to remove the lock or to cover it with masking tape. Some paint usually gets into the mechanism. To clean a paint-bound lock, follow this procedure:

- Remove the lock from the door (run a sharp knife around the edges so the new paint will not be cracked and broken).
- Disassemble the lock.
- Using a wire brush, scrape the paint from the parts. In extreme cases, you may have to resort to a small knife to do the job or else soak the individual parts in paint remover. Dry each part thoroughly.
- Check for rust and worn parts and replace as needed.
- Assemble and mount the lock.

Use paint remover only as a last resort, since it leaves a residue that attracts dust and lint. When you use paint remover, you must clean each part before assembly.

Locks that are difficult to operate usually have not been lubricated in a long time. if ever. Never use oil to lubricate a lock. The professional approach is to use a flake or powdered graphite. Apply the lubricant sparingly. Remember. a little bit goes a long way. and this is especially true of graphite. Should you overuse it. you may have to explain why there is a dark patch on the carpet that cannot be cleaned...graphite stains are almost impossible to remove.

Chapter 5
Cutting Warded Keys

Warded keys are made of iron, steel, brass, and aluminum. Iron and aluminum keys have a tendency to break or bend within a relatively short time; steel and brass keys can outlast the lock. The warded key has seven parts, as shown in Fig. 5-1. The configuration of the box, length of the shank, and the relative thickness of the shoulder are not critical to the selection or the cutting of the key.

KINDS OF WARDED KEYS

Warded keys come in various types: the simple warded key, the standard warded key, the multicut key, and the antique key. Simple warded keys are often factory made precut keys which fit several different keyholes. Multicut keys, on the other hand, are designed for specific locks. The standard warded key is usually mass-produced, but it has more precut ward and end cuts than the simple warded key. Standard warded keys can be easily converted into master keys by cutting. The antique key may have several kinds of cuts: ward cuts, end cuts, and even side (or bullet) groove cuts extending the length of the bit. Antique keys usually go to older locks, but these keys are still manufactured.

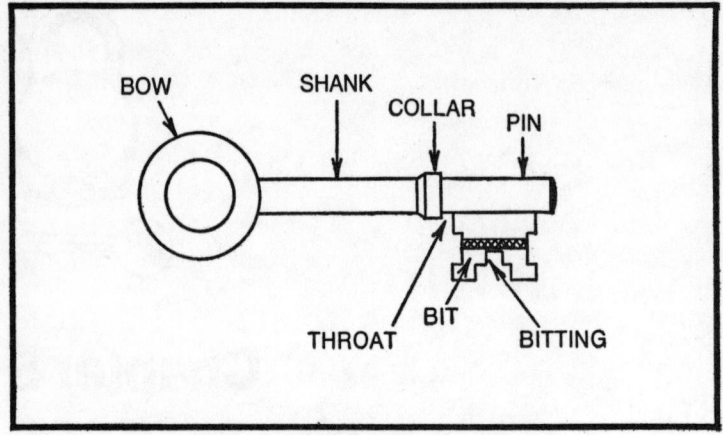

Fig. 5-1. The parts of a warded key.

SELECTION OF KEY BLANKS

A key blank is a key that has not been cut or shaped to fit a specific locking mechanism. When selecting a blank for a duplicate warded key, the following should be considered:

Pin Size—The pins of both keys (original and duplicate) must be the same diameter. If your eyesight cannot correctly determine whether or not they are the same size, you can use either calipers or a paper clip. Use the calipers to compare the diameter of the original key with the diameter of the duplicate. You can use a paper clip that has been wrapped around the original key to check the diameter of the duplicate.

Length—From the collar to the end of the pin both keys should be approximately the same length. This is important because this portion of the key enters the keyhole and operates the lock.

Height—The height of the bitting (cuts in the bit) must be the same on both keys. If the bitting in the duplicate is higher, it should be filed down; but if it is lower, another blank should be selected.

Bow—The bows need not be identical but generally should be closely matched.

Width—The width of the bittings should be approximately the same. If the bitting on the duplicate is too wide the extra thickness may prevent the duplicate from entering the lock.

Thickness—The thickness of the bits should be the same. If the original key bit is tapered the bit of the duplicate should also be tapered. You may have to select a blank with a thick bit that you can file down to the correct taper.

If you don't have a micrometer, the following table of standard wire diameters can help you determine the approximate dimeter of warded key blanks. You can take a standard piece of wire with a known diameter and compare it with any key blank.

Table of Standard Wire Gages

(in inches and millimeters)

1 mm = 0.03937 in.
1 in. = 25.4 mm

Standard Wire Gage Number	Inches	Millimeters
4/0	0.400	10.16
3/0	0.372	9.45
2/0	0.348	8.84
0	0.324	8.23
1	0.300	7.60
2	0.276	7.01
2½	0.264	6.71
3	0.252	6.40
3½	0.242	6.15
4	0.232	5.89
4½	0.222	5.64
5	0.212	5.38
5½	0.202	5.13
6	0.192	4.88
6½	0.184	4.67
7	0.176	4.47
7½	0.168	4.27
8	0.160	4.06
9	0.144	3.66

Table of Standard Wire Gages

(in inches and millimeters)

1 mm = 0.03937 in.

1 in. = 25.4 mm

Standard Wire Gage Number	Inches	Millimeters
10	0.128	3.25
11	0.116	2.95
12	0.104	2.64
13	0.092	2.34
14	0.080	2.03
15	0.072	1.83
16	0.064	1.63
17	0.056	1.42
18	0.048	1.22
19	0.040	1.02
20	0.036	0.91

You can also use a drill to determine the approximate diameter of warded key blanks. Insert the pin of a blank into the hole that matches the blank's diameter.

DUPLICATING A WARDED KEY BY HAND

1. Select the proper key blank.
2. Wrap a strip of aluminum, approximately 1½ × 2¼ or 2½ in. about the pin and bit of the original key with one edge against the collar (Fig. 5-2).
3. Clamp the original key (wrapped in the aluminum) into a vise, bitting edge up. Insure that the aluminum fits snugly about the key bit and pin.
4. Cut off excess aluminum around the bit of the key. Remove the aluminum strip and smoke the key with a candle.
5. After properly smoking the key, place the strip back on the bit and reclamp.
6. Using a warding file, cut down the aluminum strip—*not the original key*—until it is in the shape of the

original key. Since the aluminum is easily bent, use the file only in one direction—away from you. The stroke should be firm and steady at first. As you file closer to the cuts of the original key, the strokes should be shorter and lighter. When the file just barely touches the original key and starts to remove the candle black, *stop and go no further*. If the candle black is removed and the shiny surface of the original key is revealed, you know that you have filed too deep.

7. Fit the aluminum strip onto the key blank.

8. File down the exposed areas on the bit until it matches the outline of the aluminum strip. Be careful not to cut into the strip. Again, use shorter and ligher strokes as you get closer to finishing each cut.

9. If the original key has a side groove that matches a keyhole ward, this too must be cut. Use another strip of aluminum. If the original key has two grooves you must wrap the strip around the bit so that both grooves are covered.

10. Using a scriber, scratch the metal strip to indicate the top and bottom of the groove(s). Fit the strip onto the key blank and mark the positions of the groove(s) on both ends of the bit. By connecting the marks with lines, you know exactly where to file.

11. To determine the depth of a groove cut, put one edge of a metal strip into the groove of the original key and scribe the depth of the groove on it. This mark will be

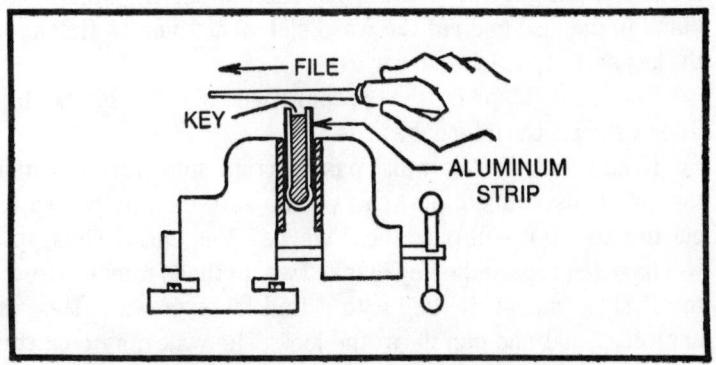

Fig. 5-2. File into the aluminum strip in the direction shown.

your depth guide when filing the groove on the duplicate key.

THE IMPRESSION METHOD

The term *impression* implies a duplicate key that is made from the lock itself as opposed to copying the original key. Impressioning is a skill that sets the locksmith apart from the individual who merely duplicates keys in the local five-and-dime store. It is a skill that is an extremely important and valuable asset.

Impressioning is the ability to decipher small marks made on a smoked key blank that has been inserted into a lock and turned. Interpretation of the marks tells the locksmith what cuts to make, where to make them, and how deep they must be. The advantage of impressioning is that you need not disassemble a lock or remove it from a door to make a key.

The first cut allows that blank to enter the keyhole. Refer to "Duplicating a Warded Key By Hand" for information on what is required to properly make such a cut. Since you do not have the original key, smoke the end edge of the blank and insert it into the keyhole so the edge comes into contact with the case ward. Scribe mark the top and bottom of the ward on the blank. Remove the key; the candle black that was removed indicates the depth of the cut and the scribed marks show the position of the cut on the blank. Transfer the scribe marks to the near end of the bit and draw lines connecting the two pairs of marks. As in the previous section, use a small piece of metal to make a depth gage so you will not file too deeply. Mount the blank in the vise and cut the ward slot. When you're finished, the key should pass the case ward.

The next step is the preparation of the blank for impressioning the internal wards.

Recall that the key must pass certain side wards. When the lock is assembled, how can you be sure exactly where to cut the key so it will pass these wards? You can't. Thus, you will have to prepare the key blank. Two methods may be used; the first is to coat the key with a thin layer of wax. This is unprofessional and can harm the lock. The wax may clog the mechanism and require that the lock be disassembled and cleaned—at your expense—you can't charge a customer for your mistake. The professional method is to smoke the key. The smoking must be thick enough to form a stable marking surface. If for example, the blank is thicker than it should be, enough blackening must be present to give true readings, a thin coat will speckle as you turn the key and send you on wild goose chases.

The technique for impressioning the key is:

1. Insert the key into the lock and turn it with authority.
2. Remove the key. You will notice one or more bright marks where the blackening has been removed. These marks indicate obstructions and you will have to file the appropriate cut for the key to pass.
3. Blacken, insert, turn, and remove the key; if a mark is present below the point you filed, the cut is too shallow. Carefully file it deeper.
4. When the cut is deep enough, use emery paper and clean off the fine burrs left by the file. This is the professional way.
5. Reinsert the key and turn it again. It should pass the wards. If it tends to stick slightly, a quick pass over the cut lightly with the warding file will alleviate this problem.

NOTE
THE CUT SHOULD BE SQUARE ON ALL SIDES. THINK OF EACH WARD CUT AS A MINIATURE SQUARE. PERFECT AND EVEN ON ALL SIDES—THIS SHOULD BE YOU GUIDE.

6. Smoke the key, insert it, and turn it. The edges of the cuts should be shiny. The brighter the spot, the greater the pressure at that point. As the individual cuts are filed deeper, the bright spots grow dimmer. As this happens you know that you are close to the point where you should stop filing.

Each time you insert the key to test the depth of the cut, you must be more cautious. You are working "in the blind,"—a small overcut means the depth is permanently wrong on the blank and you will have to start over. This is the reason for making a few light, but firm, strokes as you near the completion of each cut. Take your time to make one perfect key instead of rushing the job and making many incorrectly cut keys.

The cuts on the key should as deep and as wide as necessary—and no more. Overly large cuts interfere with the action of the lock, and may force you to cut another blank.

Once the key is completed, it must be *dressed out*. I mentioned earlier using emery paper to remove burrs before they break off and fall into the lock. Now polish the entire key with emery paper. You might also give it a light buffing on your wheel for appearance sake. It does you no good to give the customer a dirty and smudged key. Show pride in your work!

Passkeys

Skeleton-type passkeys (masterkeys) are sold in variety stores. These keys will fit many old locks and more than a few new ones. As such, they are a convenience. But no reputable locksmith stocks them. Why? The ethics of the profession forbid it. You certainly do not want to supply someone with a key that could open his neighbor's lock. Nor should you duplicate a passkey without authorization from the owner. Be leery of a customer who wants a key duplicated, but with additional cuts. Locksmiths have lost their licenses for less. Don't let it happen to you.

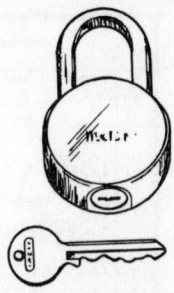

Chapter 6
The Lever Lock

Lever locks have many uses in light security roles. Available in a variety of sizes and shapes, these locks are found on desks, mailboxes, lockers, bank deposit boxes, and so on.

Figure 6-1 illustrates a popular example. The circular "window" on the back of the case is a locksmith's aid—it reveals the heights of the levers without the necessity of dismantling the lock. Thanks to the window, it is relatively easy to make a key "in the blind."

PARTS

A lever lock has six basic parts:

- Cover boss
- Cover
- Trunnion
- Lever tumblers (top, midde, and bottom)
- Bolt (bolt stop, notch, and post)
- Base

These parts are shown in Fig. 6-2.

OPERATION

This lock requires a standard flat key. When the key is turned, the various key cuts raise corresponding lever

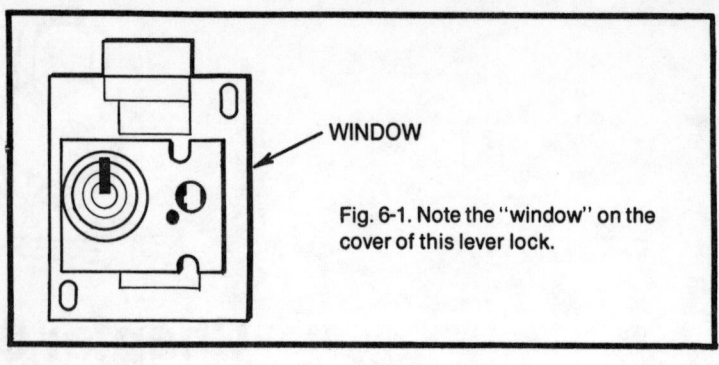

Fig. 6-1. Note the "window" on the cover of this lever lock.

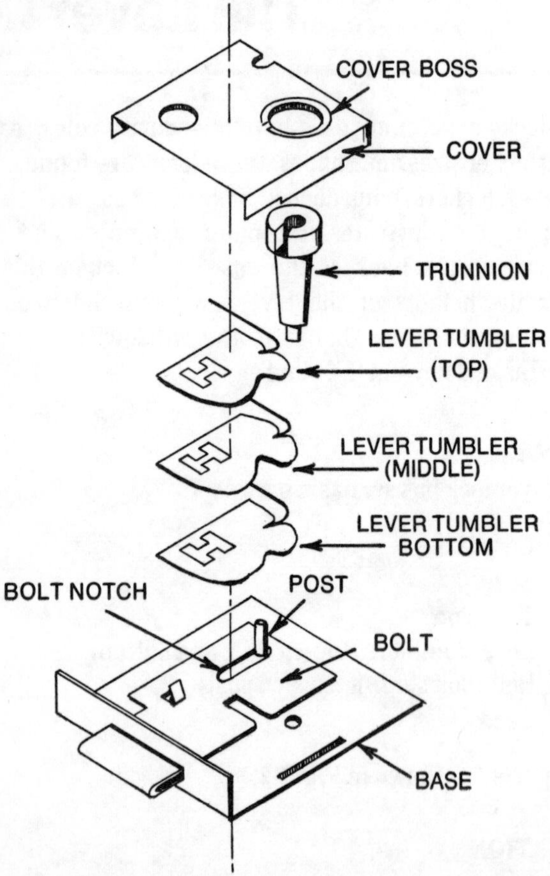

WINDOW

COVER BOSS

COVER

TRUNNION

LEVER TUMBLER (TOP)

LEVER TUMBLER (MIDDLE)

LEVER TUMBLER BOTTOM

BOLT NOTCH

POST

BOLT

BASE

Fig. 6-2. A lever lock in exploded view. This particular lock has three tumblers; others may have a dozen or more.

70

tumblers to the correct height. As the levers are raised, the gates (Fig. 6-3) of the levers align and release the bolt. The bolt stop (some call it the post) must pass through the gating from the rear to the front "trap" or vice versa. either unlocking or locking the lock.

LEVER TUMBLER

The number of lever tumblers varies. Most locks have no more than five, although deposit box locks have more. The lever tumbler consists of six parts:

- Saddle (or belly)
- Pivot hole
- Spring
- Gating slot
- Front trap
- Rear trap

Each lever is a flat plate and is held in place by a pivot pin and a flat spring. Each lever has a gate cut into it. The gates are located at various heights either with the saddle aligned for all levers or staggered. The latter approach has long since

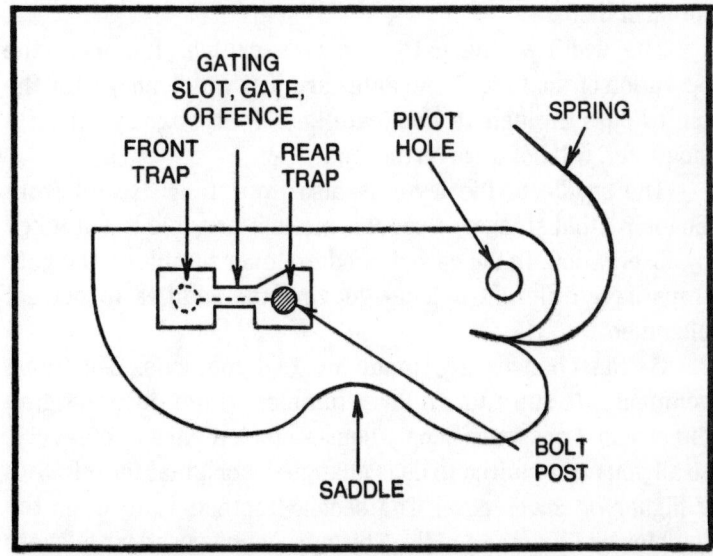

Fig. 6-3. Lever-tumbler nomenclature. The key bears against the saddle.

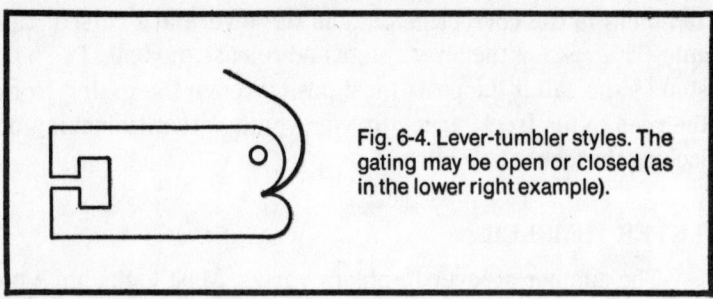

Fig. 6-4. Lever-tumbler styles. The gating may be open or closed (as in the lower right example).

been antiquated. When the levers are raised to the proper position, the gates are open and the bolt post can be shifted from one trap to another, thus locking or unlocking the lock.

Since the bolt post meets no resistance at the gating, the lock will work properly. On some designs the edge of the lever has serrated notches. The bolt post has corresponding notches. The notches on the lever and the bolt jam together if an improperly cut key is used; this effectively stops the bolt from passing through the gating and keeps the lock secure. Only a perfectly and properly cut key will open this type of lock. This feature adds immensely to the security of the lock.

Manufacturers have, over the years, developed a variety of lever types (Fig. 6-4). The operating principle is the same for all of them.

The width of the gate is also a critical factor in the operation of the lock. Some gates are just wide enough for the pin to pass through; a duplicate key, even slightly off on a single cut, will not work on the lock.

The saddle of the lever is also important. Recall from Chapter 1 that staggered saddles make it possible to cut a key by observation. In the case of modern lever tumblers, the gate traps have different heights, leaving the saddles in perfect alignment.

Gating changes are made by two methods. The most common is to substitute a lever tumbler with a different gate dimension. Locksmith supply houses stock a variety of levers, so all you are required to do is change the original for one with a higher or lower gate. The second method is to alter the tumbler by filing the saddle. This approach is used with levers whose movement is restricted at the gate. Unless the tumbler gating varies greatly. the curvature of the saddle must vary with the shape of the key.

As I mentioned earlier, the typical lever lock contains two, three. five. or possibly six levers. Bank deposit box locks may have as many as fourteen. Lever locks can be keyed individually, alike (two or more with the same key), or masterkeyed, depending upon the wishes of the buyer.

DISASSEMBLY

General-purpose lever locks come in three styles. In order of popularity these are:

- Solid case usually spot-welded or riveted.
- Pressed form with the back and sides of the case one piece—small tabs from the sides bend to hold the cover in place.
- Cover plate secured by a screw.

Riveted or spot-welded locks should be discarded when they fail. It is cheaper to purchase a new lock; the time required to drill out the rivets or chisel through the spot welds costs the customer more than the lock is worth.

Locks secured by tabs can be disassembled and reassembled quite easily. To disassemble, insert a thin tool or small screwdriver under the flanges and pry upward. Remove the cover. Look for a small object jammed in the keyway; if something is found, remove it. At times like this, you may wonder if being a locksmith is really worth it; but a locksmith must have patience with the small jobs as well as with the big ones.

Another common problem with lever locks is a broken spring. If it is not the top lever, than carefully remove each lever. in turn. placing them in a logical order so that they may

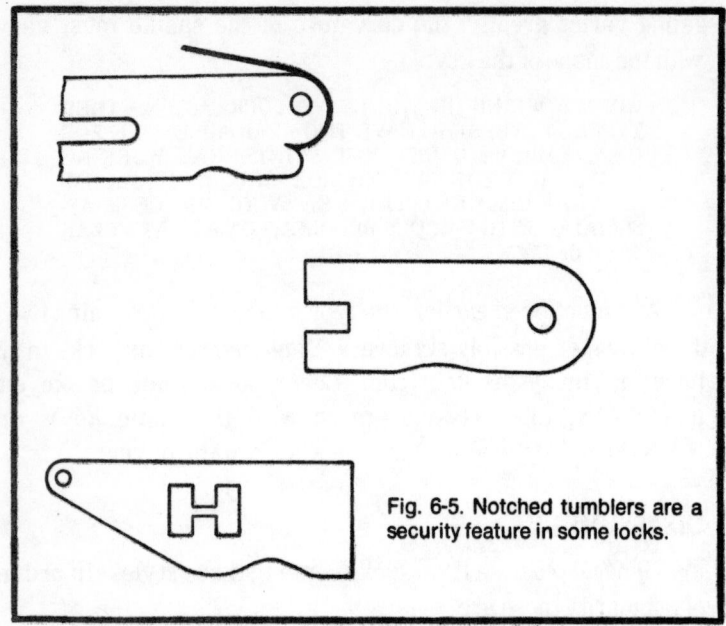

Fig. 6-5. Notched tumblers are a security feature in some locks.

be assembled as they were found. Remove the lever whose spring is broken. Select a piece of spring steel from your inventory. Cut and bend it to shape. If you have purchased an assortment or ready-made springs, select the proper one and replace the broken spring with a new one. Replace the levers and reassemble the lock.

SAFE DEPOSIT BOX LOCKS

Lever safe deposit box locks normally have a minimum of 6 and upwards of 12 or even 14 levers. Two keys are required. One key is assigned to the individual who rents the box; the second or guard key is held by the bank. Both keys are needed to open the lock.

Disassembly is simple. Remove the screws and lift off the plate. You will notice that the lock is constructed differently than previously discussed lever locks. There are two sets of lever tumblers; two bolt pins must pass through the lever gates at the same time. Note the unique lever shape (Fig. 6-5).

Many safe deposit locks have a compression spring bearing against the upper lever that forces the lever stack down.

allowing no play between them. This spring is an integral part of the lock's security mechanism. Without the spring, the levers would be able to move a fraction of an inch when a key or pick "irritates" them. This movement is enough to provide clues for the lockpick artist.

OPERATION OF THE SAFE DEPOSIT LOCK

This lock, as stated earlier, requires two keys to open it. If one key could be turned to the *open* position by itself, it would mean that the key was faulty, the lock mechanism was in some way evaded, or the bolt post was bent or broken.

The levers have a V-shaped ridge that matches a similar V cut in the bolt post. Another key, even one that is 0.001 in. off on any cut, will mesh the notches in the post and lever. This is another built-in security feature of these locks.

Broken springs are the main difficulty. Or the levers may be at fault. The saddles may wear enough to affect the gate. In cases such as this, replace the lever with a new one. Do not file the gate cut wider to compensate. Sometimes a pivot post will work loose. Repair by rapping the post with a light hammer or by brazing. Never use solder. Slightly bent parts can be straightened out, but it is best to replace the entire damaged part.

Safe deposit locks are serviced by locksmiths who enjoy high standings within the community. Unless an emergency arises, a beginning locksmith or one who is a newcomer to the area would not be given the job. But this reluctance is only a matter of the conservatism of bankers. It does not reflect upon the skills of students and those working locksmiths who have

Fig. 6-6. Typical lever-lock keys.

had only a few months experience. Many of these men and women could repair a safe deposit box lock without difficulty.

SUITCASE LOCKS

Suitcase locks are essentially simple but have been built to accept a staggering variety of keys. Ninety-nine locks out of a hundred are warded, with a primitive bolt mechanism to keep the case closed. Only a few, such as the Yale luggage lock, use a lever-type mechanism. These locks offer better security than warded locks.

The lock size, the depth the key is inserted, and the number of cuts are important clues to the lock type. A key for a lever lock will go ¼ in. or deeper into the lock before turning. Warded locks are shallower.

Opening suicase locks is relatively simple. Many times one key will open several different suitcase and luggage locks. Many of these locks on the market are not designed for security. By cutting down almost any suitcase key, it is possible to make a skeleton key for emergencies.

Chapter 7
Cutting the
Lever Lock Key

Unlike the warded or cylinder key, the lever lock key used by the average person almost never contains a keyway groove running along its side; it is a flat key. A number of different flat keys exist. Figure 7-1 illustrates some of them. Figure 7-2 identifies the various parts of a typically cut lever lock key.

The lever lock key can be cut by hand or machine. In order to make the key, you must have the proper key blank. The three critical dimensions of the lever lock key are the thickness, length, and height. If the key blank is slightly higher or wider than the original, the blank should be filed down to the proper size. If it is thicker, select another blank. Filing down the thickness of the blank weakens it structurally.

Laying the keys (original and blank) side by side, run your finger across them lightly. If the blank is thicker than the original, you will feel your finger catch as it passes from one to the other. When you insert the key into the keyway, it should not bind or be tight fitting.

The first cut that must be made is the throat cut, which enables the key to turn within the keyway. To do this, insert the blank in the keyway and scribe each side of the blank where it comes in contact with the cover boss. Determine the point where the trunnion of the lock turns. Draw a vertical line there

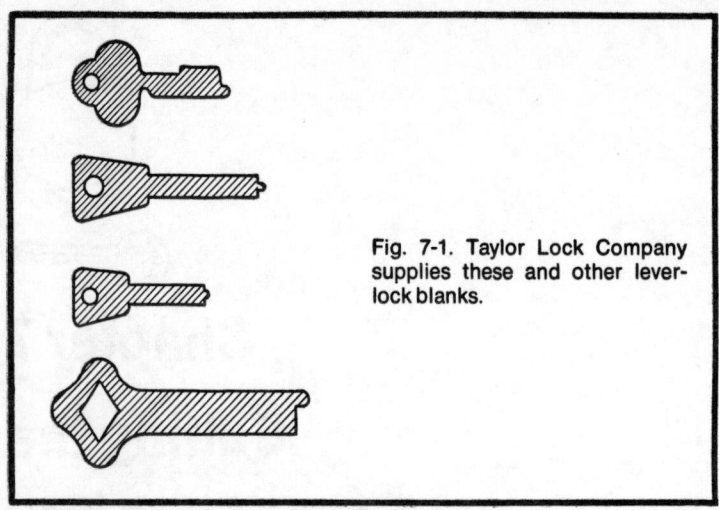

Fig. 7-1. Taylor Lock Company supplies these and other lever-lock blanks.

to indicate the depth of the throat cut. Remove the key from the lock, place it in a vise, and use a 4 in. warding file to cut alongside the vertical line to the proper depth. Cut on the side of the line towards the tip of the key.

As stated earlier, there is a small round window in the back of most lever lockcases. The window is positioned so you can see where the bolt pin meets the lever gates. You can get a general idea of the proper cuts to be made on a key blank by observing the lever action through the window.

After the throat cut is filed, the other cuts must be made. To do this, follow these steps:

1. Smoke the blank, insert it in the keyway, and turn. Remove it. The lever locations will be marked on the blank.

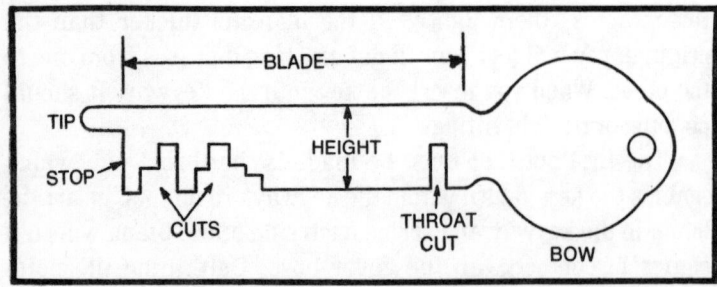

Fig. 7-2. Lever-lock key nomenclature.

2. File the marks slightly, starting with the marks closest to the throat cut.
3. Insert the key and turn it. Notice (through the window) the height that each lever comes up and the position of the pin in relation to the lever gates. The distance each gate is from the pin indicates the depth the key should be cut for each particular lever. File each key cut a little at a time, periodically inserting and turning the key to check the gate/pin relationship. Be sure to file *thin* cuts. Continue until you can insert the key and have the gate and pin line up exactly.
4. If the key in the keyway binds, observe the levers. One or more may not be correctly aligned. If the pin is too high, you have cut too deeply; if the pin is too low, you have not cut deeply enough. It may be necessary to resmoke the blank and reinsert it. The point where the key binds hardest will be indicated by the shiniest spot on the key. Just a touch with the file will usually alleviate the problem. Insure that each filed cut is directly under its own lever.

At this point, you should have corrected any variations between the original key and the blank. The dimensions must be identical: the key height, thickness, and the length.

You will need a vise for holding the two keys, a small C-clamp, a warding file, a candle, and a pair of pliers. Follow this procedure:

1. Holding the original key over a candle flame, smoke it thoroughly.
2. Allow a few minutes for the key to cool and clamp the original and the blank at the bows. Most locksmiths use a C-clamp for this initial alignment.
3. Once aligned, place the key and blank in a vise. If you wish, you can leave the C-clamp in position.
4. Using the warding file, make the tip cut first. File in even and steady strokes, bearing down in the forward, or cutting, stroke. Keep a careful eye on the original. Stop when the file just disturbs the blackening.
5. Once the tip cut is completed, move to the next cut.

6. Remove the keys from the vise and inspect the cuts. Each should be rectangular and flat.
7. Using emery paper. lightly sand away the burrs on the edges of the cuts. Wipe off the candle black from the original key.
8. Test the duplicate in the lock. Should it stick, blacken the duplicate, and try it again in the lock. Breaks in the blackening will show where the key is binding. A light stroke with the file should correct this.

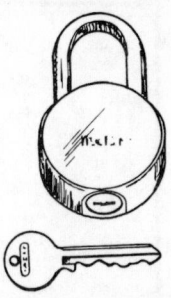

Chapter 8

The Disc-Tumbler Lock

The disc-tumbler lock gets its name from the shape of the tumblers. These are about as secure as lever locks and, as such, are superior to warded and other simple locks. However, disc-tumbler locks are far less secure than pin-tumbler locks.

Disc-tumbler locks are found in automobiles. desks, and in a variety of coin-operated machines. Some padlocks are built on this principle. Because the cost of manufacture is very low. replacement is cheaper than repair.

While similar to the pin-tumbler lock in outside appearance and in the broad principle of operation, the internal design is somewhat unique.

The disc tumblers are steel stampings, arranged in slots in the cylinder core. Figure 8-1 shows a typical disc tumbler. The rectangular hold, or cutout, in the center of the disc matches a notch on the key bit. The protrusion on the side, known as the "hook," locates the spring. The disc stack is arranged with alternating hooks. one on the right of the cylinder, one on the left.

The complete lock is illustrated in the next drawing.

OPERATION

The disc lock employs a rotating core, as does the more familiar pin-tumbler design. However. the disc-tumbler core is

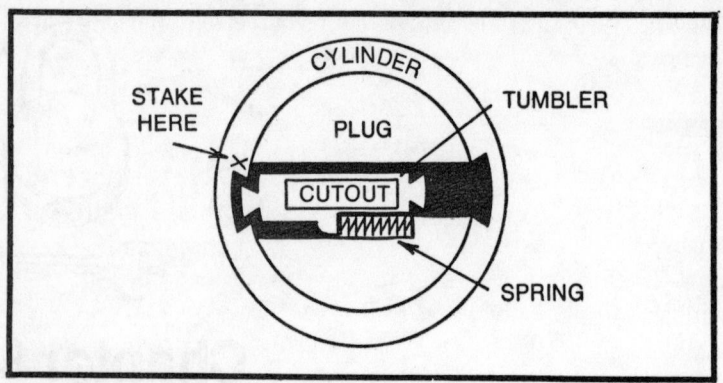

Fig. 8-1. Disc-tumbler lock in cross section. The position of the cutout determines the depth of the key cut.

cast so that the tumblers protrude through the core and into slots on the inner diameter of the cylinder. The core is locked to the cylinder so long as the tumblers are in place.

The key has cuts that align with the cutouts in each tumbler. The key should raise the tumblers high enough to clear the lower cylinder slot. but not so high as to enter the upper cylinder slot. In other words. the correct key will arrange the tumblers along the upper and lower shear lines (Fig. 8-2). The plug is free to rotate and. in the process. throw the bolt.

KEYS

The key resembles a cylinder pin-tumbler key. except that it is generally smaller and always has five cuts. A cylinder pin-tumbler key may have six or seven cuts.

DISASSEMBLY

Good quality disc locks feature a small hole on the face of the plug. usually just to the right of the keyhole. Insert a length of piano wire into the hole and press the retainer clip. Turn the plug slightly to release. The key gives enough purchase to withdraw the plug: if a key is not available. you can extract the plug with the help of a second length of piano wire inserted into the keyhole. Bend the end of the wire into a small hook. Other locks attach the plug to the cylinder with the same screw that secures the bolt-actuating cam. Others. fortunately a minority, have the plug and cylinder brazed together. File off the brass.

KEYING

Manufacturers have agree upon five possible positions for the cutouts relative to the tumblers. Keying is a matter of arranging the tumblers in a sequence that matches the key cuts. Once the sequence has been identified, install the tumbler springs over their respective hooks and mount the tumblers in the plug. The tumblers are spring-loaded and, until the plug is installed in the cylinder, are free to pop out. Lightly stake them in place with a punch or the corner of a small screwdriver blade. One or two pips are enough since the tumblers will have to be broken free once the assembly is inside the cylinder. Inserting the key is enough to release the tumblers.

SECURITY

These locks have no more than five tumblers and each tumbler cutout has five possible positions. These variations allow 3125, or 5^3, key changes, at least in theory. In practice the manufacturer will discard some combinations as inappropriate and may further simplify matters by limiting the key changes to 500 or less. Obviously, disc-tumbler locks are not high security devices.

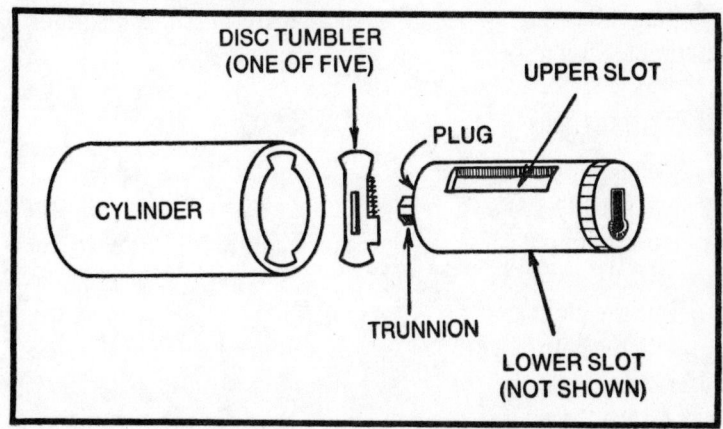

Fig. 8-2. A disc-tumbler lock in exploded view.

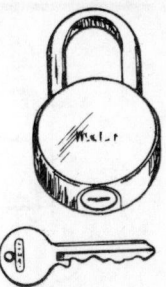

Chapter 9
Reading Disc
and Lever Locks

It's not unusual for a locksmith to be asked to make a key for a lock when the original key has been lost or misplaced. If the lock does not have a code number on its face or if the owner neglected to write down the number on the key, the locksmith has three choices. He can pick the lock, impression a key, or "read" the lock.

Like impressioning, reading a lock is a skill that must be developed through patient practice. You cannot expect to master this skill quickly, nor can you expect to remain proficient in it without constant practice. At first, practice daily then weekly or twice weekly to maintain your skill.

When called upon to fit a key, either by impressioning or reading of the lock, the locksmith invariably looks into the lock keyway. A quick glance determines whether it is a lever, disc, or pin-tumbler lock.

DISC LOCKS

The view down the keyway of a disc-tumbler lock will show a row of discs with their centers cut away and staggered. The cutaways are at different heights; the discs themselves are the same diameter. The cutaway looks like a small staircase with a surrounding wall (formed by the vertical edges) about the

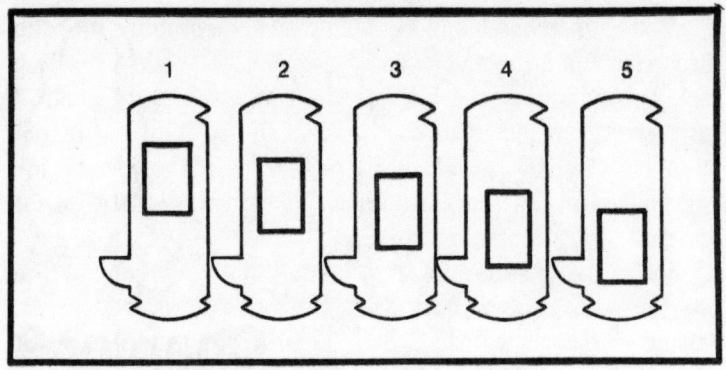

Fig. 9-1. Disc tumblers have five variations.

steps. Since each disc is the same size, only the height of the "steps" varies, and this variation is predetermined. A No. 1 disc has its cutaway toward the top of the tumbler; a No. 5 has its cutaway situated low on the tumbler (Fig. 9-1).

Here is where the skill of reading a disc lock is learned. Study the discs through the keyway. Notice the relationship of the discs to each other and to the keyhole. Through constant study and practice, which includes mixing the various discs, you will be able to determine which disc is which (Fig. 9-2).

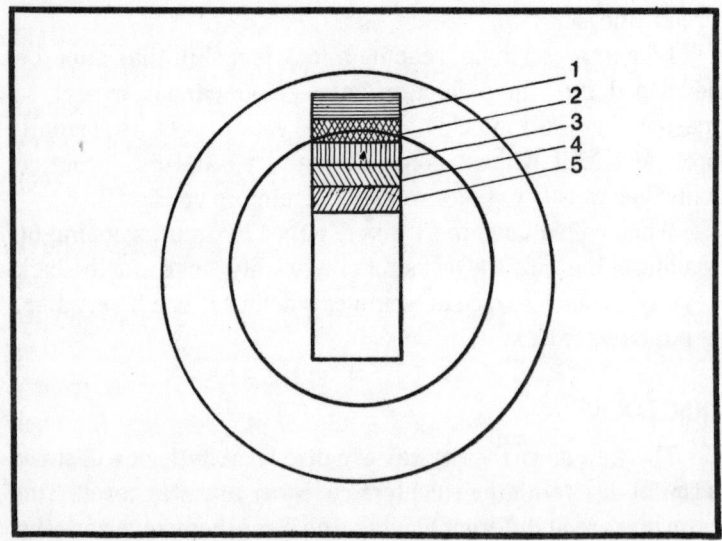

Fig. 9-2. The position of the tumblers will give you a general idea of the key profile.

You have to be able to lift each disc and compare it to the disc in front of or behind it. To do this, you will need a reading tool. This is nothing more than a stiff length of wire about 3 inches long, mounted in a small dowel handle. Think of the tool as a long hairpin attached to a short piece of wood for convenience in holding. The wire should be bent slightly so that you can see the tumblers.

Insert the tool into the lock, holding it so you can see the interior and observe the discs. By shifting the tool about, raising and lowering each disc, you can see the relationship of each disc cutaway to the next disc and to the keyway. Using your knowledge about the cutaway relationships, you will be able to decode the tumblers and cut a key for the lock.

You might ask yourself, "How do I know where to cut the key—and how deep?" Recall that when impressioning a key, you blackened the key and determined the cutting point by the pressure of the levers upon the key as you tried to turn it. The same technique applies here. Insert a blackened key and give it a slight turn. This brings the key in contact with the sides of each tumbler cutout. The cutout will leave a mark on the blank, indicating the portion of each cut.

Determining the depth of the cuts requires experience. You have already learned that cutting a key is slow, patient work. As you get closer to the proper depth, the file is moved with less pressure than before. The same care and precision is needed here. A disc with a No. 1 cut hole requires a deep key cut, as compared to a No. 4 or 5 disc, which requires a very shallow cut.

Reference aids include various extra keys that you have collected. Since the depths of the cuts are standardized in the industry, you can have a key with a 13354 cut and, by observing the differences in the depths of each cut, know exactly how deep a No. 2 cut should be. You can obtain a depth key set from a locksmith supply house. This set has a different key for each depth, with the same cut in all five tumbler positions. Thus, a No. 2 key has five No. 2 cuts; a No. 3 key has five No. 3 cuts; and so on throughout the series. With the help of these keys and your trained eyesight, you can place a blank in a vise and cut the key by hand.

Without a depth key set you can use a variety of disc keys as guides. Select a key with the proper cut and align your blank to it. Another approach is to make your own set of depth keys. Making the set teaches you how deep to make each cut at any given position on the blank.

LEVER LOCKS

As you already know, a lever-lock keyway is narrow and the view of the tumblers is further restricted by the trunnion. While this is a handicap, it can, in part, be overcome with the help of an appropriate reading tool.

The positions of the lever saddles is one clue to reading the lock. The saddle of the lock can tell you quite a bit. The wider the saddle, the deeper the cut; the narrower the saddle, the shallower the cut. Using the reading tool, you can feel the different saddle widths to determine the cuts and develop some idea the cuts and develop some idea of the key shape.

In order to do this properly, you must have an appreciation of the internal workings of the lock. Depth key sets can be useful when you are ready to cut the keys. Lever cuts are usually in 0.015−0.025 in. increments.

Practice is required. Begin with a lock that has a window in it: and, if possible, obtain extra levers so you can change the keying of the lock at will.

Raise each lever with the reading tool and try to determine the proper position for each one. Remember the general rule: a wide saddle requires more movement than a narrow saddle. Once you are satisfied that you have read the lock, disassemble it and examine the lever tumblers. You may have misread the tumblers, but do not be disheartened. The only way to achieve competence in this skill is practice and more practice.

Other tumblers are designed with uniform saddle widths. Keying is determined by the positions of the gates in the tumblers. These locks can be quite difficult to read.

Working with the reading tool, raise one of the levers as high as it will go. While this movement is not a direct indication of the depth of the key cut, it is important. The amount of upward movement establishes the minimum key cut; a shallower cut would jam the levers against the top of the lockcase. In addition, the individual gates and traps are usually in some rough alignment. You will usually find that two are in almost perfect alignment.

A cut for a lever that has the post on the upper half of the trap will be shallow; a cut for a lever that has its post on the lower half will be deep. A lever that has its gate in the intermediate position will require a key cut between these extremes.

Once you have established the general topography of the cuts, refer to your set of depth keys for the exact dimensions. Established locksmiths have reference manuals that may simplify the work.

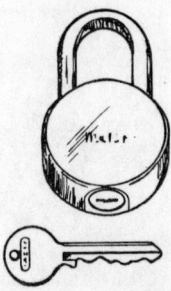

Chapter 10
Schlage Wafer-
Tumbler Locks and Keys

The wafer-tumbler design, by Schlage, offers the greatest improvement ever over the disc-tumbler lock. The wafer-tumbler cylinder is not die cast but is made of tubular steel, manufactured to stringent tolerances for minimal slippage and the greatest exactness in alignment. Wafer tumblers are made of heavier metal than disc tumblers. This insures a long life for them. Wafer-tumbler locks have more tumblers than most other locks: there are eight wafers in each lock.

The following material is furnished by the Schlage Lock Company. It covers everything a good locksmith needs to know about the Schlage wafer-tumbler lock and key.

DISASSEMBLING AND REMOVING THE KEYWAY UNIT

1. Refer to Fig. 10-1. Depress the catch (A1) through the hole in the shank of the inside knob (A) with a screwdriver to release the inside knob from the spindle (E).
2. With the knob removed (Fig. 10-2), position the screwdriver into the small notch (B1) usually located on the bottom edge of the inside rosette (B) and, with a prying motion, snap off the inside rosette.

Fig. 10-1.

3. Refer to Fig. 10-3. Remove the two machine screws (C) so the inside mounting plate (D) will slip off over the inside spindle (E). The lock will now slip out of the hole.

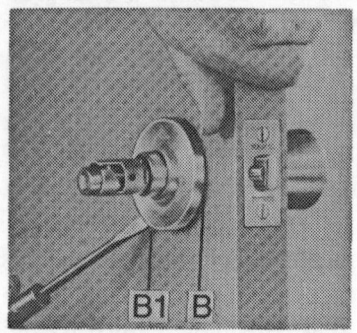

Fig. 10-2.

4. Refer to Fig. 10-4. The lock housing (F) is attached to the lock by a small cotter pin (G) or by twisted lugs. Remove the cotter pin or straighten out the lugs. Lift the housing above the lugs and, with a slight turn, rotate the housing ¼ in. and remove it.

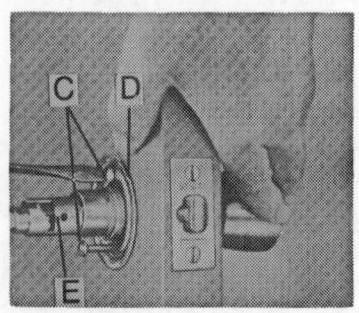

Fig. 10-3.

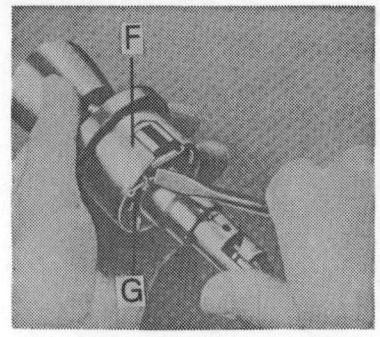

Fig. 10-4.

5. With the housing removed (Fig. 10-5), the lock frame is now exposed. Hold the lock in both hands, positioning your fingers as shown. When performing this operation, hold your palms carefully around lock to prevent springs from escaping the retractor slide. To remove the thrust plate (I), press forward with your thumbs against the frame tabs (H1) and push upward with your index fingers against the thrust plate. This will disengage plate.

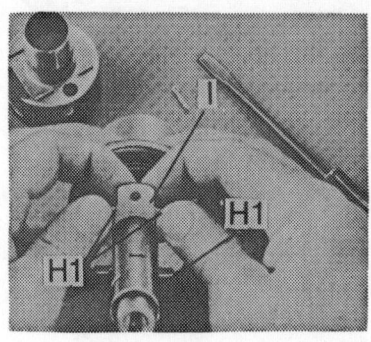

Fig. 10-5.

6. Refer to Figs. 10-6 and 10-7. In order to free the plunger unit (L) as you remove this assembly, it is necessary to push the slide (J) all the way to the rear against the compression of the two slide springs (K) with your thumb.

7. After the inside spindle (E), thrust plate (I), and plunger assembly (L) have been removed, let the slide (J) and the two slide springs (K) ease forward gradually and remove them from the lock frame (H).

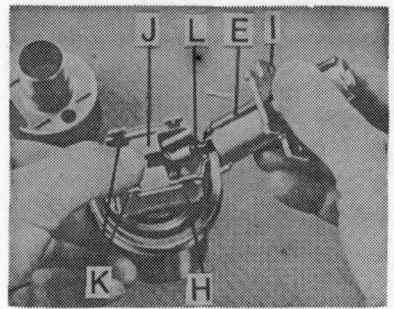

Fig. 10-6.

8. Refer to Fig. 10-8. To remove the wafer keyway unit (N) from the lock, push in on the face of the wafer keyway unit from the outside knob (M). The unit will then slide inward, where it can be removed from the knob assembly.

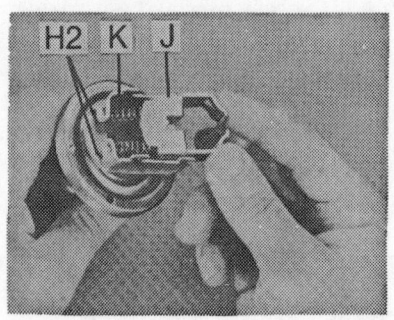

Fig. 10-7.

9. To facilitate reassembly (Fig. 10-9), remove the outside knob (M) from the lock frame by rotating the knob ¾ of a turn while pulling out.

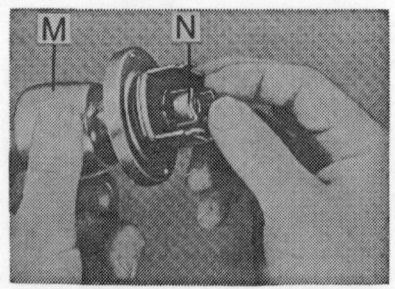

Fig. 10-8.

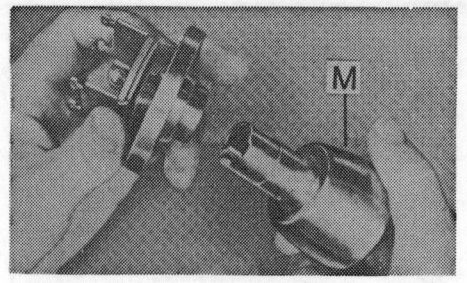

Fig. 10-9.

TYPES OF KEYWAY UNITS

Schlage wafer keyway units are made in two distinct types, type 1 and type 2. To distinguish between these two types, look first at the master wafer column. In type 1 (Fig. 10-10) the elongated slot will be at the top. Type 2 keyway has the elongated slot of the master wafer column below the V-groove.

TYPES OF KEYS

As there are different types of keyways, there are also different types of keys (Fig. 10-11). These can be recognized by

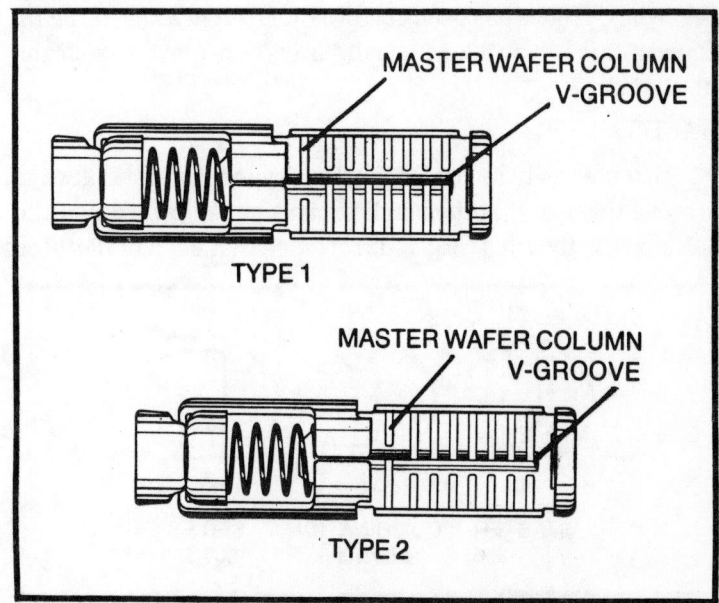

Fig. 10-10. The two distinct types of Schlage wafer keyway units.

95

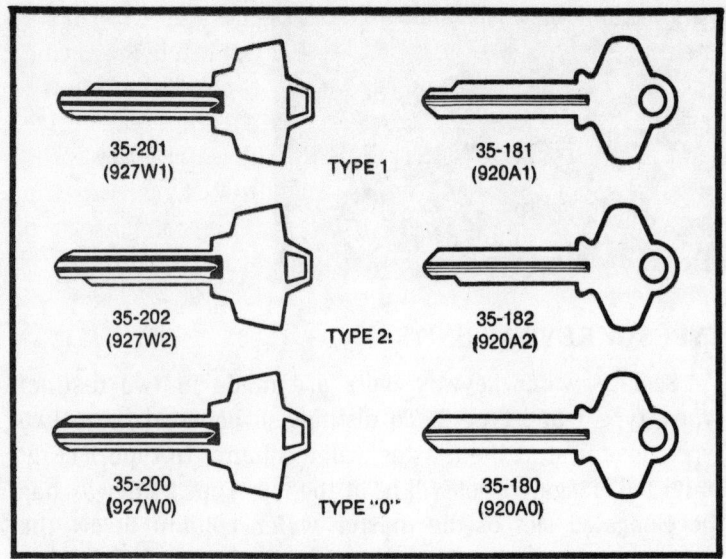

Fig. 10-11. The three different types of keys for wafer keyway units.

looking at the tip sideways to see which portion above or below the V-groove has been cut away. If the portion above the V-groove has been cut away, this is a type 1 key. If the portion below the V-groove has been cut away it is a type 2 key. If the tip is uncut it is a type 0 key, usually used as a master or grand masterkey.

WAFERS

To more easily recognize the three types of wafers, always arrange them so that the small protrusion is upward and the opening is to the right (Fig. 10-12). Notice that each of the three

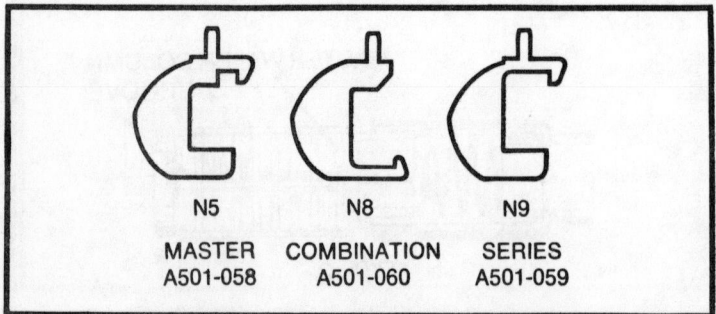

Fig. 10-12. Silhouettes of the three kinds of wafers.

types has a definite silhouette. The master wafer (N5) has a notch cut out at the base of the protrusion just inside the spring seat. The combination wafer (N8) has a protrusion on the rounded shoulder opposite the spring seat location. The series wafer (N9) has a protrusion at the top of the wafer close to the spring seat but does not have the small notch, as does the master wafer (N5). Each of these wafers performs in a different manner, and it is most important to recognize each type before it is inserted in the keyway unit.

OPERATION OF THE KEYWAY UNIT

Figure 10-13 shows two wafer keyway units set to the same combination. A is in the relaxed position (with the key out of the keyway). B has the proper key inserted. Notice in A there are four protrusions of the wafers, one at the bottom and three at the top. The first protrusion to the right of the plunger spring (N3) is the master wafer (N5). This remains out except when it is retracted by the uncut portion of the tip of the key. The cut portion of the tip is necessary to allow full insertion of the key into the keyway. The three protrusions (N9) at the top of the

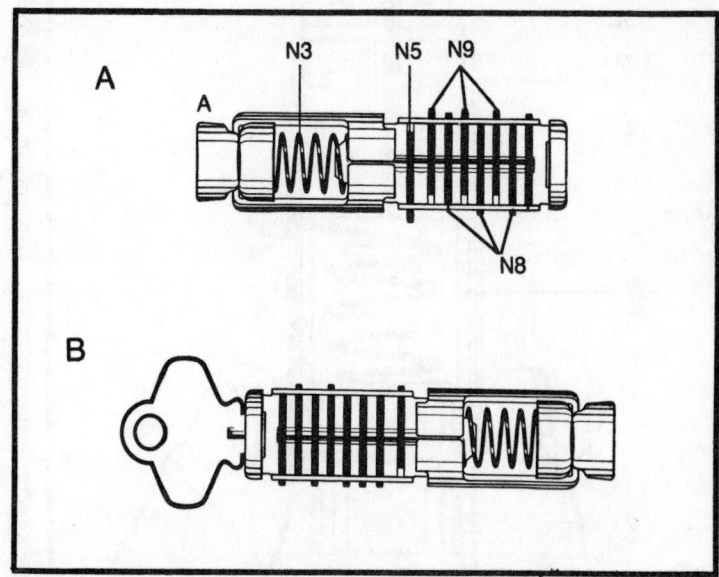

Fig. 10-13. Two wafer keyway units set to the same combination, one (A) in the relaxed position, the other (B) with the proper key inserted.

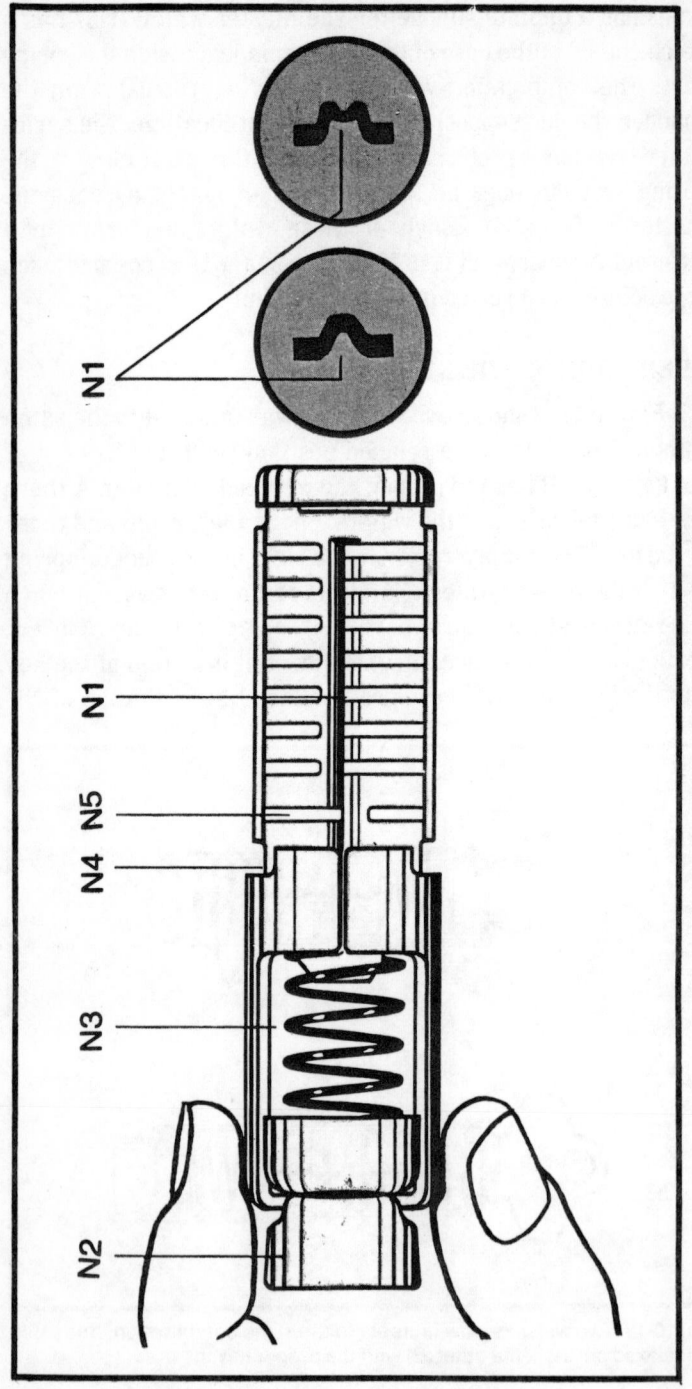

Fig 10-14. A wafer keyway unit (a view of the tumbler side).

keyway are the series wafers. When the key is inserted, the uncut portion opposite the protrusion acts upon the series wafers to pull them into the keyway.

The remaining four wafers (N8) in the keyway are combination wafers. Both in the relaxed position (A) and with the key inserted (B) these wafers lie within the confines of the keyway unit. Therefore, cuts on the key adjacent to their protrusion are required to prevent them from being pushed out into the locking position. An improper key will fail to draw back all the protrusions of the master and series wafers and will extend some or all of the protrusions on the combination wafers.

DESCRIPTION OF THE WAFER KEYWAY UNIT

In working with the wafer keyway unit (Fig. 10-14), hold it in your left hand with the V-grooved dividing strip (N1) facing you. In this position the spring comb is on the underside and cannot be seen. The protrusion at the extreme left is the keyway cam (N2). To the right is the plunger spring (N3). Next to this is the keyway frame (N4), which includes the entire steel area from the plunger spring to the finished cap of the keyway unit. In this steel framework are the wafers which are activated by the insertion of the key.

The first column (N5) to the right of the plunger spring is the location of the master wafer. Notice that the slots in this column have a different proportion than do the other columns in the keyway frame. In the other columns the top slot is shorter than the bottom slot in this particular unit. In some units the relationship of these slots is reversed, the long on top and the short on bottom.

On the reverse side of the keyway unit (Fig. 10-15), notice that there is a metal spring rack (N6) which looks like a comb and upon which are seated the wafer springs (N7).

KEYWAY CODING

As explained before, the first column to the right of the plunger spring is the master wafer column. Its proportions give a clue as to the type of keyway with which we are dealing.

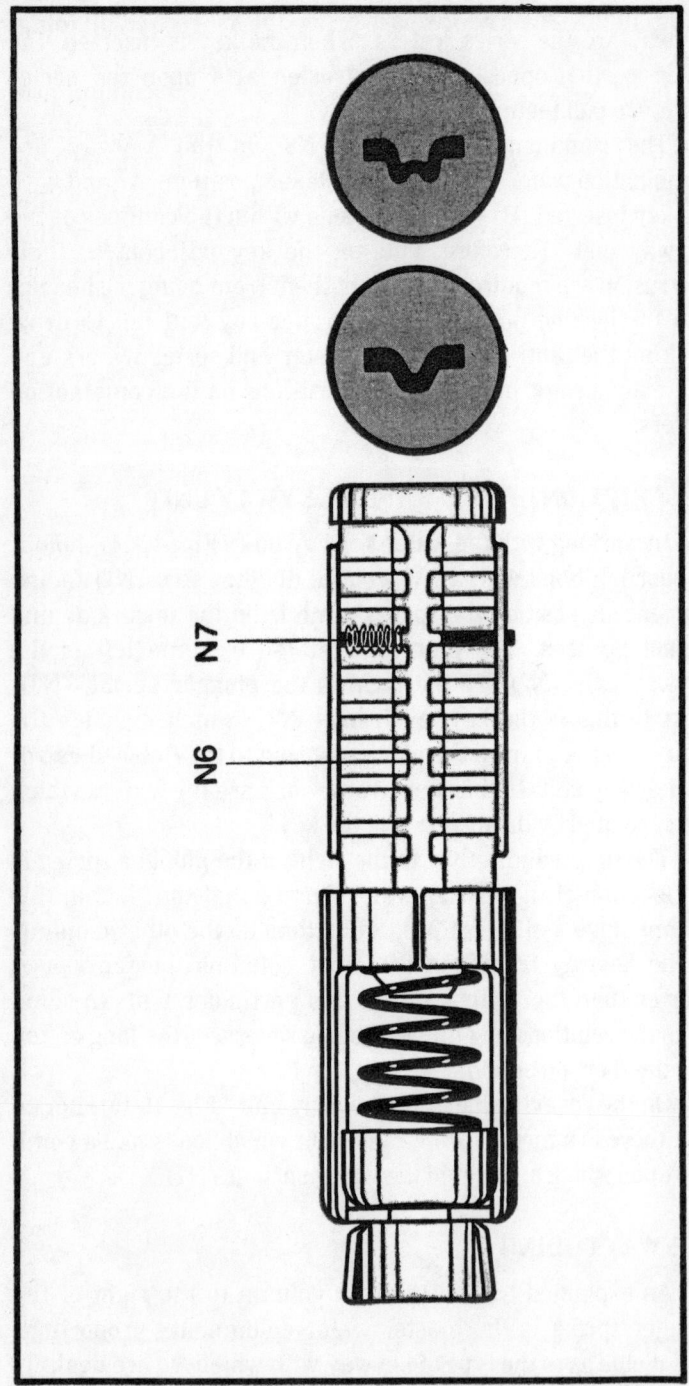

Fig. 10-15. A wafer keyway unit (the reverse side).

In Fig. 10-16A, the longer slot is below the V-grooved dividing strip, indicating a type 2 keyway.

The 14 slots to the right of the master wafer column are assigned code numbers corresponding to the placement of the combination wafers (N8). The first slot to the right of the master wafer column above the dividing strip is given the designation 1. The slot directly below this is given the designation 2. The code numbers alternate between odd and even, continuing to the right of the keyway. All the odd numbers are on top—1, 3, 5, 7, 9, 1′, 3′ (1′ and 3′ are read as 1 prime and 3 prime). All the even numbers are below the dividing strip—2, 4, 6, 8, 0, 2′, 4′.

All factory cut keys have a combination number stamped on the bow which indicates the notching on the key. In Fig. 10-16B the key has the number 203823. The first digit indicates a type 2 key with its tip cut away. The second digit, in this case 0, indicates a stock key not related to any masterkeyed system. The last four digits indicate the location of the notches

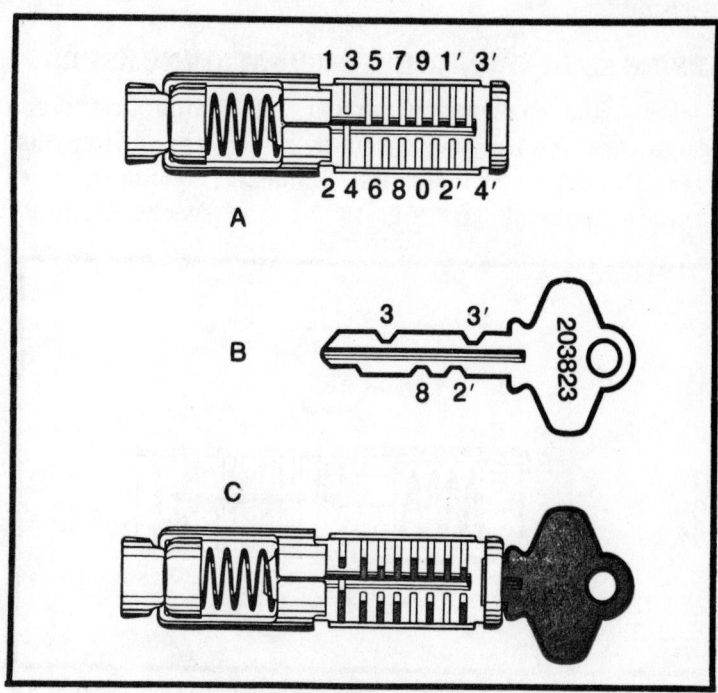

Fig. 10-16. The keyway coding system.

cut in the key. These same four digits (3823) also indicate the position of the combination wafers in the keyway since these wafers must rest within the cutaway portions of the key.

If a key is not stamped with a factory combination number, take an empty wafer keyway and insert the questionable key to determine its combination number. In Fig. 10-16C, the shank of the key may be seen through the slots in the keyway except at those code locations where the key has been notched. Looking first at the master wafer column, the slot below the dividing strip is unobstructed, indicating the tip of the key has been cut away at the bottom. This key is therefore a type 2; the first digit of the combination number would be 2. The second digit of the code number is not related to any cuts on the key and is always 0 for stock (nonmasterkeyed) keys. Next, look at the 14 code slots in the remaining 7 columns to locate the 4 cutaway portions of the key. These cuts occur at code positions 3, 8, 2', 3'. The complete combination number which would be stamped on this key is 203823.

KEYING KEYWAY UNITS BY COMBINATION NUMBERS

Using the combination number 101450, the first digit designates a type 1 keyway. Select such a keyway and into this insert a master wafer in the first, or master, column with the protrusion pointing up (Fig. 10-17). The second digit, 0,

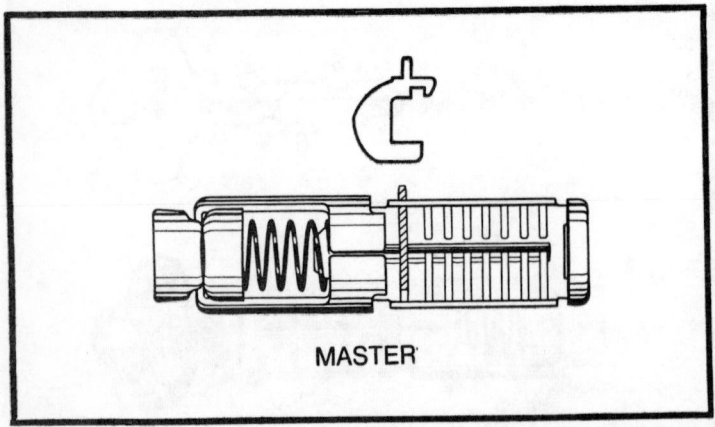

Fig. 10-17. A keyway unit showing the position of the master wafer in the master column.

indicates a stock keyway unit which should be set up in accordance with the following procedure.

Taking four combination wafers, insert them in the positions indicated by the last four digits of the combination (Fig. 10-18). The combination wafer is unique in that it may be inserted with the protrusion pointing either up or down in the seven combination columns. The code number designates in which of the 14 slots the protrusion should be inserted. For example, 1 should be inserted pointing upward in the first column after the master wafer column; 4 should point downward in the second column; 5 should be pointing upward in the third combination column; 0 should be pointing downward in the fifth combination column. After the four combination wafers have been positioned, the remaining three empty columns should be filled with the series wafers (Fig. 10-19). Note that the protrusion of the series wafers can be inserted only in the longer slots of the empty columns. The protrusion of all the series wafers, therefore, should point in the same direction within any one keyway. When springs have been properly attached to all wafers, the keyway unit is then ready to be operated by a key cut to combination 101450.

KEYING ALIKE KEYWAY UNITS

Frequently it becomes necessary to alter the combinations of one or more stock keyway units to exactly match those of

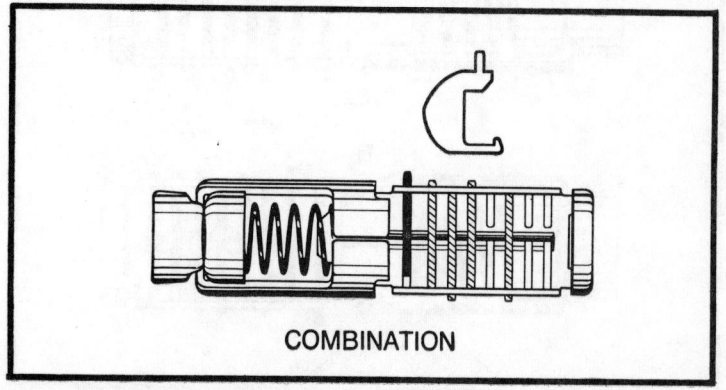

COMBINATION

Fig. 10-18. A keyway unit showing the positions of the combination wafers for the combination 101450.

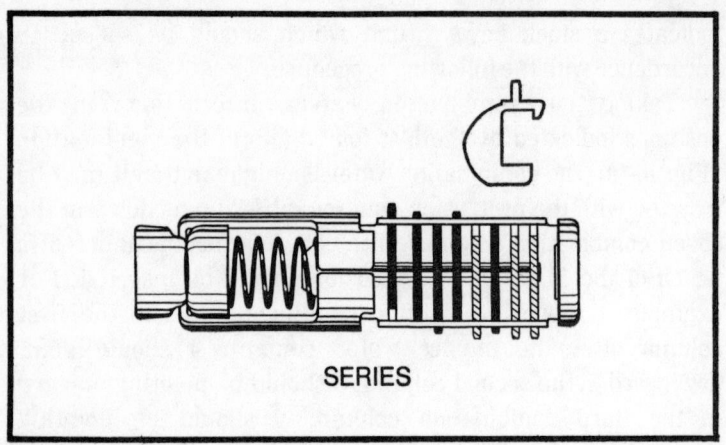

Fig. 10-19. A keyway unit showing the positions of the series wafers for the combination 101450.

another. The procedure used to accomplish this is called "keying alike" and should not be confused with "masterkeying."

One of the simplest methods of keying alike a group of stock wafer keyway units of the same type, either type 1 or type 2, is to put one aside as a control, empty the series and combination wafers from the others, then "set up" these units

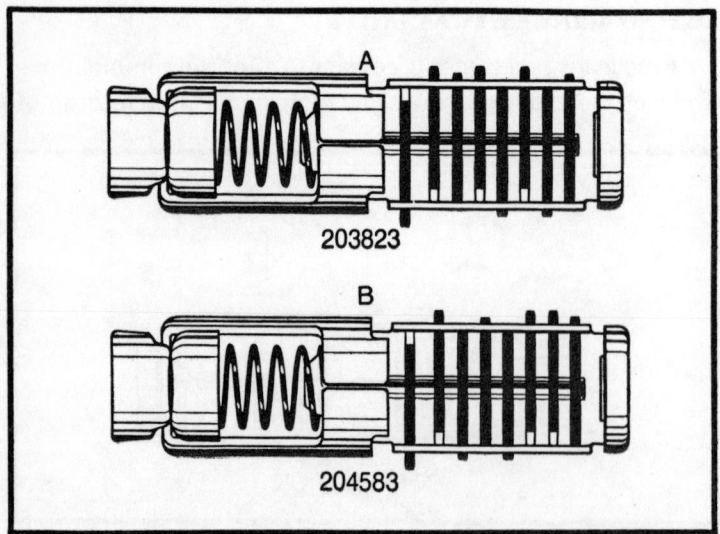

Fig. 10-20.Two keyway units of the same type but with different combinations.

to the code combination of the control keyway, using the procedure explained in the preceding section. An alternate method involving fewer operations consists of rearranging only those series and combination wafers in the random keyways which differ in position from those located in the control keyway.

As an illustration of the alternative method, let's key alike two keyway units (Fig. 10-20) of the same type (type 2). We'll use the keyway coded 203823 as the control (A). Of course, we should examine the columns one at a time, moving from the first to the seventh:

First Column No change necessary.

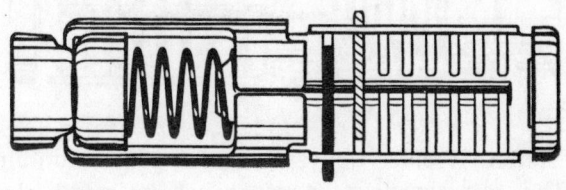

Second Column Invert the combination wafer so the protrusion extends through the code 3 slot.

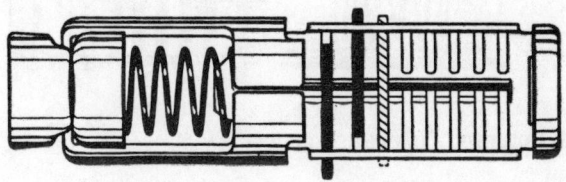

Third Column Replace the combination wafer with the series wafer.

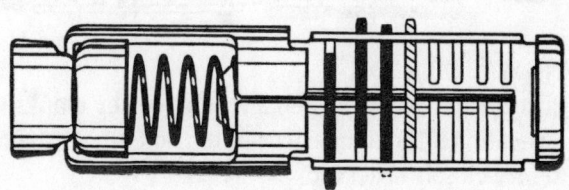

Fourth Column   No change necessary.

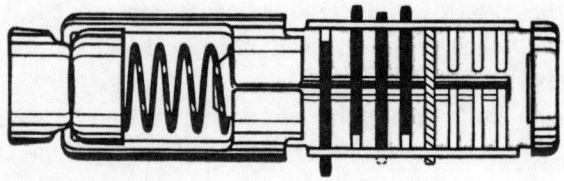

Fifth Column   No change necessary.

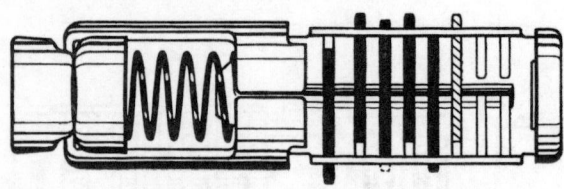

Sixth Column   Replace the series wafer with a combination wafer. The protrusion must extend through the code 2′ slot.

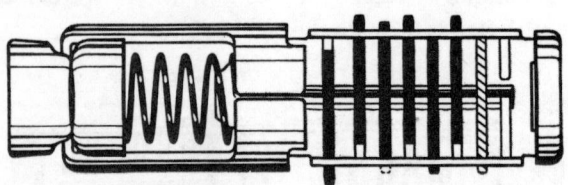

Seventh Column   No change necessary.

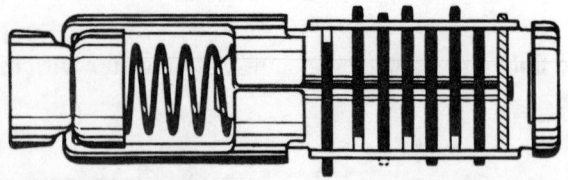

CUTTING WAFER KEYS

Keys accurately cut from genuine schlage key blanks insure smooth operation if the dimensions of the notching. as shown in Fig. 10-21. are closely observed.

The first operation normally performed on a key blank is to cut away a portion of the tip to correspond to the type of keyway unit with which it is to be used. Key blanks may be purchased with this notch already cut by specifying blanks for type 1 or type 2 keyway units. All other cuts on the key are made to the same depth (0.060 in.) and have the same width (0.060 in.) at the bottom of the notch. All the angles in the cuts should be a minimum of 90°.

It is best to use either a factory-cut key blank or a full-cut pattern key available from the factory. With full-cut pattern keys it is necessary to select only those notches corresponding to the specific combination numbers to be cut. After the keys are cut, dress them lightly with a file to remove sharp edges and check the keys in the keyway unit to make sure they operate properly. All the protrusions on the wafers should be flush with the keyway when the proper key is inserted.

ASSEMBLING THE KEYWAY UNIT

As shown in Fig. 10-22A, always hold the keyway unit in your left hand with the keyway cam (N2) to the left and the V-grooved dividing strip facing you. Then as you insert the wafers (B), you will find the series wafers will only go in the elongated slot of the column. They should not be forced

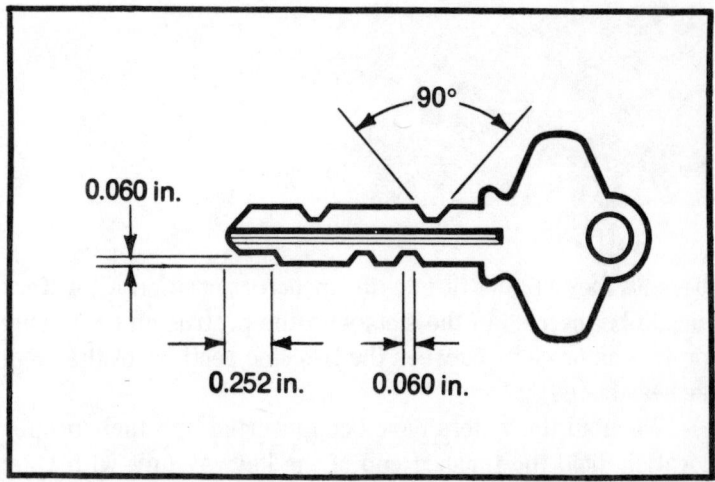

Fig. 10-21. Standard dimensions of the wafer key.

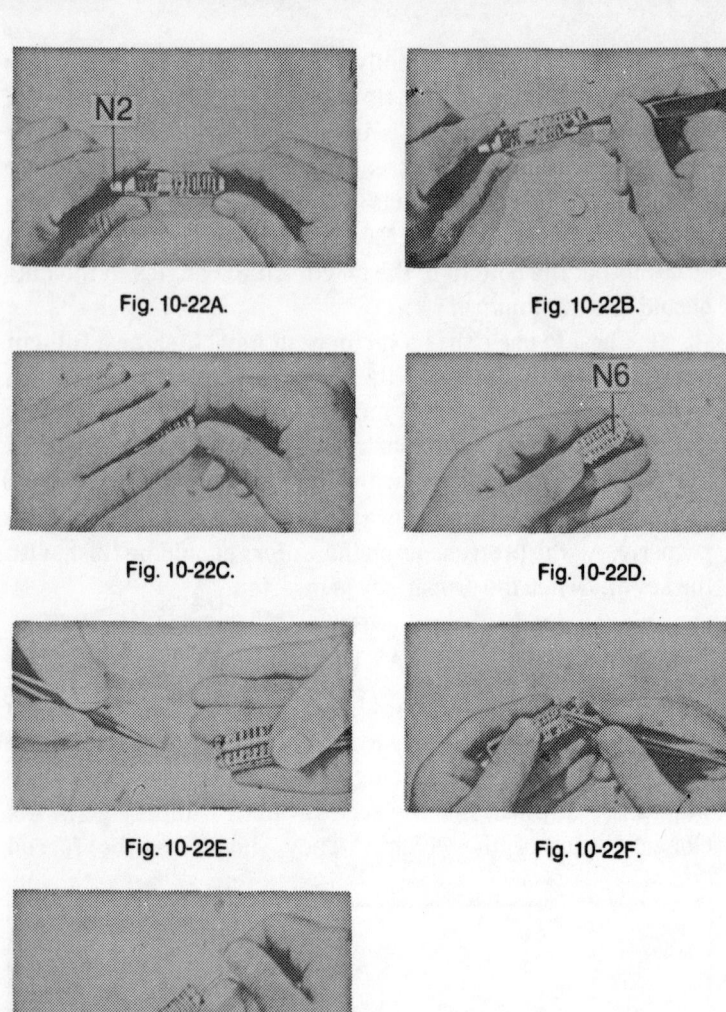

Fig. 10-22A.

Fig. 10-22B.

Fig. 10-22C.

Fig. 10-22D.

Fig. 10-22E.

Fig. 10-22F.

Fig. 10-22G.

because they will not fit into the incorrect position. All wafers should be inserted in the slots with the protrusion first. This protrusion projects between the two side sections of the steel keyway frame.

When all the wafers have been inserted into their proper location, hold the finished cap of the keyway unit with your right hand (C) and move your left hand so that your first two fingers cover the wafers, holding them in position as you rotate the keyway unit (D) to expose the spring rack (N6). If the wafers are not held in position as you rotate the keyway, they may drop out and cause you to rework the setup.

After the keyway unit has been rotated, hold it tightly between your fingers and thumb. Exert force downward against the wafers with the index finger to open the distance between the spring rack in the center of the keyway and the spring seat on the wafers to provide space for the insertion of the wafer springs.

The easiest way to insert a wafer spring is to use a pair of fine needlenose tweezers (E) to grasp the spring at the second coil back from one end. Holding it in this position, you can guide the free end of the spring over the spring rack seat and use the needlenose tweezers to guide the other end over the spring seat on the wafer (F).

Take the key (G) and run it in and out of the keyway several times as it is important that all springs are fully seated before the keyway unit is reassembled into the lock.

MASTERKEYING

Frequently it becomes necessary to provide a group of locks with keyway units, each having a different and noninterchangeable key, but all having in common one masterkey capable of operating each of the locks in the group.

Combining locks in this manner is called "masterkeying" and is quite easily accomplished with Schlage wafer keyway units. To illustrate the principle of masterkeying, consider two stock keyway units having combination numbers 203823 and 204823. These units, like all other stock keyway units, contain one master wafer, four combination wafers, and three series wafers. Their code numbers indicate that these units are identical except for the position of the combination wafer in the second column. This difference is sufficient to prevent interchangeability of their keys, but a third key could be cut which would operate both these keyway units by providing it with notches for both the code 3 and code 4 tumbler positions. In other words, their masterkey would be cut with five notches corresponding to code positions 3, 4, 8, 2, and 3.

This principle of masterkeying can be expanded to include many more combinations. However, as the number increases, it will become necessary to eliminate one or two series wafers from the keyway units. Series wafers cannot be used in any column for which the key has been notched since it is the unnotched portion of the key which retracts the series wafer. At no time should a keyway unit be set up without any series wafers because this decreases the security of the masterkey system.

One method of laying out a masterkeyed system involves five basic steps:

1. Select the required number of keyway units, all of which should be either type 1 or type 2.
2. Of the seven columns available, arbitrarily select either one, two, or three columns which will be reserved in all keyway units for the placement of series wafers. Sixteen masterkeyed units are available with three columns reserved for series wafers. Eighty masterkeyed units are available when two columns are reserved for series wafers. Two hundred and forty masterkeyed units are available when one column is reserved for a series wafer.
3. List the code numbers of all the slots in the remaining columns. These slots are available for the placement of the four combination wafers. These code numbers also represent the cuts in the masterkey which will operate all the keyway units having combinations determined by the procedure in step 4.
4. Using the digits obtained in step 3, tabulate a list of four digit combination numbers. Although each column contains two code slots, it can accommodate only one wafer. Therefore, be careful to select not more than one digit from any available column.
5. Using the code numbers determined above, key the keyway units by the combination numbers, placing series wafers only in those columns reserved in step 2.

There is an alternate method of masterkeying which is particularly useful when there is at hand a number of assembled stock keyway units or a supply of factory-numbered stock keys. The following basic steps illustrate this method:

1. Select approximately one-third more keys or keyway units (all of one type) than your anticipated needs and list their combination code numbers as shown in Fig. 10-23.

2. In this method, only one series wafer is used. So in order to determine which column should be reserved for its placement to best use the maximum number of listed keys or keyway units, represent the keyway unit by drawing a rectangle and indicate the 14 slots by writing in their appropriate code numbers (Fig. 10-24).

3. Above and below the rectangle (opposite each slot code number) write the figure which indicates the frequency of occurrence of the particular code number in the list of combinations (Fig. 10-24).

4. Add the two figures in each column (outside the rectangle) obtained in step 3 and determine the column having the lowest sum. This column will be the best location of the placement of a series wafer in all keyways selected to be masterkeyed. In Fig. 10-24 the

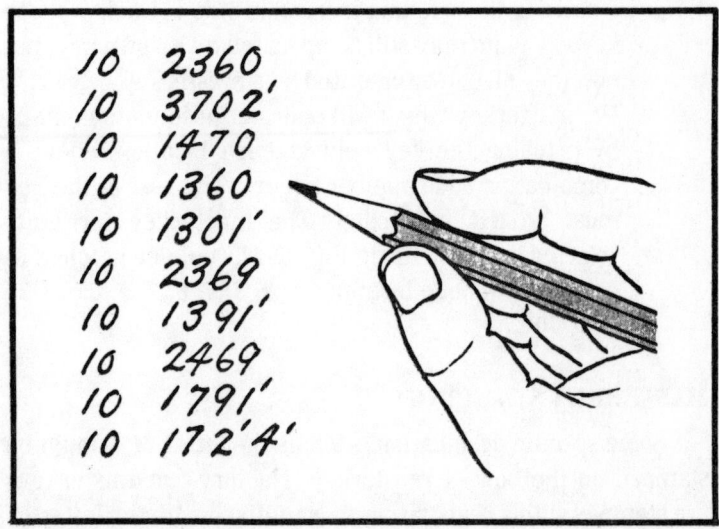

Fig. 10-23. A list of combination numbers for keys or keyway units.

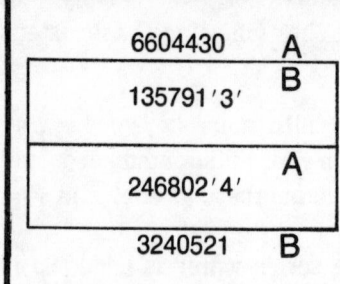

Fig. 10-24. A representation of the keyway unit, showing the 14 code slots (A) and the frequency of occurrence (B) of each code number in the list of combinations (Fig. 10-23).

seventh column with a sum of one is the most desirable location for the series wafer.

5. Review the list of combination code numbers and eliminate as impractical all combinations which contain digits corresponding to code slots which are located in the column selected in step 4. In the illustration, combination 101724 would be eliminated since it contains the digit 4′, a code slot in the seventh column.

6. The required quantity of keyway units to be masterkeyed should be selected from the remaining group of combination code numbers. Remove from each of these selected units the two series wafers not located in the column reserved in step 4. Each of these keyway units may still be operated by its own key, but now they all can be operated by a masterkey.

7. The masterkey which will operate these units is made by notching the key only at each location where a combination code number occurs. The rest of the key must be left unnotched. The masterkey for units selected from the list in Fig. 10-22 would be notched at combination code locations 1, 2, 3, 4, 6, 7, 9, 0, 1′, 2′ (Fig. 10-25).

MASTERKEY STAMPING

Some specific designation, such as the letter M, should be stamped on the bow of masterkey. Factory-cut masterkeys are stamped with a registry number prefixed with the letter R. This number is indexed in the factory records, which contain

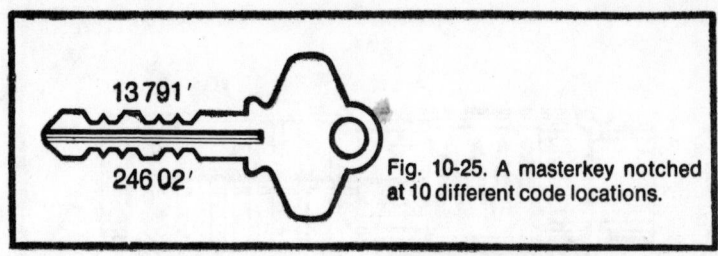

Fig. 10-25. A masterkey notched at 10 different code locations.

detailed information such as the job name, the job location, and the combinations controlled by the masterkey. Factory-cut change keys, which operate the individual keyway units in a masterkeyed system, are stamped with a combination number. This number has the same meaning as that stamped on stock keys. The second digit of the masterkey code number is never 0 as it is on stock keys. It represents the number of the column in the keyway which contains the series wafer. Stock keys converted into masterkeys should be overstamped with the appropriate second digit.

Occasionally it becomes necessary to provide two or more groups of masterkeyed keyway units (each group having its own masterkey) with a key called a grand masterkey which would operate the keyway units in all the groups. One method of laying out a grand masterkey system involves seven basic steps:

1. Select either type 1 or type 2 keyway with which to build the system (Fig. 10-26).
2. Of the seven columns available, arbitrarily select one column which will be reserved in all keyway units for a series wafer (Fig. 10-27).

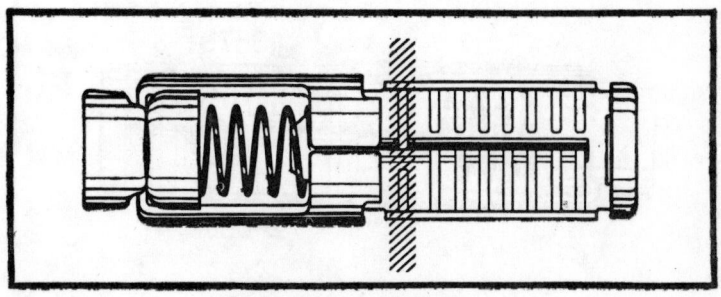

Fig. 10-26. A type 1 keyway.

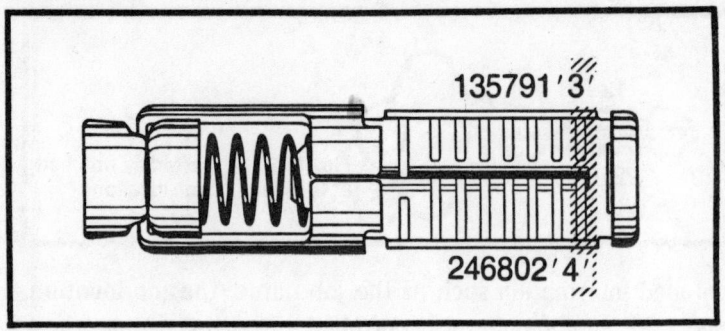

Fig. 10-27. In this keyway the seventh column is reserved for a series wafer.

3. If only two masterkeyed groups are desired under the grand masterkey, arbitrarily select another column which will provide two slots, each of which will be assigned exclusively to a masterkeyed group and will be the means for separating the two groups, thus preventing interchangeability of their keys (Fig. 10-29).

4. List the code numbers of the slots in the remaining five columns.

5. Tabulate a list of four digit combination numbers, each of which must contain one of the unique slot code numbers selected in step 3. The remaining three digits may be drawn from the available code numbers obtained in step 4, but remember to select not more than one digit from any one available column. This list of combination numbers represents one masterkeyed group (Fig. 10-29).

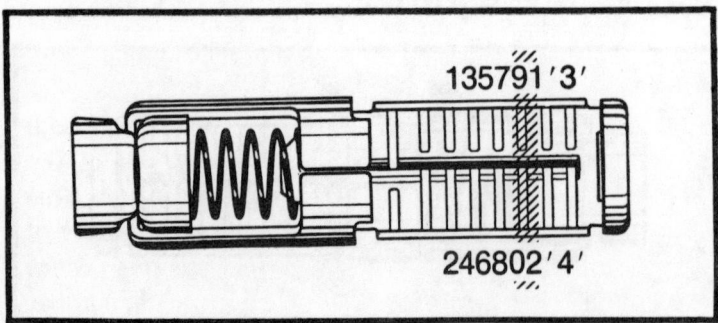

Fig. 10-28. One slot in this column is reserved for each masterkeyed group.

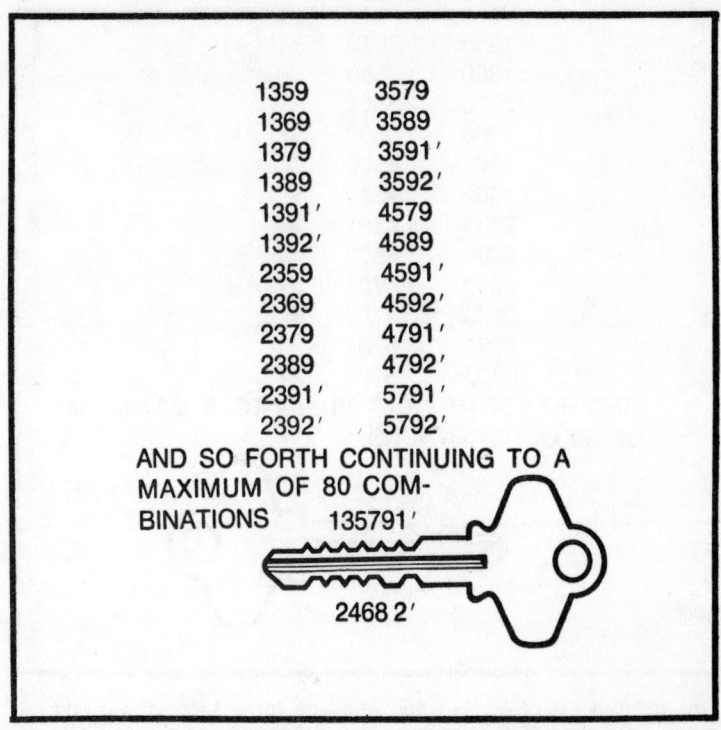

1359	3579
1369	3589
1379	3591'
1389	3592'
1391'	4579
1392'	4589
2359	4591'
2369	4592'
2379	4791'
2389	4792'
2391'	5791'
2392'	5792'

AND SO FORTH CONTINUING TO A MAXIMUM OF 80 COMBINATIONS   135791'

2468 2'

Fig. 10-29. A list of combination numbers for a masterkeyed group. Each number contains the unique slot code number 9.

6. Repeat step 5. However, substitute the other unique slot code number for the one previously used. This list of combination numbers represents the second masterkeyed group. Though these two groups may appear similar, their masterkeys are not interchangeable since one will not have the notch which is necessary to operate that combination wafer which is used exclusively in the other group (Fig. 10-30).

7. The grand masterkey has all the notches found on both masterkeys and will, therefore, operate all of the keyway units in both groups (Fig. 10-31). This illustration represents only one of a large variety of possible grand masterkeying arrangements. More complex systems or those involving a greater number of masterkeys should be referred to the factory.

1350	3570
1360	3580
1370	3501'
1380	3502'
1301'	4570
1302'	4580
2350	4501'
2360	4502'
2370	4701'
2380	4702'
2301'	5701'
2302'	5702'

AND SO FORTH, CONTINUING TO A MAXIMUM
OF 80 COMBINATIONS

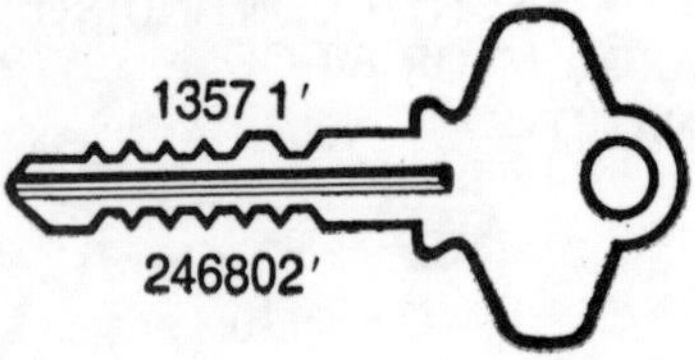

1357 1'

246802'

Fig. 10-30. A list of combination numbers for a second masterkeyed group. Each number contains the unique slot code number 0.

EXTENDED MASTERKEYED AND GRAND MASTERKEYED SYSTEMS

Masterkeyed or grand masterkeyed systems may be doubled simply by using both types of keyways (type 1 and type 2) and providing a key capable of operating both. In such an extended system, all keyways must be provided with a

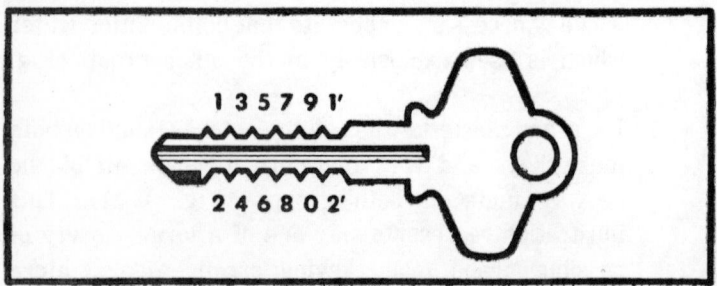

Fig. 10-31. A grand masterkey that fits all the keyway units in two masterkeyed groups (Figs. 10-29 and 10-30).

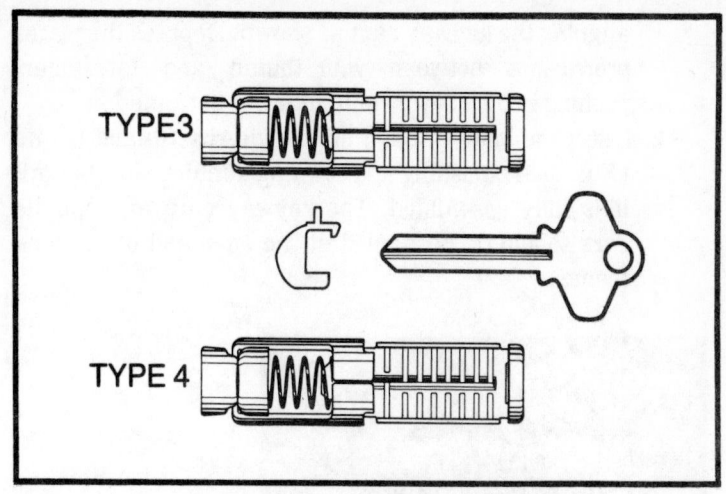

Fig. 10-32. Two keyway units (type 3 and type 4).

series wafer in the master column. This does not, in any way, impair the operation of any existing keys. However, it does make possible the use of an additional key (type 0) capable of acting as a master, grand master, or great grand masterkey. When a type 1 keyway unit is set up with a series wafer replacing the master wafer, both the keyway unit and its key are redesignated as type 3. Similarly, a type 2 keyway unit and its key are redesignated as type 4 (Fig. 10-32).

REASSEMBLING THE LOCK MECHANISM

1. Refer to Fig. 10-33. Grasp the wafer keyway unit (N) in your right hand. In your left hand, hold the outside knob (M) with the slot (M1) facing you. Insert the wafer keyway unit (N) into the outside knob spindle

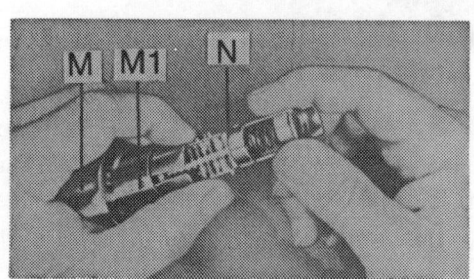

Fig. 10-33.

aligning the keyway cam as shown. Depress the wafer protrusions between your thumb and forefinger, pushing in the keyway unit until it is bottomed.

2. Insert the knob through he outside rosette and frame (Fig. 10-34), pushing and rotating simultaneously until it is fully assembled. The keyway cam and spindle ears should be positioned at the open end of the lock frame.

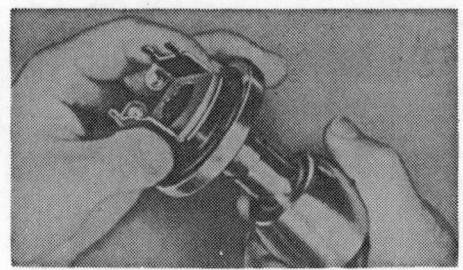

Fig. 10-34.

3. Refer to Fig. 10-35. Place the springs (K) into the slide (J) and assemble both into the frame, positioning the free ends of the springs on the ears (H2) of the lock frame. Be sure the bridge of the slide is facing you, as shown.

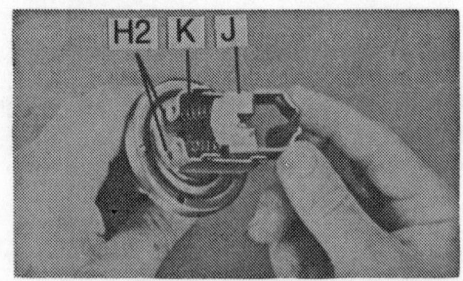

Fig. 10-35.

4. Refer to Fig 10-36. Replace the thrust plate (I) with its curved side over the springs (K) by inserting the bottom side between the bottom tabs of the lock frame and pressing the snap into position between the top tabs (H1).

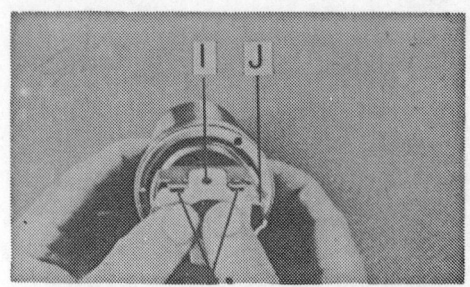

Fig. 10-36.

5. With your thumb depress the slide all the way back against the spring tension (Fig. 10-37). Insert the empty spindle (E) as shown. Release the slide so the spindle will lock into position.

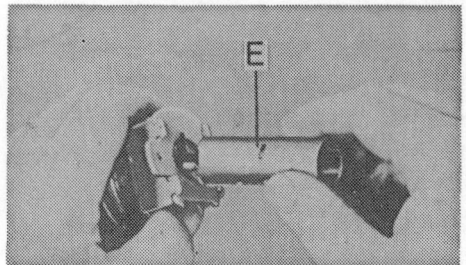

Fig. 10-37.

6. Refer to Figs. 10-38 and 10-39. Slip the plunger (L) inside the empty spindle, making sure that the tab (L1) on the turn button aligns with the locking slot (E1). Again depress the slide and work the plunger in all the way. Release the slide so the plunger will lock into position.

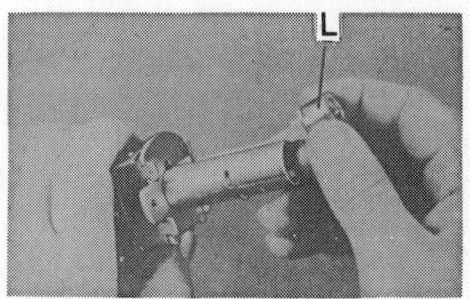

Fig. 10-38.

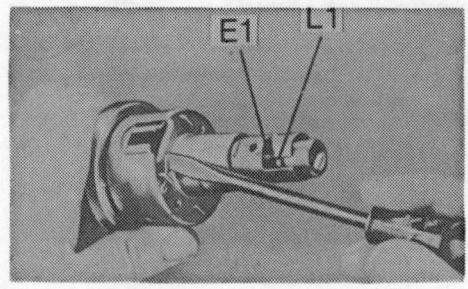

Fig. 10-39.

7. Replace the housing over the lock assembly with its opening in line with the jaws of the slide, and replace the cotter pin. The lock is now ready for installation in the door.

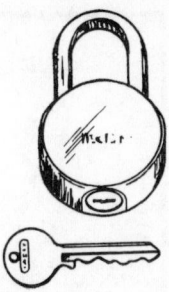

Chapter 11
Pin-Tumbler Cylinder
Locks and Locksets

Pin-tumbler cylinders are basically very simple mechanisms but can be built to give a high level of security.

CONSTRUCTION

The cylinder is a self-contained mechanism, intended to be adapted to a variety of locksets. Figure 11-1 shows a pin-tumbler cylinder in cross section.

Plug

The plug contains a keyway, usually of the paracentric, or off-center, type. It is drilled for four or, in some cases, as many as seven tumbler chambers. These chambers are made with the greatest precision. Chambers are spaced evenly along the upper surface of the plug and are aligned as closely as modern production techniques allow. The cylinder shell, sometimes called the hull, carries a matching set of chambers.

The plug may be machined with a shoulder at its forward surface; this shoulder mates with a recess in the cylinder and provides:

- A reference point for regulating the alignment of the pin chambers in the shell and the plug.

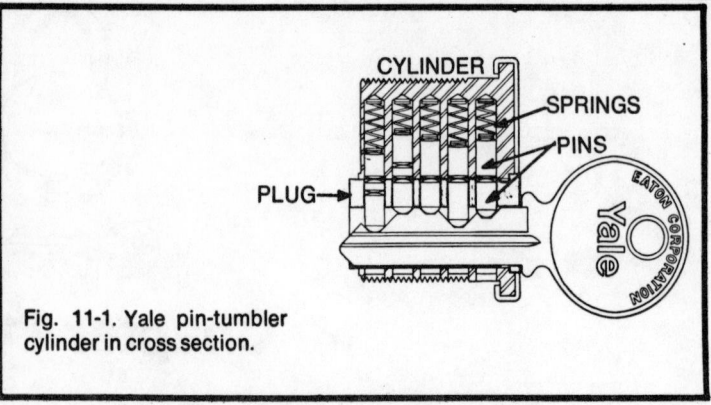

Fig. 11-1. Yale pin-tumbler
cylinder in cross section.

- A safeguard to prevent the plug from being driven
  through the cylinder, either deliberately or through
  resistance developed as the key enters the plug.
- A safeguard to discourage a thief from shimming the
  pins. Without this shoulder it would be possible to force
  the pins out of engagement with a strip of spring steel.

The plug is retained at the rear by a cam and screws, a
retainer ring, or a driver pin that locks the plug into the
cylinder. The retainer pin is mounted securely in the plug and
fits into a groove milled on the inside diameter of the cylinder
shell. This groove determines the degree of plug rotation. The
groove can be milled to stop plug travel when the bolt is
thrown; the key can be withdrawn only if the bolt is in the
locked position. Without this travel-limiting feature, the
cylinder can be rotated to release the key in the unlocked
position.

Pins

Pins are supplied in a variety of sizes and shapes, e.g.,
truncated, truncated with modified cone, hemispherical,
capped, or any of a number of other designs. Figure 11-2
illustrates some of the more popular configurations. The
drawings have been exaggerated for clarity.

The shape of the pins can foil would-be lock pickers. A
standard straight-sided cylindrical pin can easily be raised to
the shear line while the plug is turned slightly. A pin with a
broken profile may tend to "hang up" before it passes the

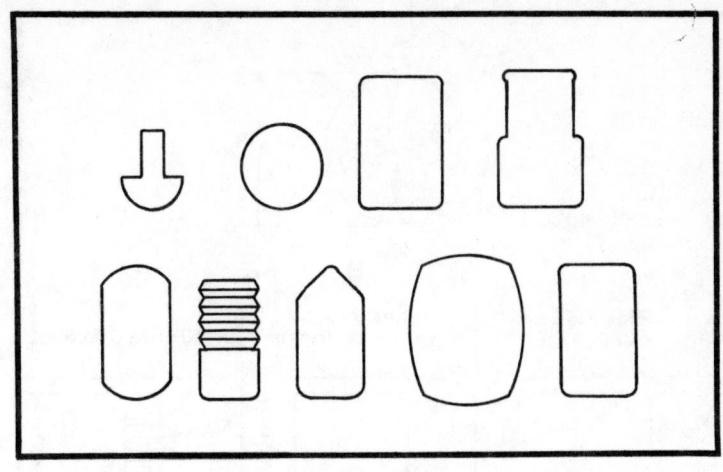

Fig. 11-2. Various pin tumblers.

shear line. In this case, only the proper key would align the pin at the shear line.

Pins also come in varying lengths and diameters, depending upon the size of the pin chamber diameter and whether the lock can be masterkeyed.

Although variations in pin length and diameter seem insignificant, these small differences can be critical. For example, a pin that is a few hundredths of an inch smaller than the chamber diameter can bind in the plug or cylinder shell. In extreme cases, the pin may jam so solidly that the cylinder will have to be broken in order to open the lock.

Pins are color-coded according to length. The code differs among manufacturers, but each of the seven pin lengths has its own distinctive color.

Cylinder Bars

The typical rim pin-tumbler cylinder has a flat cylinder bar, or tailpiece, attached at the rear of the cylinder plug (Fig. 11-3A and B). The attachment, or tailpiece, is loose to allow some flexibility in the location of the auxiliary lock on the other side of the door. Alignment, while not absolutely critical, should be as accurate as possible; under no circumstances should the tailpiece be more than ¼ in. off the axis of the plug.

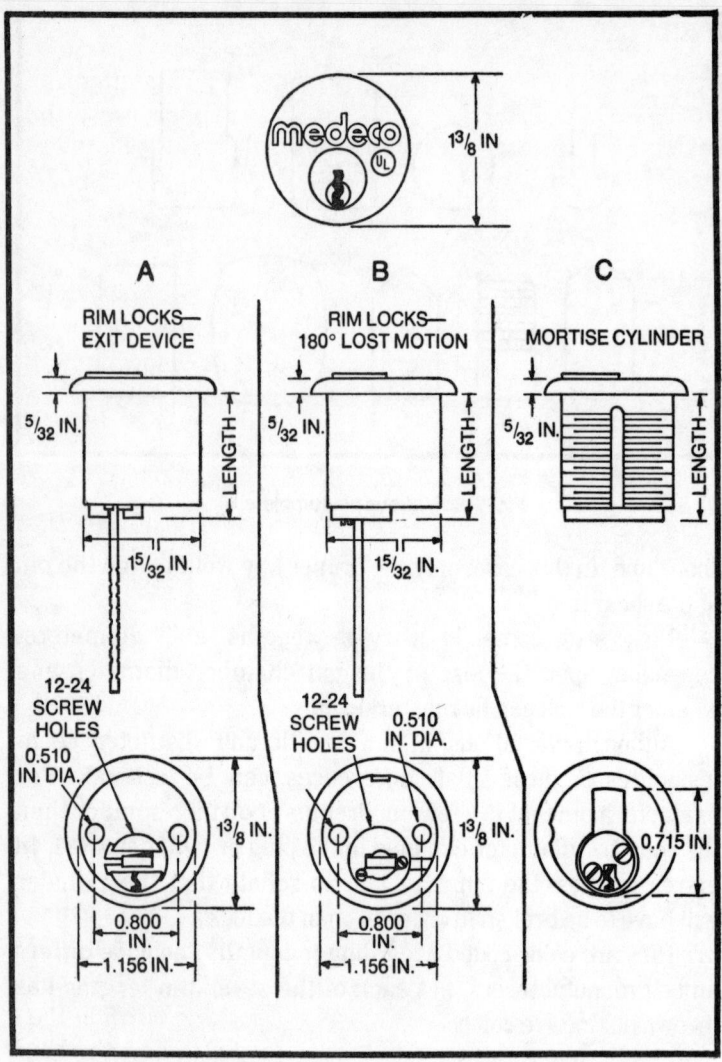

Fig. 11-3. Medeco cylinders. The tailpiece may be serrated (A), solid (B), or replaced by a cam (C).

On the other hand, mortise-lock cylinders for mass-produced doors do not have this tolerance. The best is driven by a cam on the back of the lock (Fig. 11-3C). If the lock is used on office equipment, the cam is a milled relief on the back of the plug, or else it is a yoke-like affair, secured to and turning with the plug. It is important that these locks be aligned with the bolt mechanism.

OPERATION

When the correct key is inserted, the key cuts align the tumblers as shown in Fig. 11-1. If the key is cut improperly, the break between the upper and lower pins will not align with the shear line, and the plug will not turn. Better pin-tumbler locks will require a key cut to a tolerance of ±0.025 in.

DISASSEMBLY

Disassembly of a plug cylinder is simple, requiring only a few basic tools. You will need a screwdriver, a following plug of the correct diameter, and a pin tray. Proceed as follows:

1. Remove the cam or tailpiece.
2. Turn the key in the plug about 30°.
3. Holding the plug so the pins are vertical, slip the appropriate plug follower into the cylinder from behind.
4. Push the plug follower all the way over the plug. Remove the cylinder.
5. Taking the plug in hand, slide the follower past one pin chamber; turn the plug over and allow the pin to drop into your hand or on the work table. Place the pin into the first compartment of the pin tray.
6. Uncover the next plug chamber and place the pin in the adjacent compartment of the tray.
7. This step involves removing the upper pins and springs from the shell. Invert the shell. Slide the plug follower back. The pin and spring will drop out as the follower uncovers the No. 1 chamber. Place these parts in the No. 1 pin try slot above the lower pin. Proceed to the other chamber.

ASSEMBLY

Assembly, or reassembly, of a cylinder from scratch is the reverse of this procedure. You will need pin tweezers to install the upper pins, springs, and lower pins into their chambers. You will need a plug follower to keep the parts together. A plug holder is optional; it holds the plug in an upright position and leaves your hands free while you insert the lower pins.

This procedure assumes that you have a key but do not know the pin combinations. (In normal reassembly, you already know which pins go into which chambers.)

1. Place the plug into the plug holder and insert the key.
2. Taking one lower pin at a time, insert it into the cylinder chamber. If a pin stands above or below the shear line, another must be tried. If the pin looks like it is just about at the shear line or perhaps right on it, turn the plug with the key. If the plug turns, you know that the pin is the correct one for the chamber.
3. If the pin is the wrong one, remove it and select another. Once the proper pin is located, leave it in the chamber and move on to the next one. Repeat this procedure until you have the proper pins for all the chambers.

NOTE
DO NOT REMOVE THE KEY OR TURN THE PLUG OVER.

4. Move to the cylinder shell and insert the plug follower.
5. The plug follower has a notch cut into its end; keep this notch aligned with the chambers. Start at the third chamber from the front.
6. Using yor tweezers, select a spring and lace it into the loading notch of the follower. Rotate the cylinder so that the spring slides into the chamber. Select an upper pin (all upper pins are the same size) and, using the tweezers, place it into the loading notch.
7. Push the pin against the spring and into the pin chamber. At the same time, slide the plug follower forward. This movement holds the pin in No. 3 chamber and uncovers No. 2 chamber.
8. Load No 2 chamber as before and proceed to No. 1 chamber.
9. Once the first three chambers are filled, retract the follower, leaving the last two chambers exposed (if you push the follower too far, the pins in the front chambers will pop out).
10. Load No. 4 chamber and then No.5. Once the chambers are loaded, you are ready to insert the plug into the cylinder.

11. Place the plug against the end of the plug follower with the assembly positioned in front of the cylinder. Rotate the plug to bring the pins about 30° off the vertical; press against the plug follower tool as you push the plug home. The plug forces the tool out the rear of the cylinder. Since the plug and plug follower are butted, there is no opportunity for the upper pins to fall out.

12. Once the plug is home, hold it in place and turn the key. If all is well, the plug will turn smoothly. If it binds, disassemble the lock and check the pin lengths.

13. Attach the cam or tailpiece and retest.

WHAT TO LOOK FOR IN A PIN-TUMBLER CYLINDER LOCKSET

Pin-tumbler cylinder locksets are the most popular type of locking device. Since these locksets are available in a wide variety of types and styles, some discussion of their general characteristics is in order.

Security—The pin-tumbler cylinder lockset provides better than average security by virtue of its design. However, security also depends upon the quality of the lock and upon its application. Metal doors give more protection than wood doors.

Quality—Quality depends upon the intended service of the lockset. Light, medium, and heavy-duty sets are available.

Type—Locksets are identified by their function—lavatory, classroom, residential, and so on.

Visual appeal—The lockset should match the decor of its surroundings. This is particularily important in new constructions.

Hand—The location and direction of swing of the hinges determines the *hand* of the door (Fig. 11-4). Taking the entrance side of the door as the reference point, there are four possible hands: left hand, left hand reverse (the door opens outward), right hand, and right hand reverse. It is important to match the lockset to the door hand. Failure to do so can cause bolt/striker misalignment and may require that the cylinder be rotated 180° so that the weight of the pins is on the

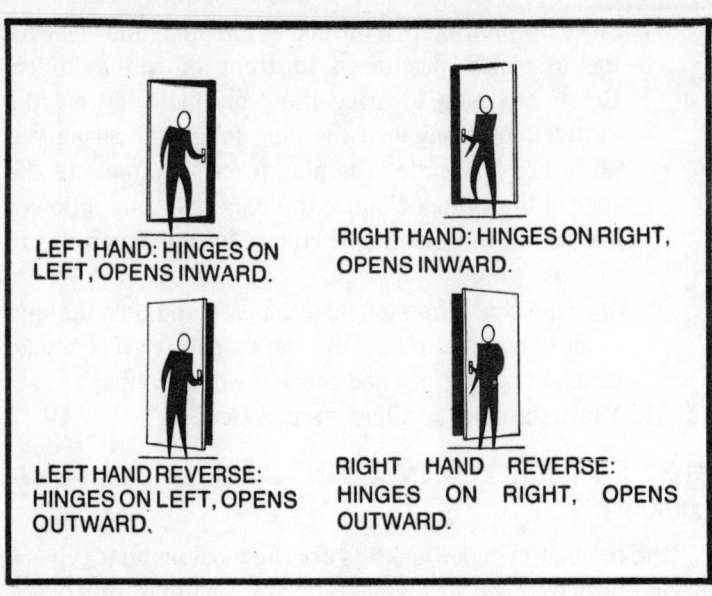

LEFT HAND: HINGES ON LEFT, OPENS INWARD.

RIGHT HAND: HINGES ON RIGHT, OPENS INWARD.

LEFT HAND REVERSE: HINGES ON LEFT, OPENS OUTWARD.

RIGHT HAND REVERSE: HINGES ON RIGHT, OPENS OUTWARD.

Fig. 11-4. The hand of a door is a term that describes the location of the hinges and the direction of swing. (Courtesy Eaton Corp.)

springs. The pins weigh only a few grams, but this is enough to collapse the springs and disable the lock. Unless stipulated in the order, manufacturers supply right-hand mortise locksets. Some of these locksets can be modified in the field; others are fixed.

SPECIFIC LOCKSETS

The information presented in the following pages has been supplied by the various manufacturers. While it does not begin to cover all locksets, it does give an overview of the current state of the art. If you need further information, contact one of the factory representatives.

Corbin

Figure 11-5 is a sectional view of the popular Corbin Model 68 lockset. This lockset gives adequate security for most homes and apartments. Model 323D is intended for medium- and heavy-duty applications and features a positive thumbturn (Fig. 11-6). Model 354 is a superior lockset intended for severe

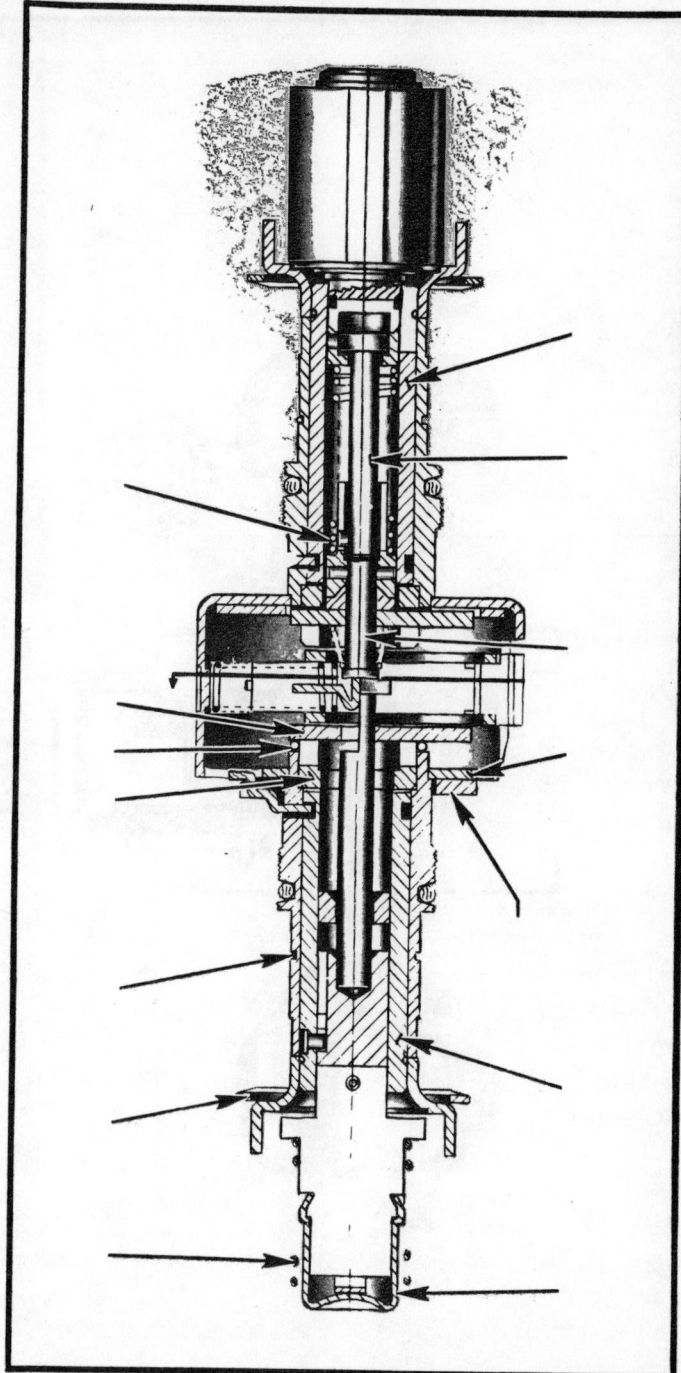

Fig. 11-5. Corbin Model 68 lockset in cutaway.

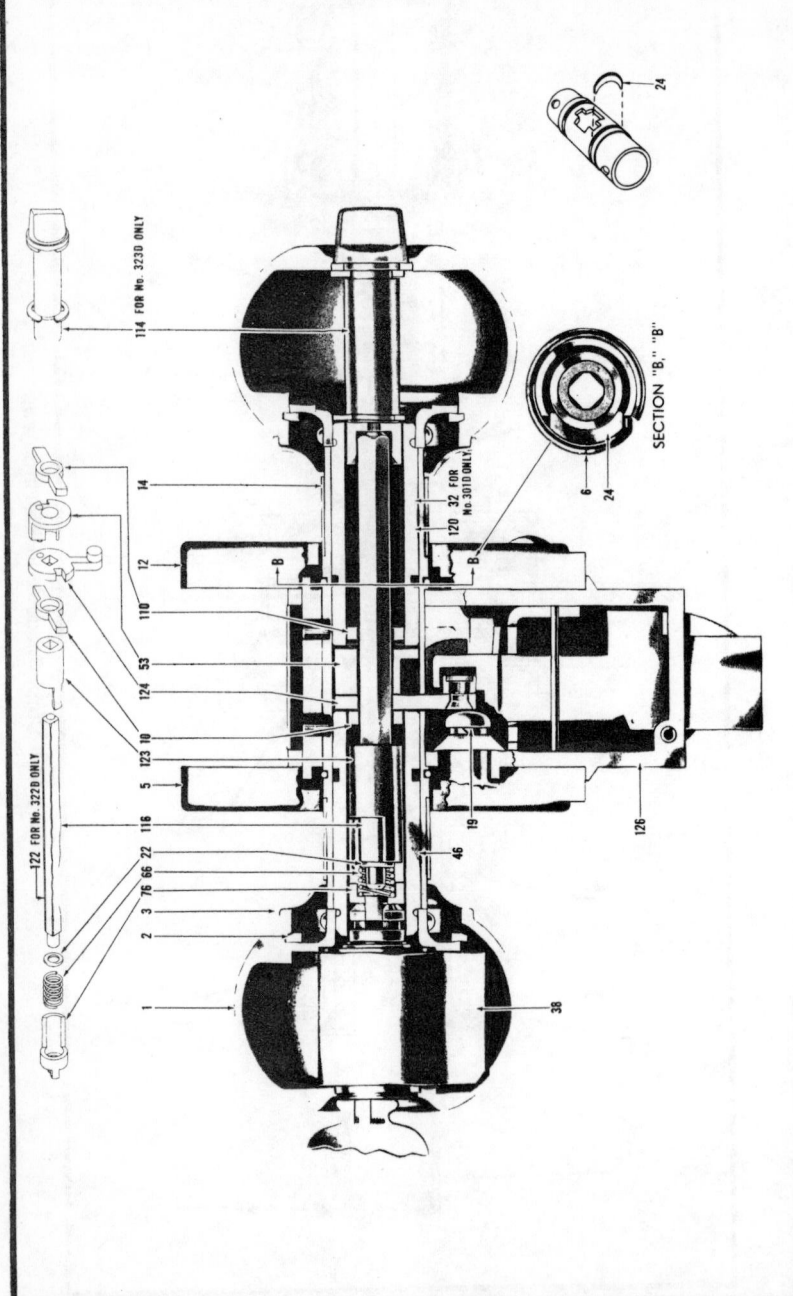

SECTION "B," "B"

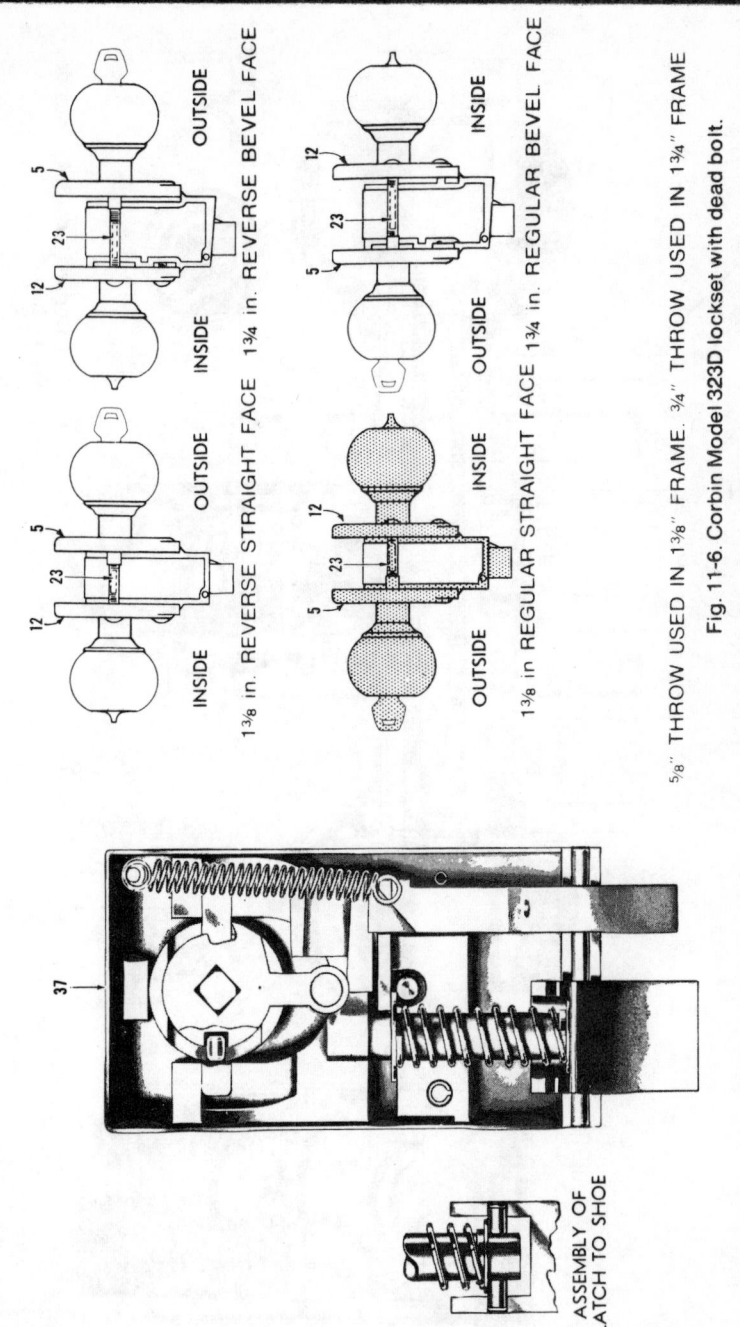

OUTSIDE / INSIDE

1⅜ in. REVERSE STRAIGHT FACE

1¾ in. REVERSE BEVEL FACE

1⅜ in REGULAR STRAIGHT FACE

1¾ in. REGULAR BEVEL FACE

⅝" THROW USED IN 1⅜" FRAME. ¾" THROW USED IN 1¾" FRAME

ASSEMBLY OF LATCH TO SHOE

Fig. 11-6. Corbin Model 323D lockset with dead bolt.

131

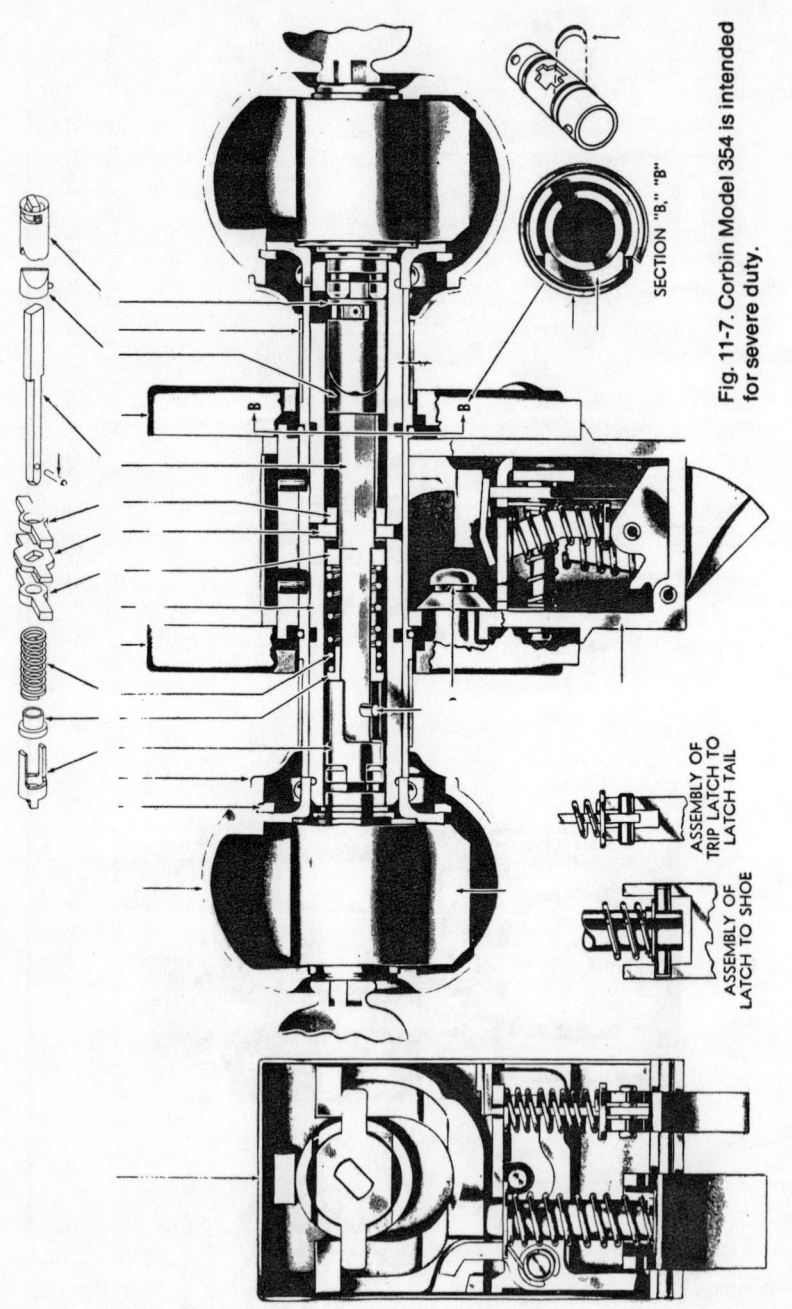

SECTION "B," "B"

Fig. 11-7. Corbin Model 354 is intended for severe duty.

ASSEMBLY OF TRIP LATCH TO LATCH TAIL

ASSEMBLY OF LATCH TO SHOE

duty (Fig. 11-7). Parts are made of high-quality steel: they are stronger than they need to be.

Corbin's unitized lockset has some unique features (Fig. 11-8). The knob is mounted on a full-width bearing for better wear characteristics; the swinging latchbolt insures positive locking action; the interior parts are bronze plated to prevent corrosion and to reduce friction; retractor springs are made of phosphor bronze and are stressed in compression to reduce fatigue. Other features include an optional extended-lip escutcheon that guards the bolt and the deadlocking latchbolt (Fig. 11-9).

Mounting the lock is simple. All that is required is a rectangular cutout from the door edge and a pair of holes for the through bolts.

The Corbin Fortune 1400 is one of the most interesting locksets on the market (Fig. 11-10.) This lockset is priced competitively but is designed to give long service. The bolt mechanism has only four moving parts and the knob is super rigid. The cylinder can be changed in a matter of minutes, a feature that should reduce maintenance costs.

Corbin's Ensign series is reversible; it can be used on left- or right-hand doors. However, these locksets do not have an auxiliary bolt (Fig. 11-11).

Schlage

Schlage engineers have worked for many years to perfect the cylinder lockset. These efforts, together with rigid quality-control standards, have made Schlage a leading name in locks.

"A" series standard duty locks are recommended for normal use applications; "G" series heavy-duty locksets combine a dead bolt with a deadlatch for additional security and durability.

"A" Series Locksets

Figure 11-12 is a cutaway view of the "A" series lockset. As shown, the cylinder has five pin tumblers; an additional pin is optional. The pins are encased in an extruded brass cylinder plug and shell. These locks are recommended for residential

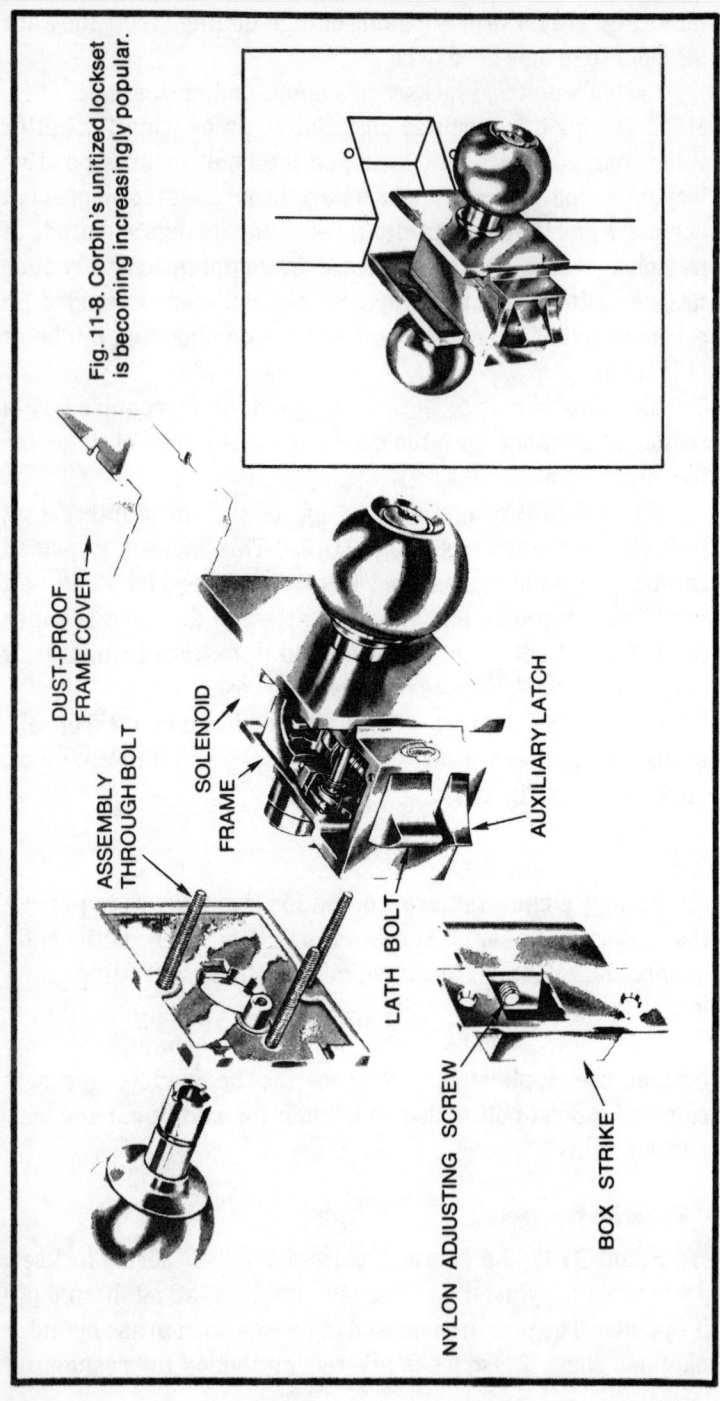

DUST-PROOF
FRAME COVER →

ASSEMBLY
THROUGH BOLT

SOLENOID
FRAME

AUXILIARY LATCH

LATH BOLT

NYLON ADJUSTING SCREW

BOX STRIKE

Fig. 11-8. Corbin's unitized lockset is becoming increasingly popular.

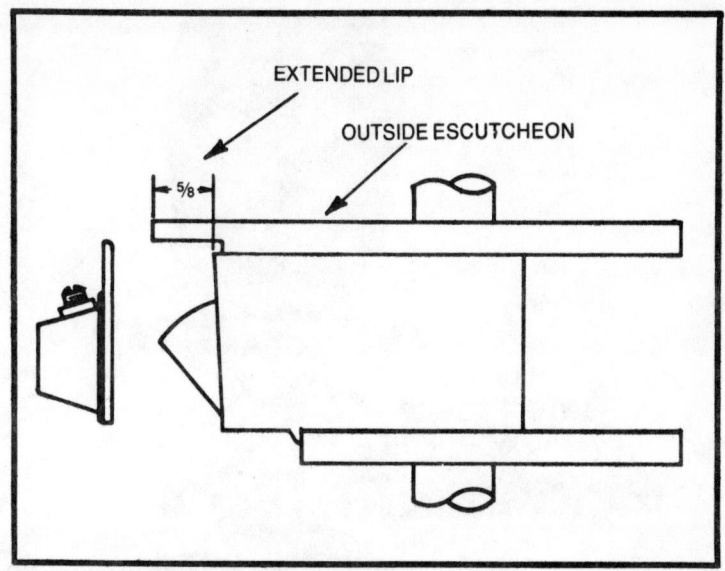

EXTENDED LIP

OUTSIDE ESCUTCHEON

⅝

Fig. 11-9. Extended lip escutcheons limit access to the latchbolt.
(Courtesy Emhart Corp.)

and commercial buildings where the normal frequency of operation is anticipated, and are used in motels, office buildings, warehouses, and retail stores, as well as in apartment complexes, residential homes, medical clinics, schools, and churches. Unlike many other locksets on the market, the Schlage "A" series can be adapted to meet almost any installation requirement. An extension link can be mounted on the latch, and the knob can be positioned as far back as 10 in. from the door edge. The backset— the distance from the front of the lock to the center of the cylinder hole or knob hub—is 5 in. (Fig. 11-13).

A51DP—The A51DP is an entrance lock (Fig. 11-14). Rotating either knob will retract the latchbolt. The turnbutton is on the inside knob; moving it to the horizontal position locks the outside knob. It must be released manually and restored to the vertical position; otherwise, only the key will unlock the lock. The inside knob is always free for immediate exit. The latchbolt automatically deadlocks the door when it is closed.

A73PD—Rotating either knob will retract the latchbolt; pushing the button on the inside knob will lock the bolt; from the

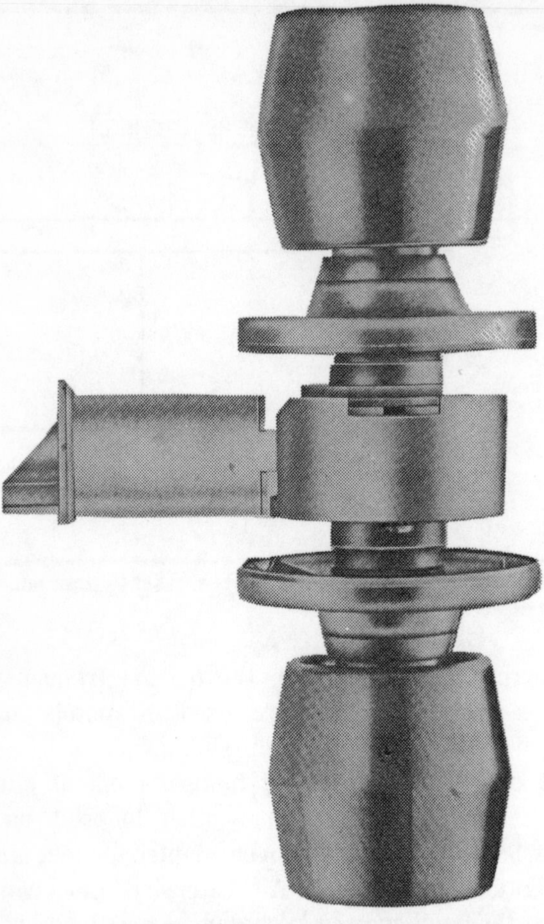

Fig. 11-10. Corbin Fortune 1400.

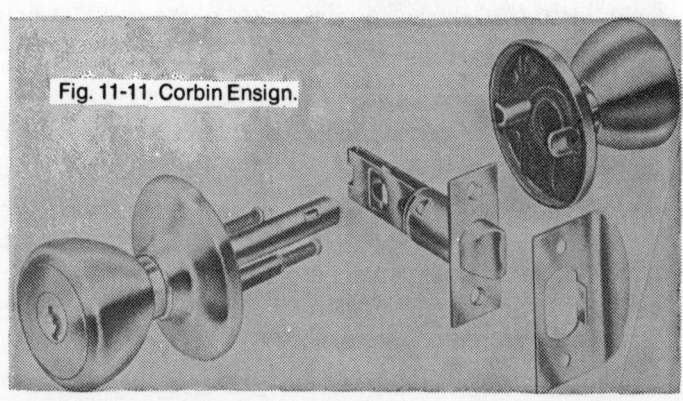

Fig. 11-11. Corbin Ensign.

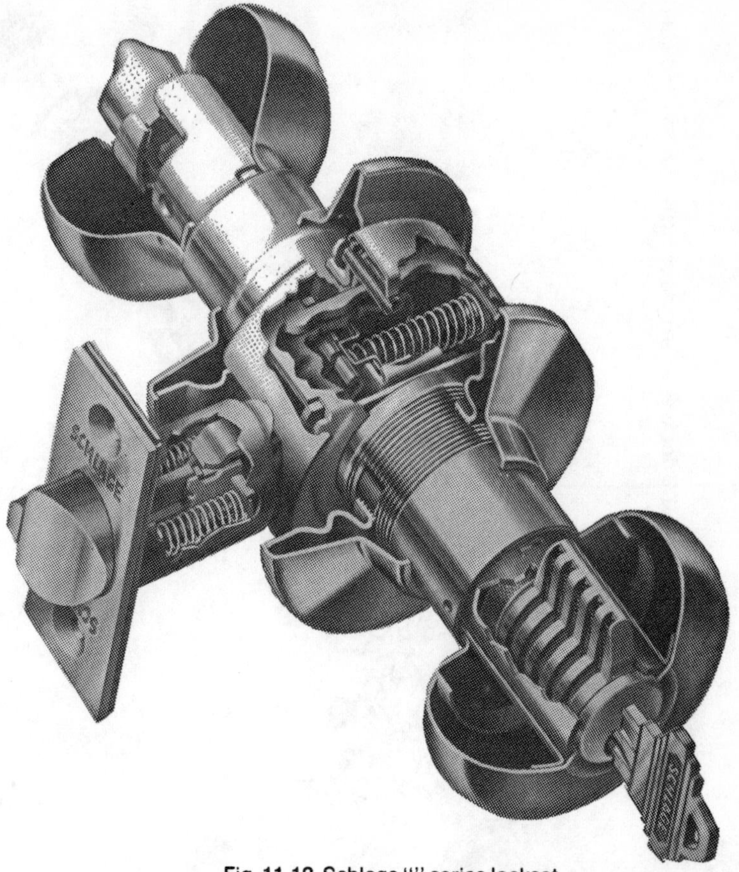

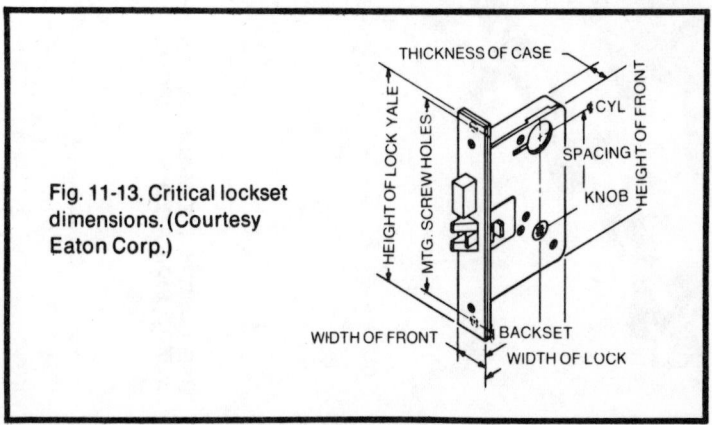

Fig. 11-12. Schlage '''' series lockset.

Fig. 11-13. Critical lockset dimensions. (Courtesy Eaton Corp.)

THICKNESS OF CASE

HEIGHT OF LOCK YALE

MTG. SCREWHOLES

CYL

SPACING

HEIGHT OF FRONT

KNOB

WIDTH OF FRONT

BACKSET

WIDTH OF LOCK

137

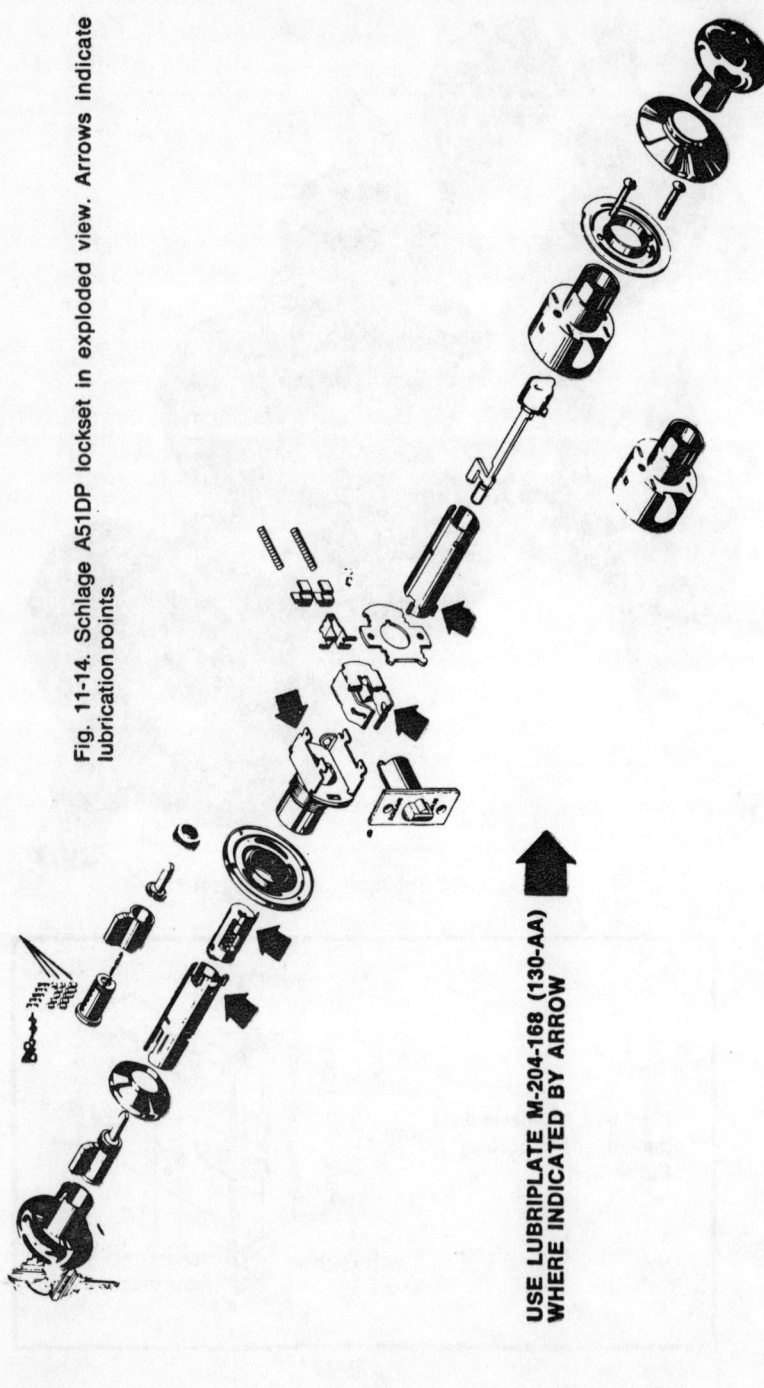

Fig. 11-14. Schlage A51DP lockset in exploded view. Arrows indicate lubrication points.

USE LUBRIPLATE M-204-168 (130-AA) WHERE INDICATED BY ARROW

outside, the bolt is key actuated (Fig. 11-15). Turning the inside knob or closing the door releases the button to prevent accidental lockout. Lubrication points are indicated in this and the previous figure by arrows. While these locks share many of the same parts the A73PD has a few more for better security.

"G" Series Locksets

"G" series double-locking security locksets are designed for high security and long wear. The "G" series combines a 1 in. throw dead bolt—with a roller tip—and a ½ in. throw deadlatch. The lock case is backed up with armor plate, and the internal mechanism is made of cold-rolled steel chemically treated to resist corrosion. When locked, the outside knob spins freely to discourage forced entry. Coil compression springs are used exclusively. In addition, these locks have *six* pin tumblers set into an extruded brass cylinder. The pins are made of nickel-silver to insure long wear.

The lockset has been designed for easy installation; all that is required is to drill five holes in the door. Friction-grip screws secure the lockset to the door. While the hand can be reversed in the field, it is simpler to order a lockset with the correct hand. The standard backset is 2¾ in. Backsets up to 5 in. can be ordered.

G51PD—The G51PD is illustrated in Fig. 11-16. It has these characteristics.

- Rotating either knob retracts the latchbolt.
- Turning the button in the inside knob to a horizontal position disengages the outside knob and makes it inoperative.
- The button does not release unless manually restored to the vertical position.
- Rotating the thumbturn from the inside, or the key from the outside, extends the dead bolt to the locked position.
- Both the dead bolt and the latchbolt can always be retracted to the unlocked position by rotating the knob or the thumbturn from the inside or the key from the outside.

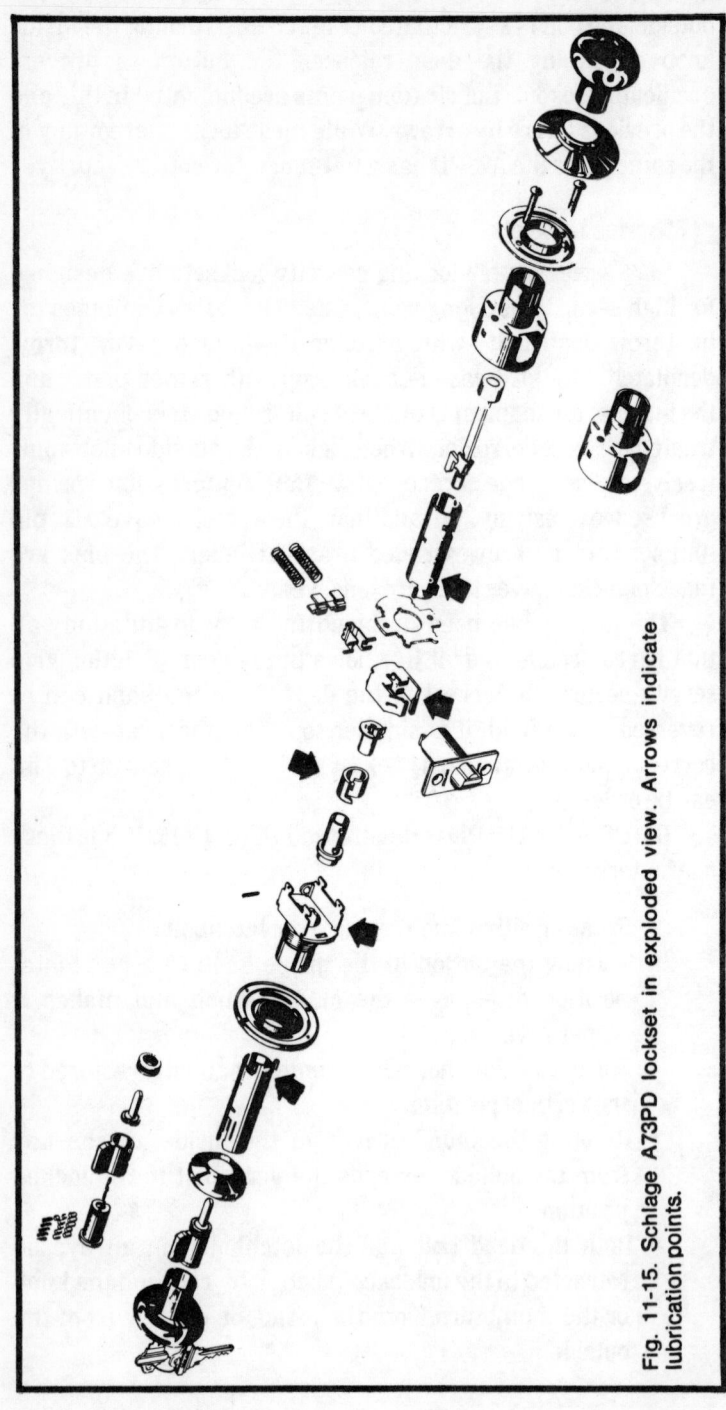

Fig. 11-15. Schlage A73PD lockset in exploded view. Arrows indicate lubrication points.

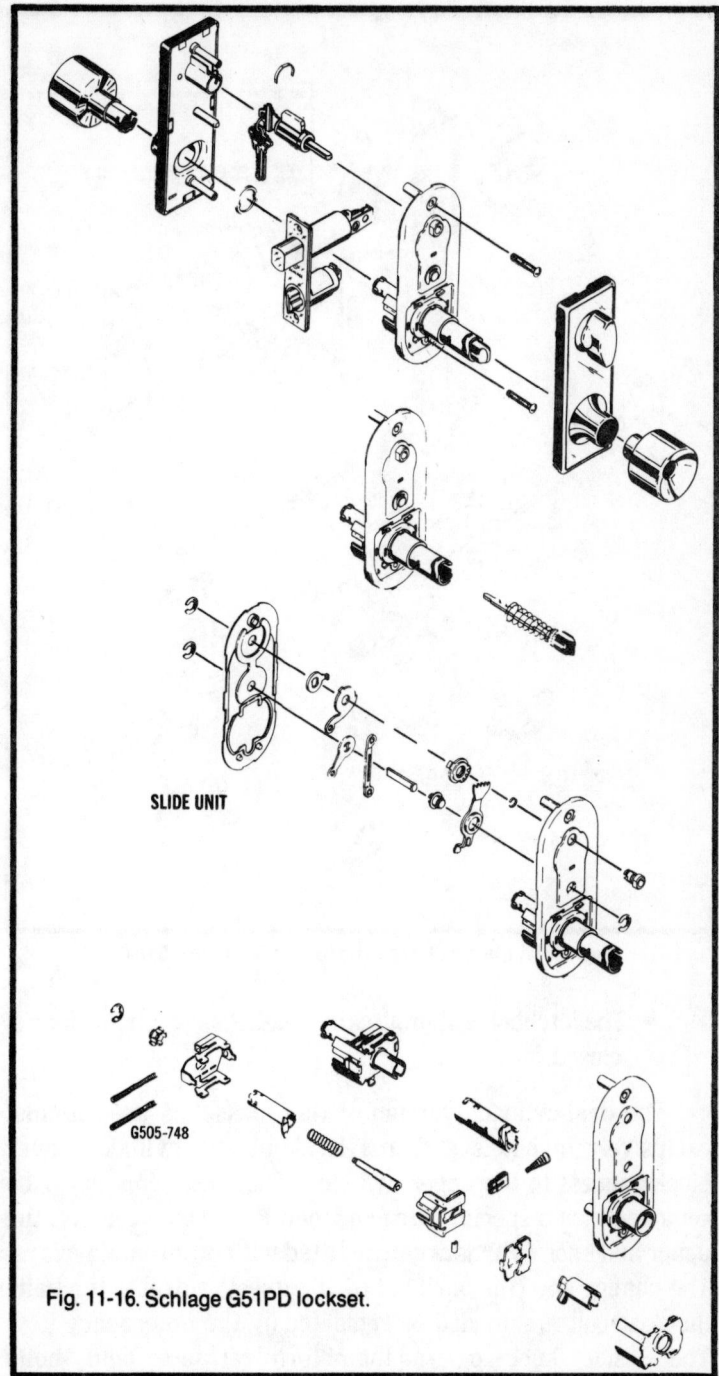

G505-748

SLIDE UNIT

Fig. 11-16. Schlage G51PD lockset.

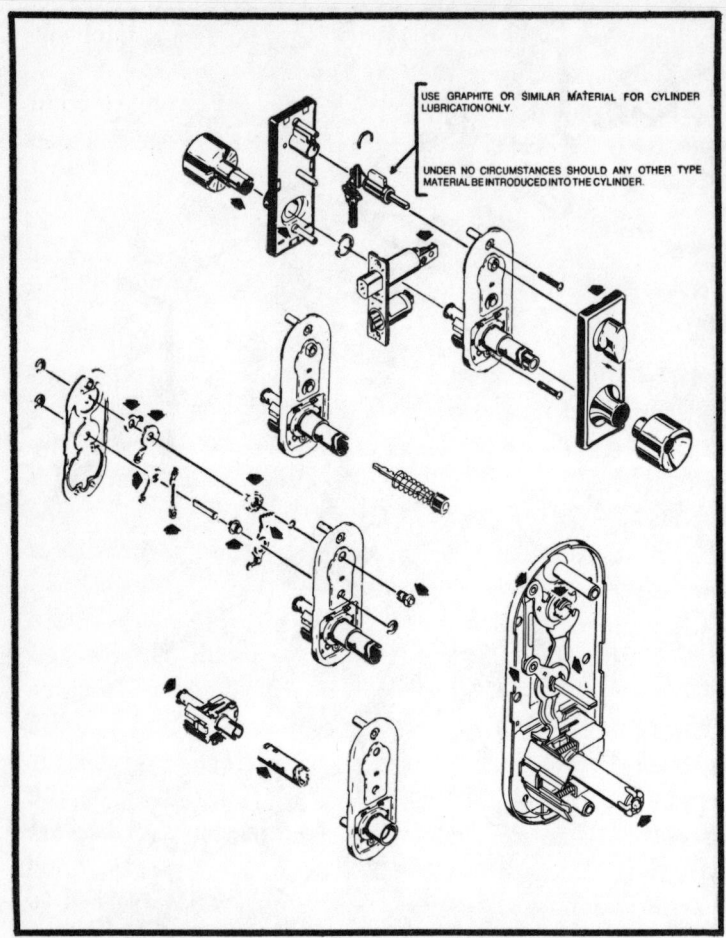

USE GRAPHITE OR SIMILAR MATERIAL FOR CYLINDER LUBRICATION ONLY.

UNDER NO CIRCUMSTANCES SHOULD ANY OTHER TYPE MATERIAL BE INTRODUCED INTO THE CYLINDER.

Fig. 11-17. Lubrication points for the Schlage G51PD.

- The latchbolt automatically deadlocks when the door is closed.

The dual-cylinder version of this lockset is used almost exclusively in hotels and motels. A plastic cylinder cover blocks access to the upper cylinder. The cover can easily be removed with a special extractor tool. For extra security, the upper and knob cylinders can be fitted with different keyways. The change key (the one held by the guest) retracts the bolt; the dead bolt is extended or retracted by the emergency key. The inside knob or the thumbturn retracts both bolts simultaneously for immediate exit. The latchbolt automatically deadlocks when the door is closed.

The next illustration, Fig. 11-17, indicates the lubrication points for the "G" series lockset. Use graphite in the cylinder and Lubriplate M204-168 on the other parts.

Medeco

Many locksmiths believe that the Medeco cylinder is the ultimate in pin-tumbler security.

Operation—Figure 11-18 is an exploded view of a typical Medeco six-tumbler cylinder. The pin tumblers must be raised to the shear line between the plug and cylinder shell by the key cuts. The depth of the individual cuts corresponds to the length of the tumblers. This principle is used on all pin-tumbler locks built on the Yale pattern. But Medeco has added another line of defense, one that makes key duplication almost impossible.

The tumblers, responding to angled bites in the key, rotate as they raise. The tumblers are slotted and, in the unlocked position, align with extrusions on the fence or sidebar. Tumbler slots and the fence are shown as parts Nos. 11 and 4 in the exploded view. The next drawing, Fig. 11-19, shows the slot for the sidebar on the plug. A matching slot is milled on the cylinder shell. In the locked position, the sidebar is athwart these slots, effectively blocking plug rotation (Figs. 11-20 and 11-21). In the unlocked position, the tumblers have turned so that their slots are at right angles to the blade of the key (Fig. 11-22). When this condition is met, the sidebar is free to move into the plug. If the tumblers are at the correct height, the lock will open (Figs. 11-23 and 11-24). Figure 11-25 is a closeup of the engagement mechanism.

The key is a very special item (Fig. 11-26). Angles milled on the sides of the bites determine the direction and amount of pin rotation and add another level of complexity. There are some 23 million usable combinations.

An ordinary key machine cannot duplicate a Medeco key. Machines are available through the factory, but they are expensive and distribution is rigidly controlled. Only well established locksmiths need apply.

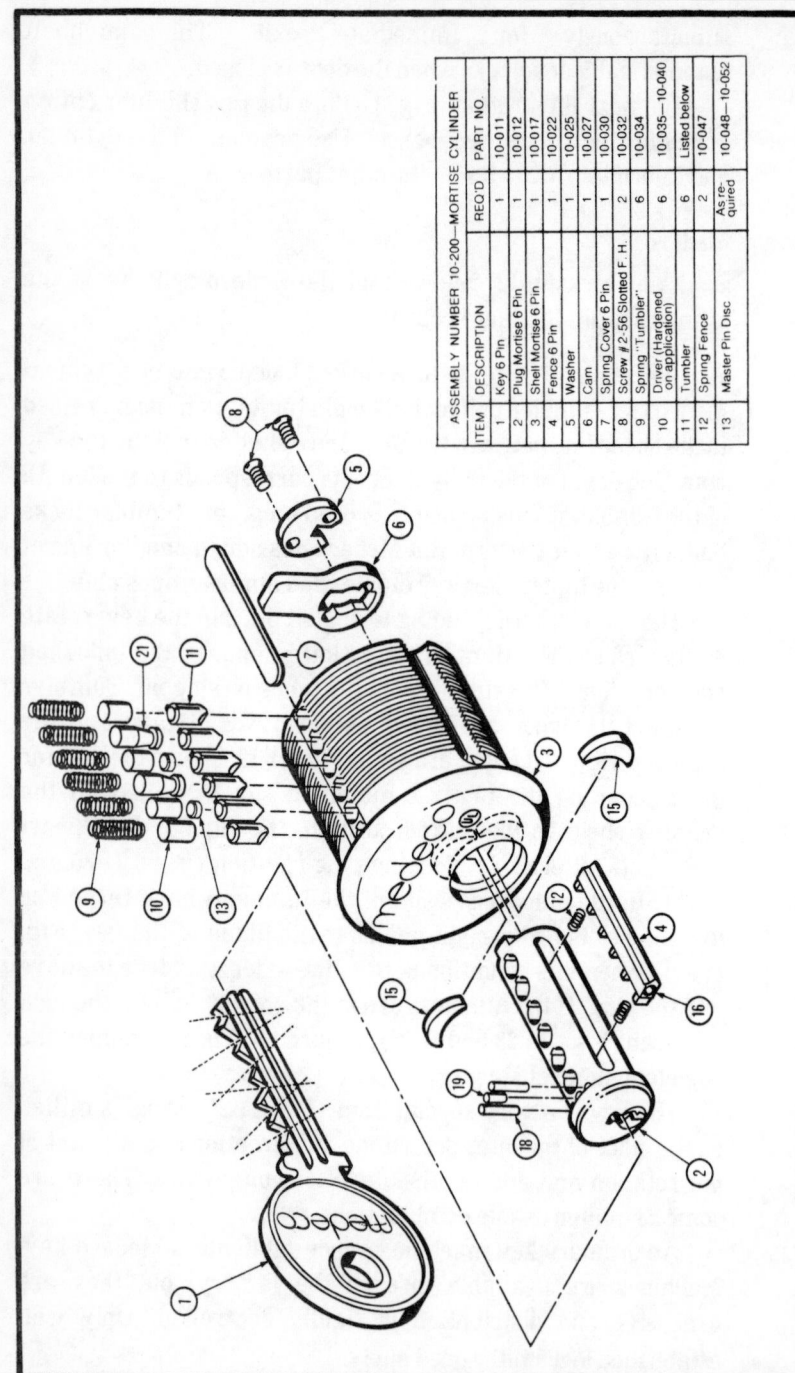

ASSEMBLY NUMBER 10-200—MORTISE CYLINDER		
ITEM	DESCRIPTION	REQ'D	PART NO
1	Key 6 Pin	1	10-011
2	Plug Mortise 6 Pin	1	10-012
3	Shell Mortise 6 Pin	1	10-017
4	Fence 6 Pin	1	10-022
5	Washer	1	10-025
6	Cam	1	10-027
7	Spring Cover 6 Pin	1	10-030
8	Screw # 2-56 Slotted F. H.	2	10-032
9	Spring "Tumbler"	6	10-034
10	Driver (Hardened on application)	6	10-035—10-040
11	Tumbler	6	Listed below
12	Spring Fence	2	10-047
13	Master Pin Disc	As re-quired	10-048—10-052

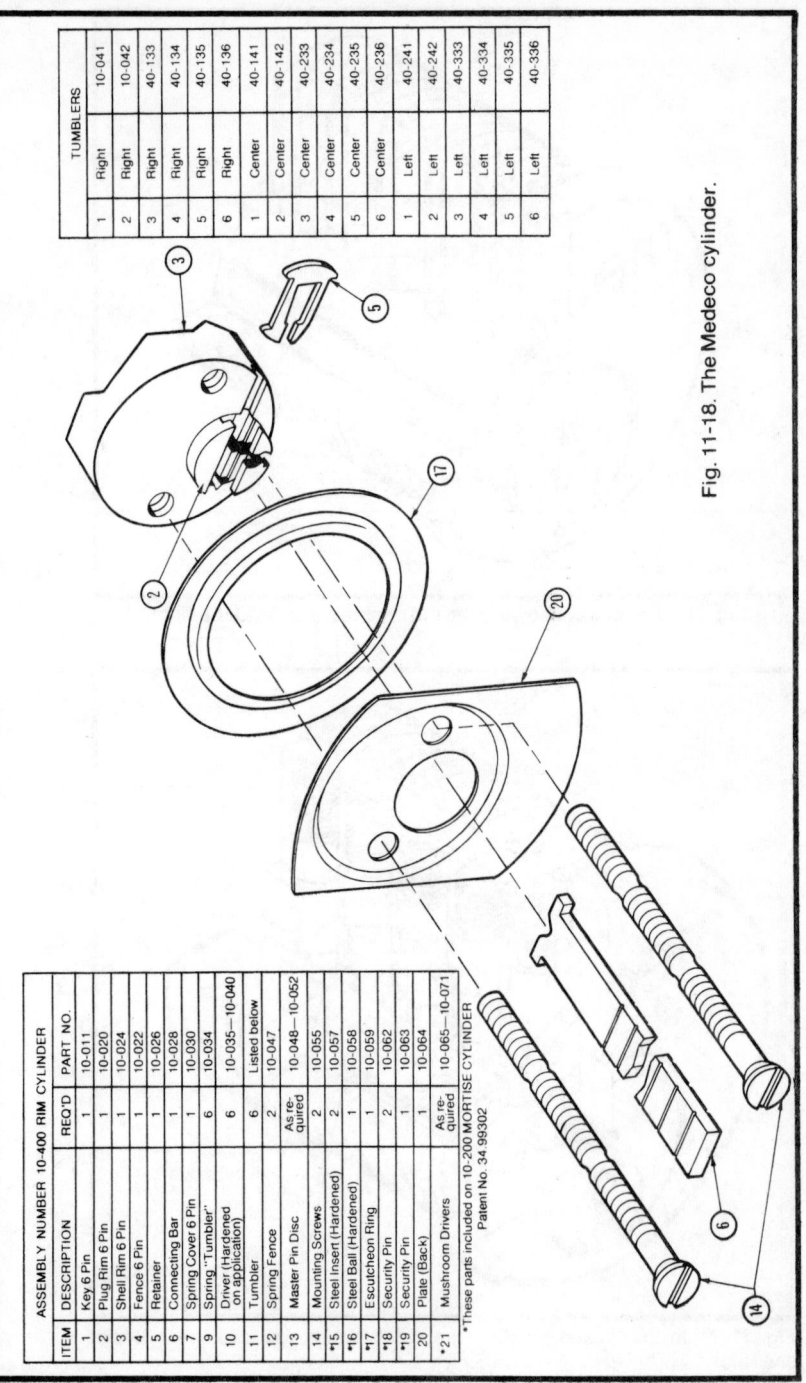

ASSEMBLY NUMBER 10-400 RIM CYLINDER		
ITEM	DESCRIPTION	REQ'D	PART NO.
1	Key 6 Pin	1	10-011
2	Plug Rim 6 Pin	1	10-020
3	Shell Rim 6 Pin	1	10-024
4	Fence 6 Pin	1	10-022
5	Retainer	1	10-026
6	Connecting Bar	1	10-028
7	Spring Cover 6 Pin	1	10-030
9	Spring "Tumbler"	6	10-034
10	Driver (Hardened on application)	6	10-035—10-040
11	Tumbler	6	Listed below
12	Spring Fence	2	10-047
13	Master Pin Disc	As required	10-048—10-052
*14	Mounting Screws	2	10-055
*15	Steel Insert (Hardened)	2	10-057
*16	Steel Ball (Hardened)	1	10-058
*17	Escutcheon Ring	1	10-059
*18	Security Pin	2	10-062
*19	Security Pin	1	10-063
20	Plate (Back)	1	10-064
*21	Mushroom Drivers	As required	10-065—10-071

*These parts included on 10-200 MORTISE CYLINDER
Patent No. 34 99302

		TUMBLERS	
1	Right	10-041
2	Right	10-042
3	Right	40-133
4	Right	40-134
5	Right	40-135
6	Right	40-136
1	Center	40-141
2	Center	40-142
3	Center	40-233
4	Center	40-234
5	Center	40-235
6	Center	40-236
1	Left	40-241
2	Left	40-242
3	Left	40-333
4	Left	40-334
5	Left	40-335
6	Left	40-336

Fig. 11-18. The Medeco cylinder.

145

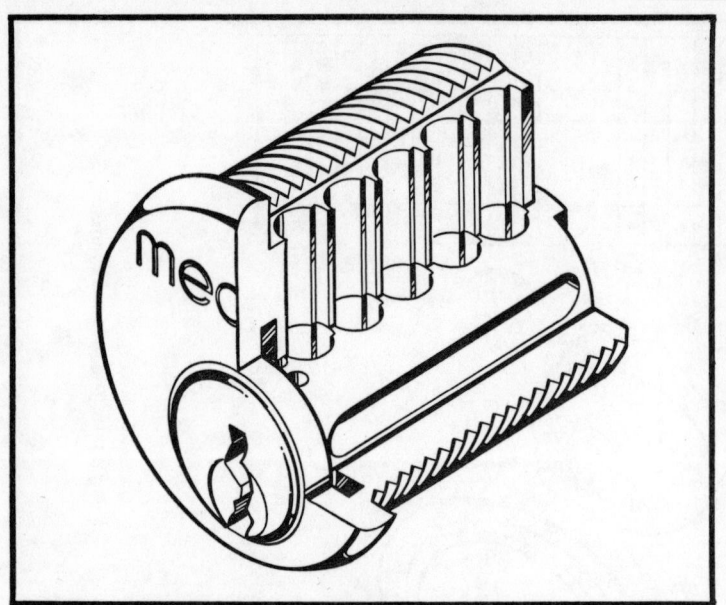

Fig. 11-19. Cutaway showing the pin chambers and sidebar slot.

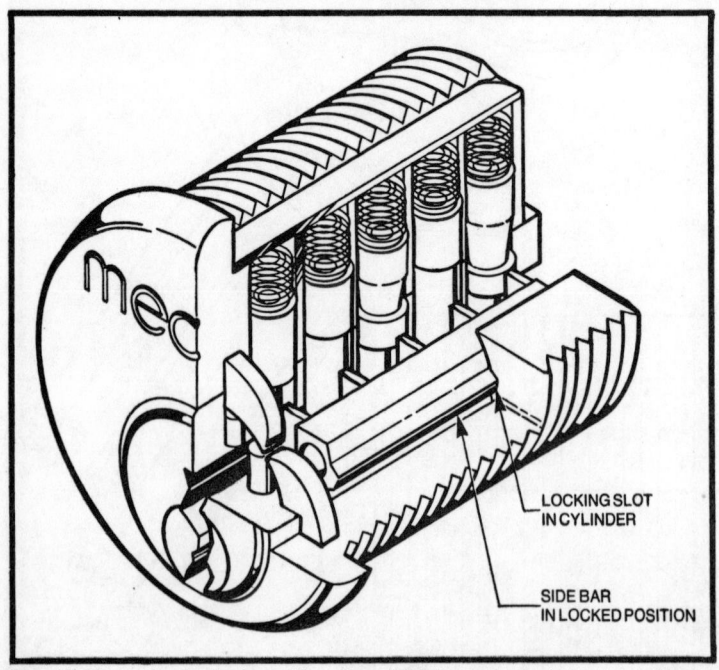

LOCKING SLOT
IN CYLINDER

SIDE BAR
IN LOCKED POSITION

Fig. 11-20. In the locked position the sidebar extends into the cylinder shell slot.

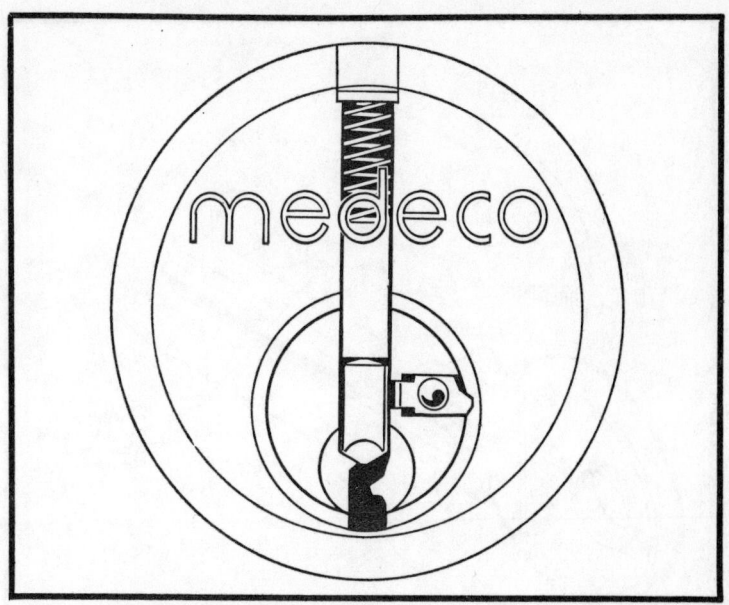

Fig. 11-21. This drawing shows both locking arrangements: the sidebar and the pin tumblers lie across the shear line.

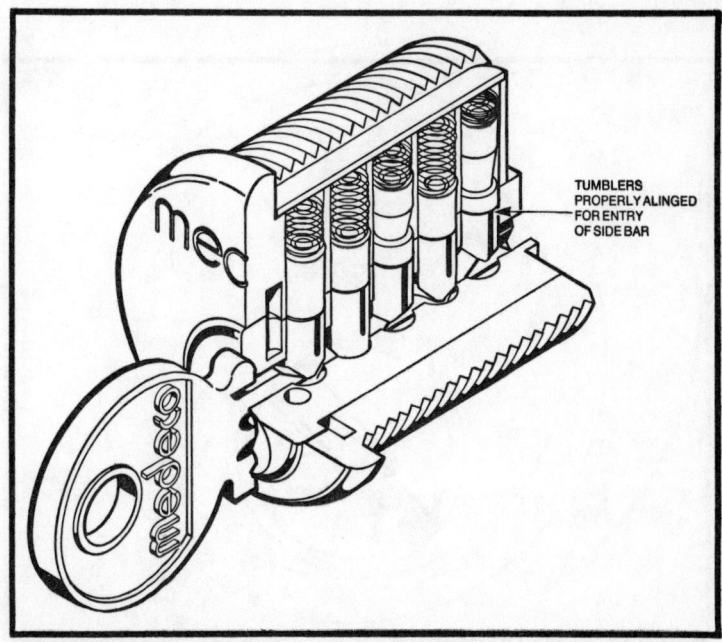

TUMBLERS PROPERLY ALINGED FOR ENTRY OF SIDE BAR

Fig. 11-22. The key rotates the tumblers, bringing the slots into alignment with the sidebar.

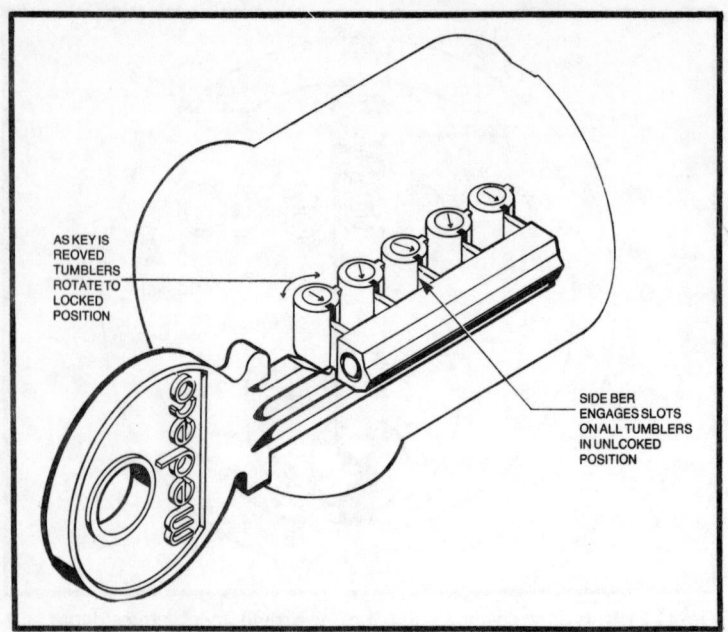

Fig. 11-23. Once the pins are in alignment, the sidebar retracts into the plug.

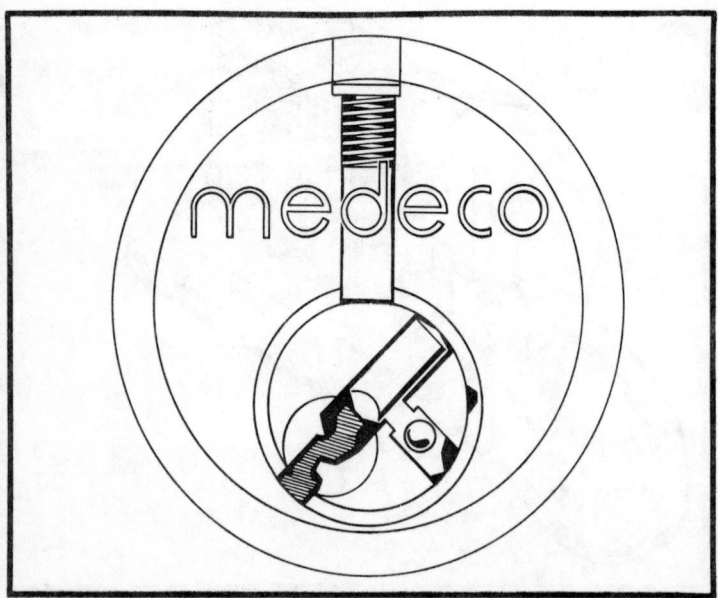

Fig. 11-24. In the unlocked position the pins and sidebar are below the shear line.

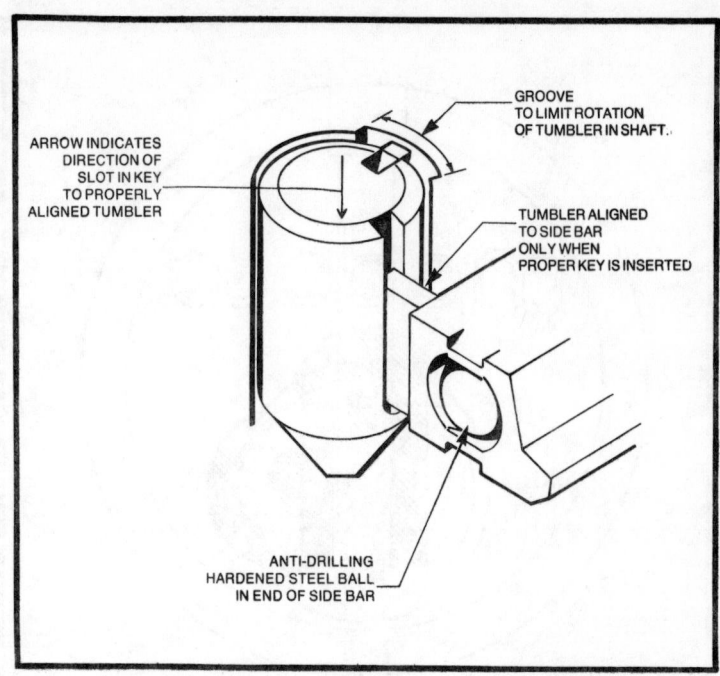

Fig. 11-25. A closeup of the sidebar and pin mechanism.

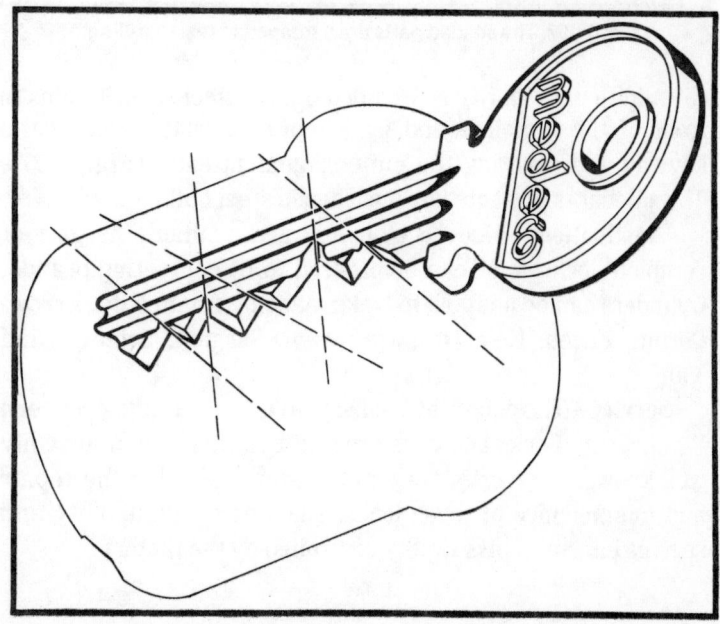

Fig. 11-26. Carefully milled angles on the key bitting rotate the pins.

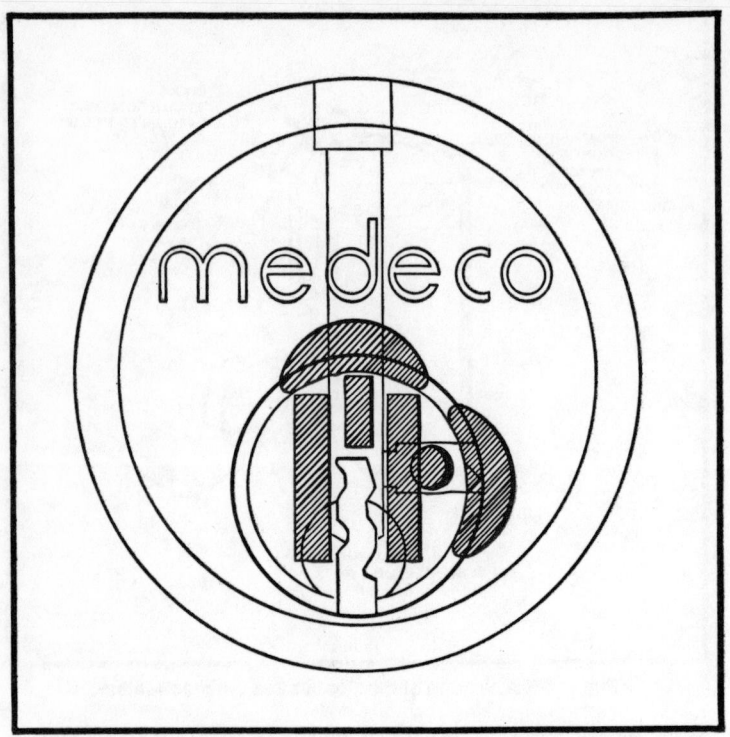

Fig. 11-27. The shaded parts are hardened to resist drilling.

Additional security is provided by tool-steel security pins in front of the tumblers and by armored inserts flanking the keyway and covering the pin engagement slots (Fig. 11-27). The sidebar is protected by a hardened steel ball.

Applications—Medeco cylinders are available as part of complete locksets or can be ordered to fit competing brands. Cylinders can be adapted to locks made by Adam Rite, Arrow, Corbin, Facon, Keil, Russwin, Segal, Sargent, Schlage, and Yale.

Service—The people at Medeco have an overriding concern for security. Locks and keys are sent by registered mail. Only well-known and trusted locksmiths are involved in the repair and maintenance of these locks, since parts availability and service information is rigidly controlled by the factory.

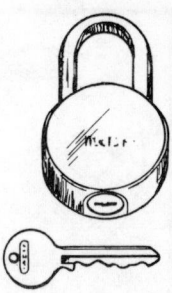

Chapter 12

The Cylinder Key

Figure 12-1 illustrates basic key nomenclature. The drawings in Fig. 12-1 give some indication of the variety of blanks available. Notice the differences in the bow, blade length, and in the width, number, and spacing of the grooves.

The bow usually has a specific shape that identifies the lock maker. Traditionally, key blank bows were unmarked; today many of these bows mimic the original design of the key. The blade length is indicative of the number of pins, e.g., a five-pin key is shorter than a six-pin key, and a sever-pin key of the same make is longer than either.

The height of the blade implies something about the depth of the cuts in the series; the higher the blade the deeper the cuts.

MATERIALS

Key blanks are made of brass, nickel-brass, nickel-steel, steel, aluminum, or a combination of these and other alloys The metal determines the strength of the key and its resistance to wear.

KEY BOARD

Earlier, I mentioned the need for a key board. The same board can be easily expanded to include cylinder

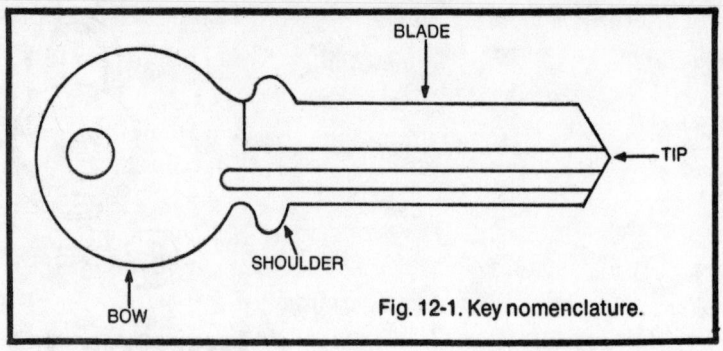

Fig. 12-1. Key nomenclature.

keys. Key-blank interchange charts can be purchases from the local locksmith supply house. These charts are sometimes published in house catalogs. You might also consider labeling each blank.

SELECTION

Determine if you have the proper key blank with reference to the manufacturer of the lock. Suppose you need a No. 14 Yale blank and do not have it. Star Key Company's interchange chart shows that Star blank No. 5YA11 will substitute. This blank will fit the lock with the same precision as the original Yale product.

Sometimes you will not know the key-blank number. In this case, follow these steps:

1. Examine the bow of the key—its shape is a clue. Look for other blanks that have approximately the same shape.
2. Once you have found several candidates that might be acceptable, compare the key blades. The key section (the shape of the blade when viewed from the front), the blade length, and shoulder configuration should be identical.

DUPLICATING BY HAND

Duplicating a pin-tumbler key by hand is not much different than duplicating a lever or warded key. The prime requirement is patience.

1. Smoke the original and mount it together with the blank in a vise. The blank must be in perfect alignment with the original.
2. Using Swiss round and warding files, begin at the tip of the key and work towards the bow. Start in the center of each cut.
3. Go slow near the end of each cut, but continue to cut with firm and steady strokes. Stop when the file touches the black on the original key.
4. Once you have finished, remove the keys from the vise. Wire-brush the duplicate to remove any burrs and polish the key.
5. Hold the duplicate up to the light and place the original in front of it and then behind it. Is the duplicate a faithful copy of the original? Shallow cuts can be dressed out with a file; cuts that are too deep mean that you will have to start over with a new blank. In extreme cases, the original may be damaged.

Making a substitute for the original is tricky. Use the original as a model, but do not file to the bottom of the grooves. When you have cut as far as you dare, check your work in the

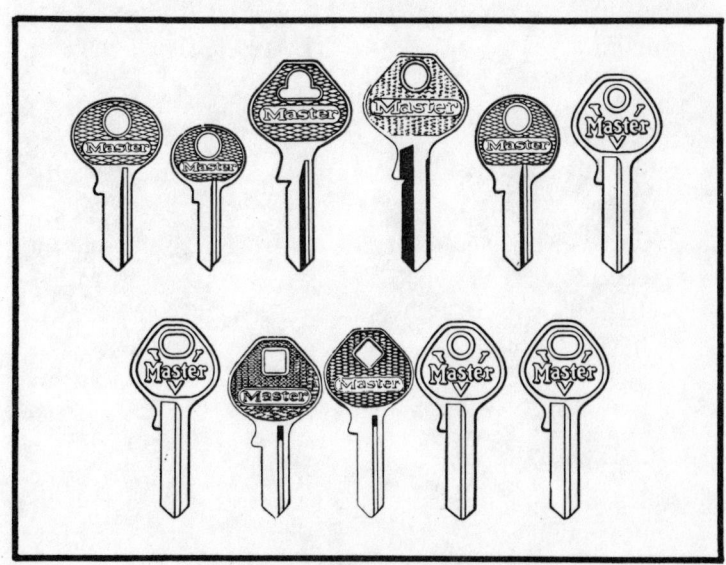

Fig. 12-2. A collection of Master blanks.

Fig. 12-3. Two key-duplicating machines.

lock. File as needed. but with circumspection. Be content to remove some metal and test again. When the key turns. stop.

Now you have a working duplicate key and you can leave it at that. But a professional will take time to disassemble the lock and check the pin heights. He may replace a tumbler or modify the key.

DUPLICATING BY MACHINE

A key-duplicating machine deserves care and respect since it is vital to your livelihood (Fig. 12-3). The machine's instruction sheet details service procedures and schedules. Once the machine is aligned and the cutter tolerances are in order. proceed as follows:

1. Place the original in one vise and the blank in the vise next to the cutter wheel. The keys should rest on ledges at the base of each vise.
2. Check the tip and the shoulder alignment with the cutter and cutter guide.
3. Put on eye goggles—particles thrown off by the cutter can blind you—and start the machine.
4. Move the blank into the cutter. The guide regulates the cuts so that the blank takes on the same profile as the original. "Rough out" the cuts on the first few passes.
5. Finish the job with one or two light passes. Be thorough: a less-than-perfect duplicate will not open the lock.
6. Wire-brush away any burrs and polish the key. Test it in the lock. It should work as smoothly as the original; if not. compare the original and duplicate under a light. File off any rough spots.

IMPRESSIONING

The technique for impressioning cylinder keys is different than for warded keys. The cylinder key is not smoked since inserting the key would wipe the black off the top of the blade. Instead. one must depend upon marks left on the blank by the tumblers. These marks indicate the position of the tumblers relative to the blank and the depth of each cut. Proceed as follows:

1. Examine the blank. The marks left by the tumblers are difficult to see in the best of circumstances; scratches across the top of the blade will make them impossible to see. Polish the top of the blade with emery paper.
2. Place the bow of the key in a small C-clamp and insert the blade into the keyway.
3. Twist the blank hard. If done correctly, the tumblers will leave their "footprints" on the blade.
4. Examine the blank under a light. Tip it slightly to give the marks definition.
5. Identify each mark with a light file cut.
6. Make a shallow cut at the mark closest to the tip with a Swiss No. 4 file.
7. Insert the key, twist it back and forth, and remove it. Check the mark an file a little more. Continue until the tumbler no longer leaves a mark. When this happens the pin is resting on the shear line.
8. When you finish one cut, move to the next and repeat the procedure. Keep the slopes even and smooth.
9. When the last cut is finished, all pins should be at the shear line.
10. Remove the key and clean it. This operation includes smoothing the various angles and removing burrs.

Rock the blank up and down as well as from side to side. It may be useful to work the key with the help of a steel rod. Insert the rod through the hole in the bow of the key and tap the bow with a mallet.

An impressioning tool can be purchased from a locksmith supply house. Squeezing the handle impressions the key.

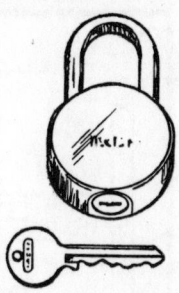

Chapter 13

Masterkeying

Masterkeying can provide immediate and long range benefits that a beginning locksmith can find most desirable. This chapter covers the principles of masterkeying, as well as techniques for developing masterkey systems.

CODING SYSTEMS

Coding systems help the locksmith to distinguish various key cuts and tumbler arrangements. Without coding systems, masterkeying would be nearly impossible.

Most coding systems (those for disc-, pin-, and lever-tumbler locks) are based upon depth differentiation: Each key cut is coded according to its depth; likewise each matching tumbler receives the same code. Depths for key cuts and tumblers are standardized for two reasons: (1) It is more economical to standardize these depths. Mass production would be impossible without some kind of standardization. (2) Depths, to some extent, are determined by production.

MASTERKEY SYSTEMS

In most key coding systems, tumblers can be set to any one of five possible depths. These depths are usually numbered

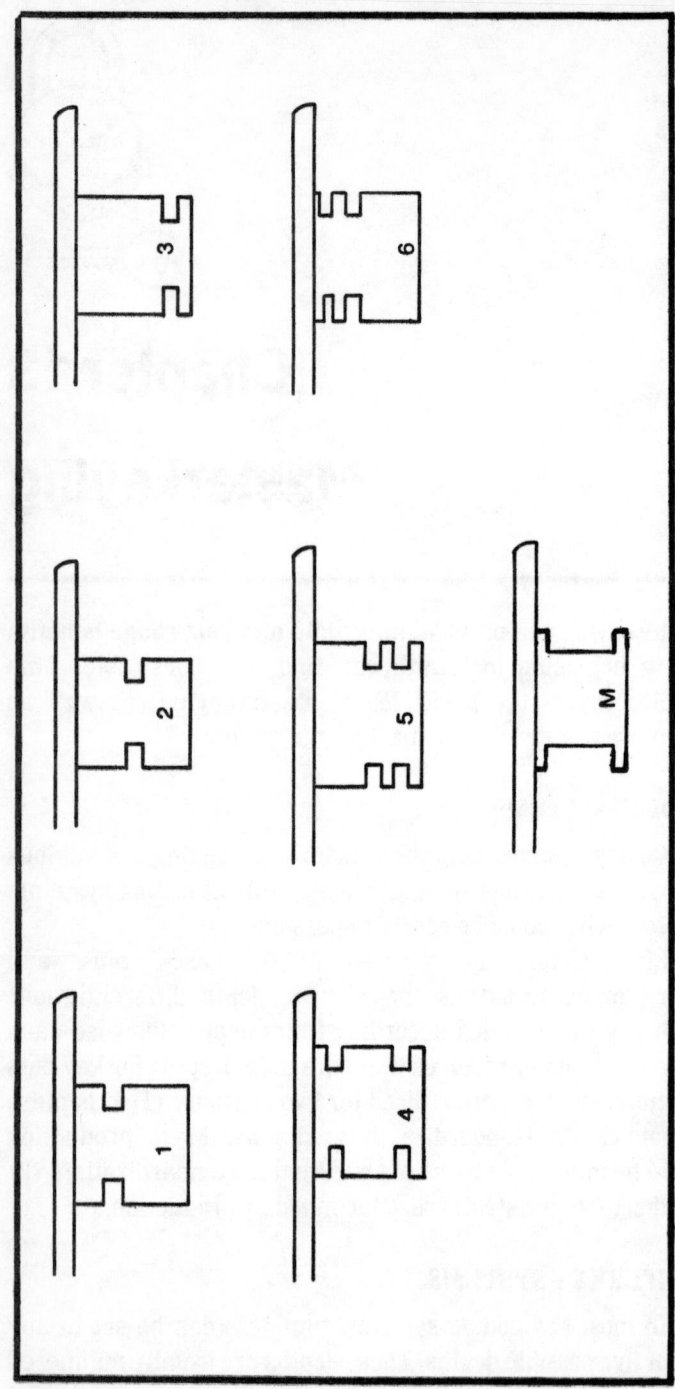

Fig. 13-1. The masterkey at the bottom of the drawing replaces the six change keys above it

consecutively 1 through 5. Since most locks have five tumblers, each one having five possible settings, there can be thousands of combinations. Masterkeys are possible because a single key can be cut to match several lock combinations.

In developing codes, there are certain undesirable combinations which cannot be used. The variation in depths between adjoining tumblers cannot be too great. For example, a pin-tumbler key cannot be cut to the combination 21919 because the cuts for the 9's would rule out the cuts for the 1's. Likewise a pin-tumbler lock with the combination 99999 would be too easy to pick. The undesirable code combinations vary depending upon the type of tumblers, the coding system, and the number of possible key variations. The more complex the system, the greater the possibility of undesirable combinations.

MASTERKEYING WARDED LOCKS

Since a ward is an obstruction within a lock that keeps out certain keys not designed for the lock, a masterkey for warded locks must be capable of bypassing the wards. Figure 13-1 shows a variety of side ward cuts that are possible on warded keys. The masterkey (marked *M*) is cut to bypass all the wards in a lock admitting the other six keys.

As explained earlier, cuts are also made along the length of the bit of a warded key. These cuts correspond to wards in the lock. To bypass such wards, a masterkey must be narrowed.

Because of the limited spaces on a warded key, masterkeying is limited in the warded lock. The warded lock, because it offers only a very limited degree of security, uses only the simplest of masterkey systems. Figure 13-2 shows some of the standard masterkeys that are available from factories.

MASTERKEYING LEVER LOCKS

Individual lever locks may be masterkeyed locally; but any system that requires a wide division of keys would have to be set up at the factory. A large selection of tumblers is required. The time involved in assembling a large number would make the job prohibitive for the average locksmith.

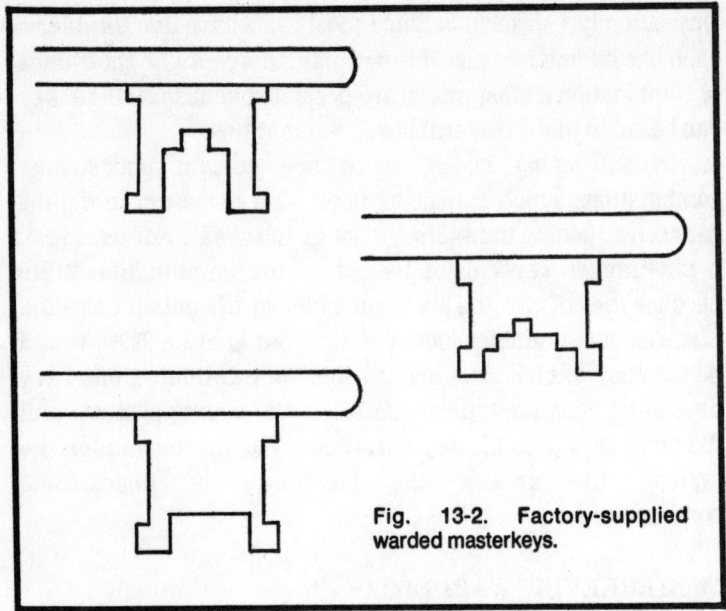

Fig. 13-2. Factory-supplied warded masterkeys.

There are occasions when you will be asked to masterkey small lever locks. There are two systems: the first is the double-gate system (Fig. 13-3); the other, the wide-gate system (Fig. 13-4). Double gating is insecure. As the number of gates in the system increases, care must be taken to prevent cross-operation between the change keys. For example, you may find a change key for one lock acting as the masterkey.

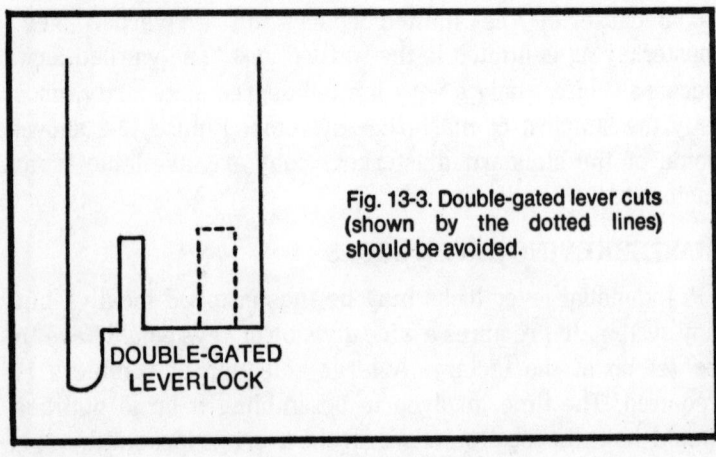

Fig. 13-3. Double-gated lever cuts (shown by the dotted lines) should be avoided.

DOUBLE-GATED
LEVERLOCK

With either system, begin by determining the tumbler variations for the lock series in question. If the keys to all the locks are available, read the numbers stamped on the keys. Otherwise the locks must be disassembled so that you can note the tumbler depths for each one. Next, make a chart listing the tumbler variations (Fig. 13-5).

The masterkey combination can be set up fairly easily now. The chart in Fig. 13-5 is for 10 lever locks, each having 5 levers with 5 possible key depths per lever. The masterkey for these locks will have a 21244 cutting code.

Suppose the tumblers in the first position have depths of 1, 2, 3, and 5. Depths 1 and 3 must be filed wider to allow the depth 2 cut of the masterkey to enter. The tumbler with the depth 5 cut requires a separate cut, or double gating. It will have a cut that will align it properly at two positions.

Moving to position 2 on the chart, we see that four levers will have to be cut; all of these will require a double gating cut. In position 3, four will require a double gating and one will need widening. In position 5, only one will require another gating to be cut, while four will require widening at the current gate.

Another masterkeying method is to have what is known as a master-tumbler lever in each lever lock. The master tumbler has a small peg fixed to it that passes through a slot in the series tumblers. The masterkey raises the master tumbler. The peg, in turn, raises the individual change tumblers to the

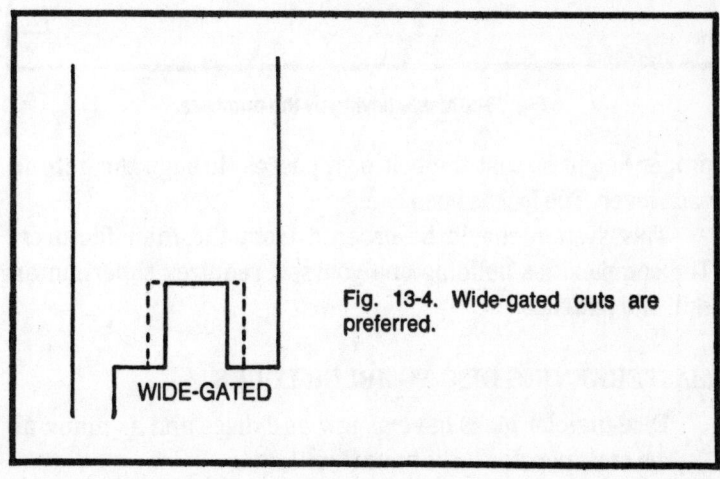

WIDE-GATED

Fig. 13-4. Wide-gated cuts are preferred.

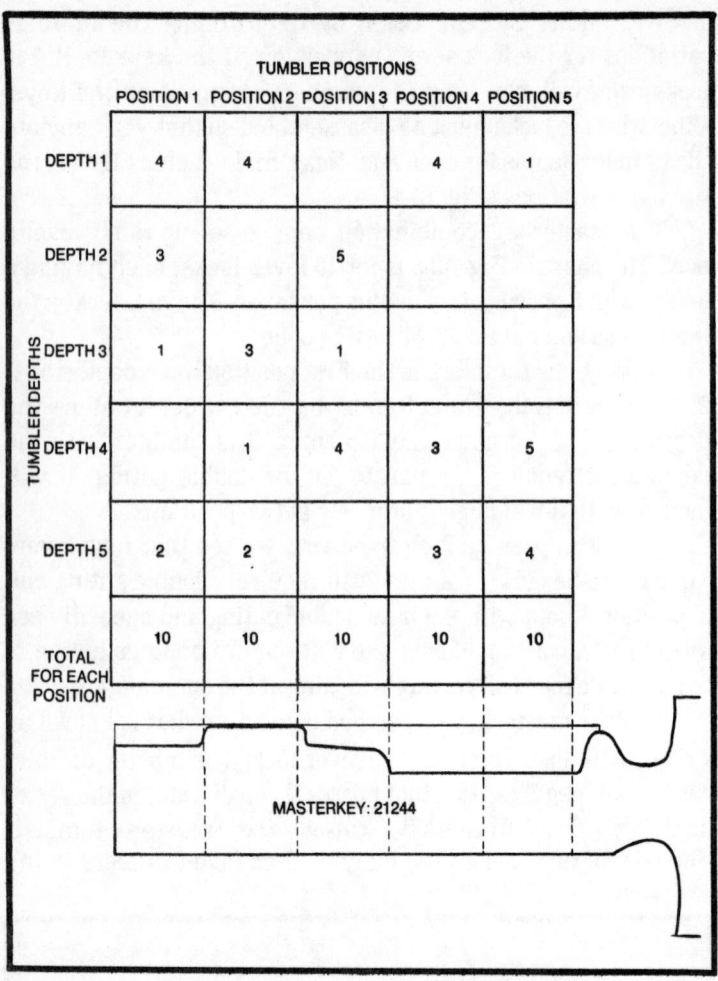

Fig. 13-5. Masterkeying by the numbers.

proper height so that the bolt post passes through the gate in each lever. The lock is open.

This system should be ordered from the manufacturer. The complexities building one yourself requires superhuman skill and patience.

MASTERKEYING DISC-TUMBLER LOCKS

Disc-tumbler locks have as few as 3 discs and as many as 12. The most popular locks have five.

Figure 13-6 shows how the tumbler is modified for masterkeying: The left side of the tumbler is cut out for the master; the right side responds to the change key. The key used for the master is distinct from the change key in that its design configuration is reversed. The cuts are, of course, different. The keyway in the plug face is patterned to accept both keys. Since the individual disc tumblers are numbered from 1 to 5 according to their depths, it is easy to think of the masterkey and disc cuts on a 1 to 5 scale but on different planes for both the key and the tumblers.

Uniform cuts are taboo. The series 11111 or 22222 would give very little security since a piece of wire could serve as the key. Other uniform cuts are out because they are susceptible to shimming. To keep the system secure, it is best to keep a two-depth interval between any two change keys. For example, 11134 is only one depth away from 11133, so 11134 should be used and 11133 omitted. The rationale is that 11134 is the more complex of the two.

The next step is to select a masterkey combination composed of odd numbers. At the top of your worksheet, mark the combination you have selected for the masterkey. Below it

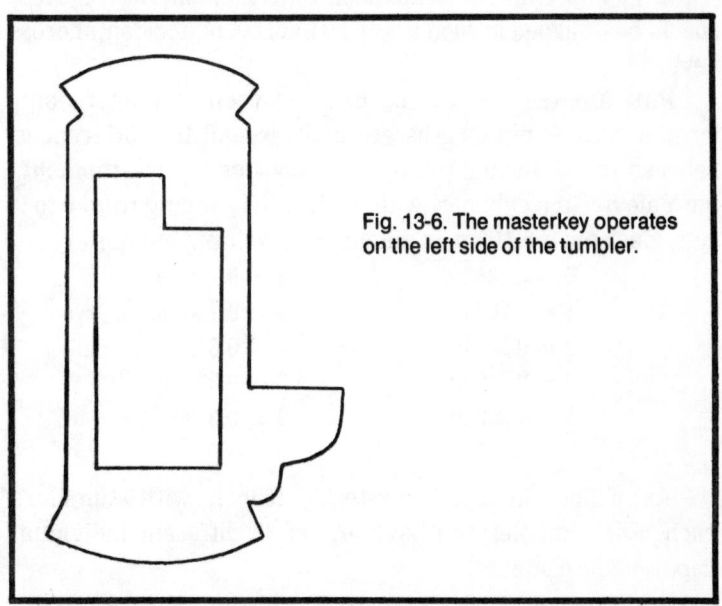

Fig. 13-6. The masterkey operates on the left side of the tumbler.

163

add a random list of possible change-key numbers. If you choose a systematic approach in developing change-key numbers, you compromise security. On the other hand, the systematic approach insures a complete list of possible combinations. You could begin systematically, then randomly select the change-keys combinations.

A single code could be used for all customers. The main point to remember is to use different keyways.

MASTERKEYING PIN-TUMBLER LOCKS

Masterkeying is more involved than modifying the cylinder. It requires the addition of another pin sandwiched between a top and bottom pin. This pin is, logically enough, called the master pin.

The master system is limited only by the cuts allowed on a key, the number of pins, and the number of pin depths available. Since this book is for beginning and advanced students, I will cover the subject on two levels: the simple masterkey system for no more than 40 locks in a series; the more complex system involving more than 200 individual locks.

It is important to remember that a masterkey system should be designed in such a way as to prevent accidental cross keying.

Pins are selected on the basis of their diameters and lengths. Master pin lengths are built around the differences between the individual pin lengths. Consider, for a moment, the Yale five-pin cylinder, with pin lengths ranging from 0 to 9 cuts. Each pin is 0.115 in. in diameter. Lower pin lengths are:

0 = 0.184 in.	5 = 0.276 in.
1 = 0.203 in.	6 = 0.296 in.
2 = 0.221 in.	7 = 0.315 in.
3 = 0.240 in.	8 = 0.334 in.
4 = 0.258 in.	9 = 0.393 in.

As an illustration, let's masterkey 10 locks with 5 tumblers each. Each tumbler can have any of 10 different individual depths in the chamber.

1. Determine the lengths of each pin in each cylinder. Mark these down on your worksheet. The masterkey selected for this system may have one or more cuts identical to the change keys.

2. Cut a masterkey to the required depths. In this instance, each cylinder plug is loaded by hand.

3. Using the known masterkey depth, subtract the depth of the change key from it. The difference is the length of the masterkey pin. If the change key is 46794 and the masterkey is 68495, the master pin combination will be as follows:

Chamber Position	Bottom	Master
1	4	2
2	6	2
3	7	3
4	9	0
5	4	1

This procedure is followed for each plug. Notice that not all the chambers have a master pin. Such complexity is not necessary and makes the lock more vulnerable to picking. Each master pin represents another opportunity to align the pins with the shear line. Figure 13-7 is an extreme instance, with five master pins and three masterkeys.

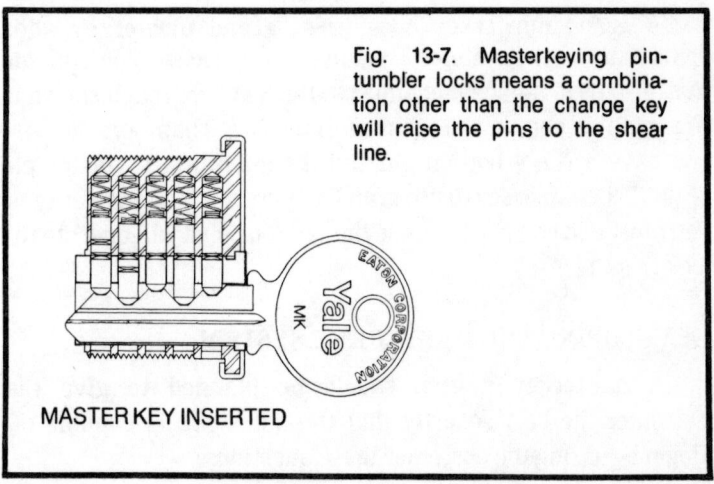

Fig. 13-7. Masterkeying pin-tumbler locks means a combination other than the change key will raise the pins to the shear line.

MASTER KEY INSERTED

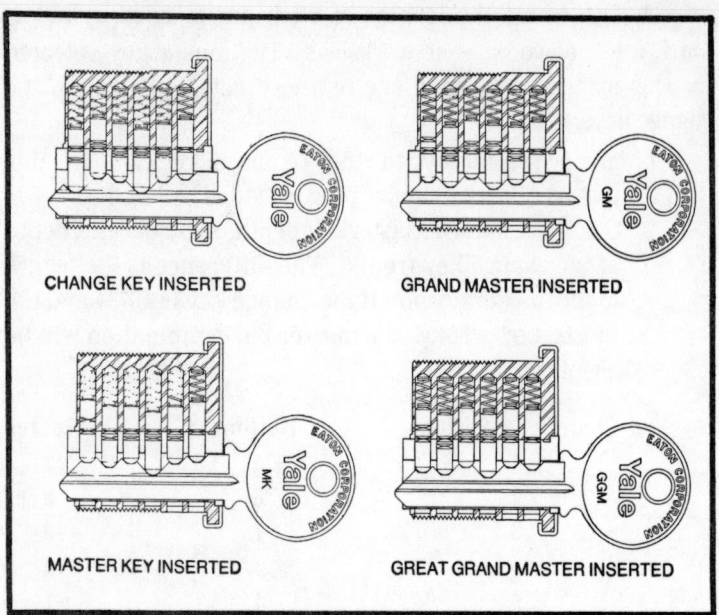

CHANGE KEY INSERTED

GRAND MASTER INSERTED

MASTER KEY INSERTED

GREAT GRAND MASTER INSERTED

Fig. 13-8. The Yale great grand master system requires five master pins.

In practice, locksmiths avoid most of this arithmetic by compensating as they go along. For example, an unmasterkeyed cylinder has double pin sets. Masterkeying means that an additional pin is added in (usually) position 1. This drives the bottom pin lower into the keyway. The bottom pin must be shortened to compensate.

A grand masterkey or a great grand masterkey adds complexity to the system. You have two choices: You can add master pins in adjacent chambers (the Yale approach shown in Fig. 13-8) or you may stack pins in the first chamber. Suppose you have a No. 4 bottom pin and the appropriate master pin (Fig. 13-9A). In order to use grand and great grand masterkeys, you must add two No. 2 pins so that all four pins will operate the lock (Fig. 13-9B).

DEVELOPING THE MASTERKEY SYSTEM

A masterkey system should be planned to give the customer the best security that the hardware is capable of. Begin by asking the customer these questions:

- Do you want a straight masterkey system with one master to open all locks? Or do you want a system that will have submasters? That is, do you want a system with several submasters of limited utility and a grand or great grand master?
- What type of organizational structure is within the business? Who should have access to the various levels?
- How many locks will be in each submaster grouping?

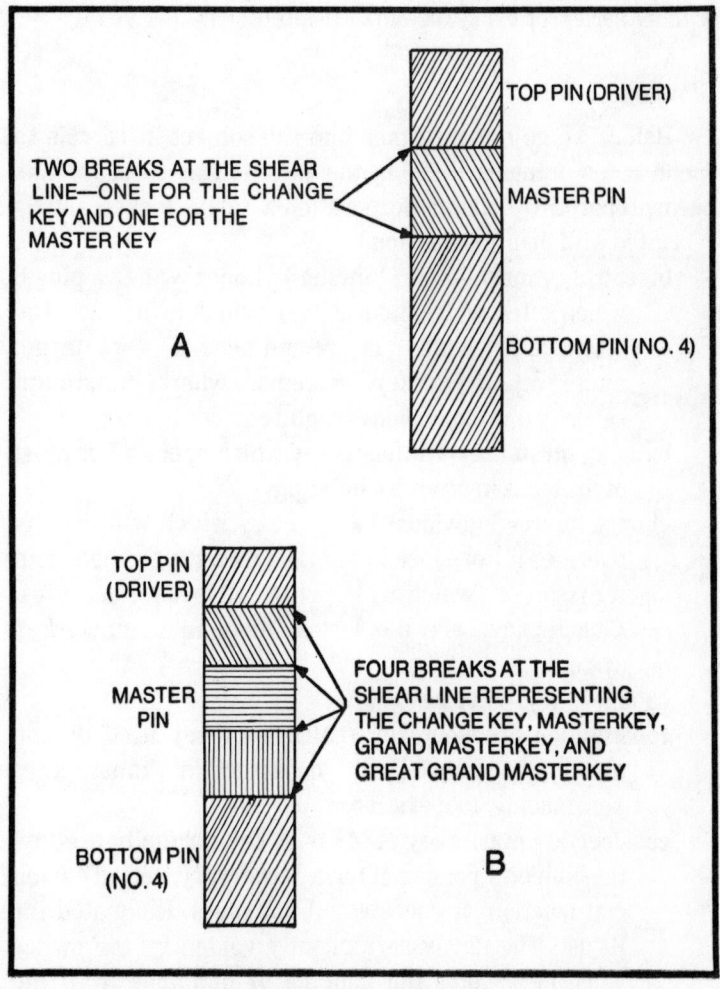

Fig. 13-9. Stacking pins is an alternative way of masterkeying.

- Is the system to be integrated into an existing system, or will the system be developed from scratch?
- What type of locks do you have?

Once you have (with the help of your catalogs, specification sheets, tecnical bulletins, and experience) digested these answers, you are ready to begin development of a masterkey system.

The purpose of masterkeying is to control access. A key may open one lock, a series of locks, a group composed of two or more series, or every lock in the system.

Glossary

Before we go much further into the subject, it is wise to spend a few minutes defining the terms. This glossary has been prepared by Eaton Corporation's (Yale locks) and is reprinted with their permission.

bicentric cylinder—A pin-tumbler cylinder with two plugs, which effectively make it two cylinders in one. The bicentric cylinder is recommended for large, multilevel masterkey systems, where maximum security and expansion is required.

building masterkey—A masterkey which opens all or most of the locks in an entire building.

change key (or individual key)—A key which will usually operate only one lock in a series, as distinguished from a masterkey which will operate all locks in a series. Change keys are the lowest level in a masterkey system.

changes (key)—See **key changes**.

construction breakout key (CBOK)—A key used by the owner to make all construction masterkeys permanently inoperative.

construction masterkey (CMK)—A key normally used by the builder's personnel for a temporary period during construction. It operates all cylinders designated for its use. The key is permanently voided by the owner when he accepts the building or buildings from the contractor.

control key—A key used to remove the central core from a removable core cylinder.

controlled cross keying—See **cross keying**.

cross keying—When two or more different change keys (usually in a masterkey system) intentionally operate the same lock.

cross keying controlled—When two or more change keys under the same masterkey operate one cylinder.

cross keying uncontrolled—When two or more change keys under *different* masterkeys operate one cylinder.

department masterkey—A masterkey that gives access to all areas under the jurisdiction of a particular department in an organization. regardless of where these areas are in a building or group of buildings.

display room key—A special hotel change key that will allow access to only one designated. even if the lock is in the shutout mode. With many types of hotel locks. this key will also act as a shutout key. making all other change and masterkeys inoperative. except the appropriate individual display room key and the emergency key.

dummy cylinder—One without an operating mechanism; used to improve the appearance of certain types of installations.

emergency key (EMK)—A special, usually top level, hotel masterkey that will operate all the locks in the hotel at all times. An emergency key will open a guestroom lock even if it is in the shutout mode. With many types of hotel locks. this key will also act as a shutout key, making all other change and masterkeys inoperative, except the appropriate individual display room key and the emergency key.

engineer's key (ENG)—A selective masterkey which is used by various maintenance personnel to gain access through many doors under different master and grand masterkeys. The key can be set to operate any lock in a masterkey system and. typically. fits building entrances, corridors, and mechanical spaces. Establishing such a key avoids issuing high level masterkeys to maintenance personnel. See also **selective masterkey**.

floor masterkey—A masterkey that opens all or most of the locks on a particular floor of a building.

grand masterkey (GM or GMK)—A key that operates a large number of keyed-different or keyed-alike locks. Each lock is usually provided with its own change key. The locks are divided into two or more groups, each operated by a different masterkey. Each group can be operated by a masterkey only, or by the grand only, or by a master and the grand.

great grand masterkey (GGM or GGMK)—A key that operates a large number of keyed-different and/or keyed-alike locks. Each lock is usually provided with its own change key. The locks are divided into two or more groups, each operated by a different grand masterkey. Each of these groups is further subdivided into two or more groups, each operated by a different masterkey. A group can be operated by the great grand only, or a grand only, or a master only, or any combination of the three.

guestroom change key—The hotel room key that is normally issued to open only the one room for which it was intended. The guestroom key cannot be used to set a hotel lock in the shutout mode from the outside of the room, nor will it open a hotel lock from the outside if the lock is in the shutout mode.

hotel-function shutout—When a hotel-function lock is in the shutout mode, regular masterkeys of all levels and the guestroom key will not open the lock from the outside of the room. Most hotel-function locks can be set in the shutout mode with the thumbturn or pushbutton from the inside, or with the emergency or display room key from the outside.

hotel keying—Keying for hotel-function locks.

great grand masterkeys, grand masterkeys, or master keys—Depending on the level of the system, these keys function as they would in a normal system except they cannot be used to set a hotel lock in the shutout mode from the outside of the room, nor can they open a hotel lock from the outside if the lock is in the shutout mode.

housekeeper's key—A grand masterkey in a great grand level hotel system that normally operates all the guestrooms and linen closets in the hotel.

interchange—See key interchange.

key bitting (pronounced like fitting)—A number that represents the depth of a cut on a tumbler-type key. A bitting is often expressed as a series of numbers or letters that designate the cuts on a key. The bittings on a key are the cuts which actually mate with the tumblers in the lock.

key bitting depth—The depth of a cut that is made in the blade of a key. See also **root depth**.

key bitting list—A list originated and updated by the lock manufacturer for every masterkey system established. This list contains the key bittings of every masterkey and change key used in the system. Each time an addition is made to the system, all new bittings used are added to the list. It is essential that a complete copy of this list be furnished to any personnel servicing a masterkey system locally. The lock manufacturer should be informed of any changes made locally to a keying system.

key bitting or cut position (also called spacing)—The location of each cut along the length of a key blade. It is determined by the location of each tumbler in the lock. Bitting position is measured from a reference point to the center of each cut on the key. The most common reference point is the key stop, but the tip of the key is sometimes used.

key change number—A recorded number, usually stamped on the key for identification. A key change number can be either the direct bitting on the key or a code number.

key changes (chges)—The total possible number of different keys available for a given type of tumbler mechanism. In masterkey work, the number of different change keys available in a given masterkey system.

key interchange—An undesirable situation, usually in a masterkey system, whereby the change key for one lock unintentionally fits other locks in the system.

key section (KS)—The cross-sectional shape of a key blade that can restrict its insertion into the lock mechanism through the keyway. Each key section is assigned a designation or code by the manufacturer. A key section is usually shown as a cross section viewed from the bow towards the tip of the key. See also **keyway**.

key set (or set)—A group of locks keyed exactly the same way. A key set is usually identified with a key symbol. See also **standard key symbols**.

keyed alike (KA)—A group of locks operated by the same change key. Not to be confused with masterkeying.

keyed different (KD)—A group of locks each operated by a different change key.

keying—A term used in the hardware industry that refers to the arrangement of locks and keys into groups in order to limit access.

keying levels—The stratification of a masterkey system into hierarchies of access. Keying systems are available with one or more levels. The degree of complexity of the system depends on the number of levels used. Generally, the top level masterkey can open all locks in the system. Each successive intermediate level of masterkey can open fewer locks but can open more locks than a change key.

keying system chart—A chart indicating the structure and expansion of a masterkey system, showing the key symbol and function of every masterkey of every level.

keyway (Kwy)—The shape of the hole in the lock mechanism that allows only a key with the proper key section to enter. See also **key section.**

levels—See **keying levels**.

maid's key—A hotel masterkey given to the maid which will give access only to the guestrooms and linen closets in her designated area of responsibility. A hotel is normally divided into floors or sections with a different maid's key for each floor or section. A maid's key will not open a guestroom if the lock is in the shutout mode.

masterkey (MK or Mky)—A key that operates a series of locks. each of which has its own change key.

masterkeyed lock—A lock designed so it can be operated by its own change key. as well as by a masterkey.

multiple key section system (also called **sectional key sections**)—Used to expand a masterkey system by repeating the same or similar key bittings on different key sections. Keys of one section will not enter locks with a different section, yet there is a masterkey section milled so it will enter some or all of the different keyways in the system. See also: **simplex key section.**

paracentric keyway—A keyway in a cylinder lock with one or more side wards on each side projecting beyond the vertical centerline_aof the keyway to hinder picking. See also **simplex key section**.

pin tumblers—Small sliding pins in a lock cylinder. working against drivers and springs and preventing the cylinder plug from rotating until raised to the exact height by the bitting of a key.

plug (of a lock cylinder)—The round part containing the keyway and rotated by the key to transmit motion to the bolt or other locking mechanism.

plug retainer—The part of a lock cylinder which holds the plug in the shell.

privacy key—A change key set up as part of a masterkey system but not operated by any masterkeys or grand masters of any level. This key is set up for such areas as liquor-storage rooms in hotels. narcotic cabinets in hospitals. and food storage closets where valuables are kept.

removable core cylinder—A cylinder containing an easily removable assembly which holds the entire tumbler mechanism, including the plug, tumblers, and separate shell. The cores are removable and interchangeable with other types of locks of a given manufacturer by use of a special key called the control key. See also **control key.**

root depth—Refers to the distance from the bottom of a cut on a key down to the base or bottom of the key blade. Root depth is easy to determine since it measures the amount of blade remaining, rather than the amount which was cut away (bitting depth).

selective masterkey—A special top level masterkey in a grand or great grand system that can be set to operate any lock in the entire system, in addition to the regular floor or section masterkey, without cross keying. Typical selective masterkeys include an engineer's key (ENG), nurse's key (NUR), and attendant's key (ATT). The number of selective masterkeys is normally limited to one or two and should be setup when the original system is established. See **engineer's key.**

simplex key section—A single independent key section that cannot be expanded into a multiple key section system. Simplex key sections such as the Yale "Para" are used for stock locks and small masterkey systems.

spacing—See **key bitting position**.

standard key symbols—A uniform way of designating all keys and cylinders in a masterkey system. The symbol automatically indicates the exact function of each key or cylinder in the system, without further explanation.

tailpiece—The connecting link attached to the end of a rim cylinder which transmits the rotary motion of the key through the door into the locking mechanism.

top level masterkey—The highest level masterkey in a multilevel keying system that fits most of the locks in the system.

tumbler—One or more movable obstructions in a lock mechanism which dog or prevent the motion of the bolt or rotation of the plug and that are aligned by the key to remove the obstruction during locking or unlocking.

uncontrolled crosskeying—See **cross keying**.

visual key control—A system of stamping all keys and the plug face of all lock cylinders with standard key symbols for identification purposes. Other key and cylinder stamping arrangements are available but are not considered visual key control.

Standard Key Symbol Code

Figure 13-10 illustrates keying levels of control and the rudiments of the standard key symbol code. Great grand masterkeys are identified by the letters GGM. Grand masters carry a single letter, beginning with the first in the alphabet and identifying the hierarchy of locks that the individual grand masterkeys open. Masterkeys carry two letters; the first identifies its grand master, the second identifies the series of locks under it. Thus, a masterkey labeled AA is in grand master series A and opens locks in masterkey series A. Masterkey AB is under the same grand master, but opens locks in series B. Master BA is under grand master B and opens locks in series BA. Change keys are identified by their masterkey and carry numerical suffixes to show the particular lock that they open. Any key in the series can be traced up and down in the hirarchy. Thus, if you misplace change key AB4, you know that masterkey AB, grand master A, or the great grand master will open the lock.

There are special keys that are out of series. Some of these keys are mentioned in the glossary, together with the appropriate key symbols for them.

If cross keying is introduced into the system—that is, if a key can open other locks on its level, the cylinder symbol should be prefixed with an X. If the cylinder has its own key, it is identified with the standard suffix. For example, XAA4 is a change-key cylinder that is fourth on this level. It may be cross keyed with AA3 or any other cylinder or cylinders on this level.

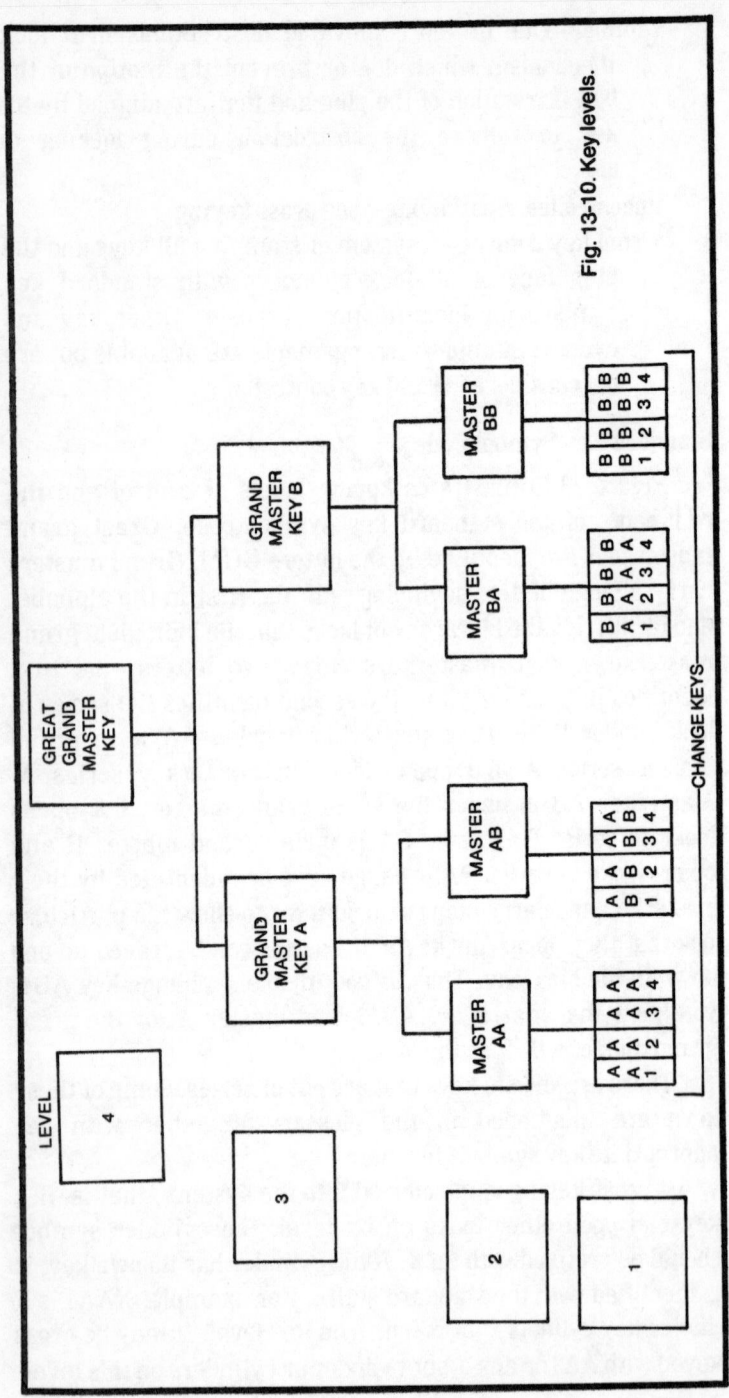

Fig. 13-10. Key levels.

By the same token, masterkey AA, grand master A, and the great grand master will open it. Elevator cylinders are often cross keyed without having an individual change key. It is no advantage to have a key that will operate the elevator cylinder and none other in the system. These cylinders are identified as X1X, X2X, and so on.

The symbols that involve cross keying apply to the cylinder only; all other symbols apply to the cylinder *and* the key. This point may seem esoteric, but ignoring it causes the factory and everybody else grief. There is, for example, no such thing as an X1X key. Nor is there an XAA4 key. Change key AA4 fits cylinder XAA4 that happens to be cross keyed with another cylinder. Key AA4 does not fit any cylinder except XAA4.

There are certain advantages to using the standard key symbol code:

- It is a standardized method for setting up the keying systems.
- It maintains continuity from one order to the next.
- It indicates the position of each key and each cylinder in the hierarchy.
- It helps to control cross keying since each cross keyed cylinder is clearly marked.
- It offers a method of projecting future keying requirements.
- It can be easily rendered on a chart.
- It allows better control of the individual keys within the system.
- It is a simple method of selling and explaining the keying system to the architect or building owner.

Selling the System

It is important to be able to communicate the advantages of the key symbol code and the implications it has for setting up an ordered, coherent, and secure keying system. This may take some selling on your part since architects and building owners tend to think of keys and locks as individual entities and not part of a larger system. Selling a comprehensive masterkeying system involves the following:

- Being able to explain the subject of masterkeying to the architect or owner.
- Reviewing the plans of the building(s).
- Choosing the proper level of control required.
- Selecting a keying system that will cover present and future expansion requirements.
- Presenting the system to the architect or owner at a meeting dealing *specifically* with keying.
- Recording all changes to the system as agreed upon at the meeting.
- Marking plans with the proper key symbol on the side of each door where the key is to operate.
- Presenting the owner with a schematic layout of the entire system, showing him the layout of the masters, grand masters, etc.

VARIATIONS IN THE MASTERKEY SYSTEM

So far I have discussed the structure of the masterkey system. Within this broad structure there are many opportunities for variations. Some of these variations involve the possible range of keyways; others involve special hardware, such as removable cylinders, master-ring cylinders, and rotating tumblers. Each of these variations can extend the range, flexibility, and security of the system. A locksmith must be conversant with all of them.

Keyway Variations

Figure 13-11 illustrates a system of control based on keyway design. General Lock's series 800 key section passes all cylinders in the system; 29, 30, 31, 32, and 33 are submasters, each passing four cylinders; the change keys are restricted to their own individual cylinders. In its fullest expansion, the system includes 35 different keyways on the change-key level and 4 masterkey levels.

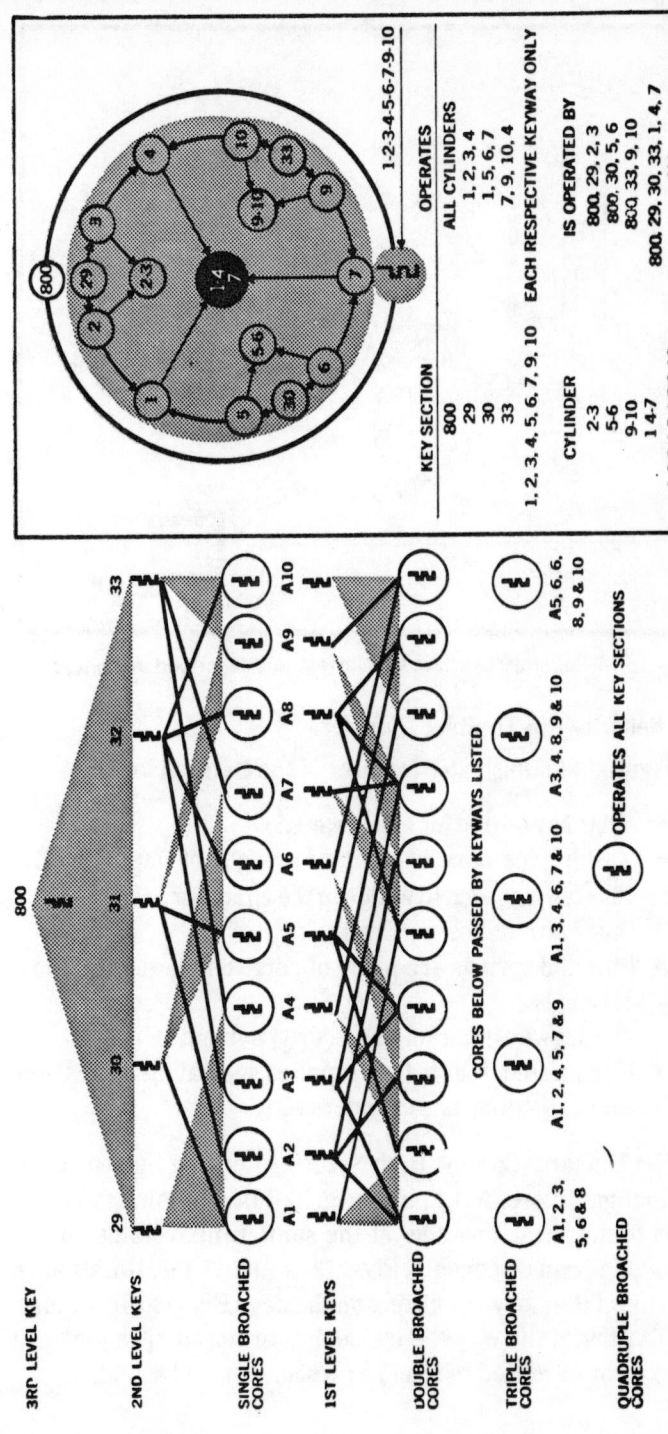

Fig. 13-11. A keyway system developed by the General Lock Company.

179

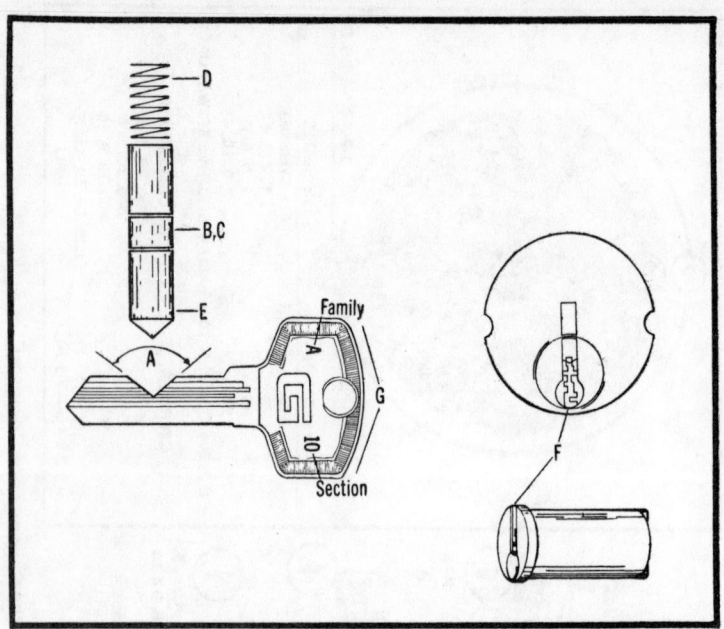

Fig. 13-12. General locks have special features described in the text.

High Security Pin-Tumbler Cylinders

Figure 13-12 illustrates features of the General lock:

- A 100° key bitting for long wear (A).
- Master pins have a minimum length of 0.040 in. (B). Shorter pins tend to wedge in the chamber.
- Only two pins are standard (C).
- Pins and springs are made of corrosion-resistant alloy (D and E).
- The keyway is part of the security system (F).
- If requested, the factory supplies special identification for keys, cores, and cylinders (G).

The Emhart (Corbin) High Security Locking System uses rotating and interlocking pins (Fig. 13-13). The pins must be raised to the shear line and, at the same time, rotated 20° so the coupling can disengage (Figs. 13-14 and 13-15). Rotation is by virtue of the skew-cut bitting on the key (Fig. 13-16). Figure 13-17 illustrates the way the cylinder is armored. The pins are protected by hardened rods and a crescent-shaped shield.

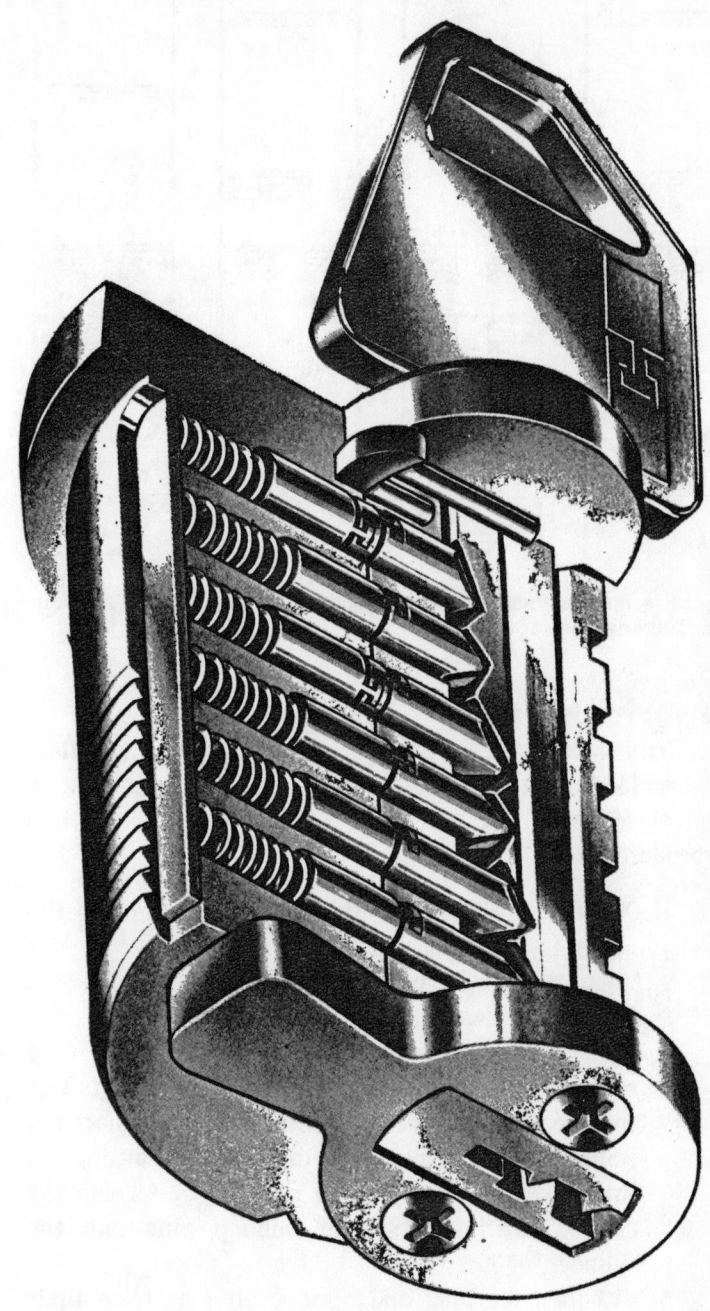

Fig. 13-13. Corbin's High Security Locking System depends upon split and rotating pins.

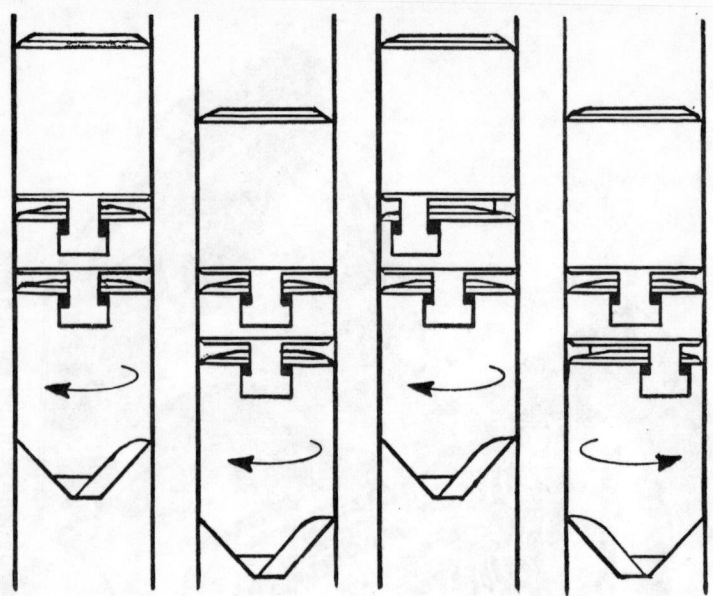

Fig. 13-14. Pins are rotated in either direction to allow the joint to uncouple. (Courtesy Emhart Corp.)

Removable-Core Cylinders

Removable-core cylinders are increasingly popular. Figure 13-18 illustrates the Corbin cylinder, a type that is typical of most. To rekey the change key, follow this procedure:

1. Obtain a Corbin rekeying kit. The kit includes the necessary pins, gages, and tools.
2. Mount the cylinder in a vise.
3. Remove the plug retainer.
4. Select the key with the deepest bitting as the plug extractor. Normally the grand masterkey meets this specification; however, there are instances where the engineer's key will have the deepest bitting. A shallow-cut key complicates matters by forcing the control pins, drivers, and buildup pins into the cylinder.
5. Withdraw the plug and remove all pins from their chambers. Figure 13-19 illustrates this procedure.

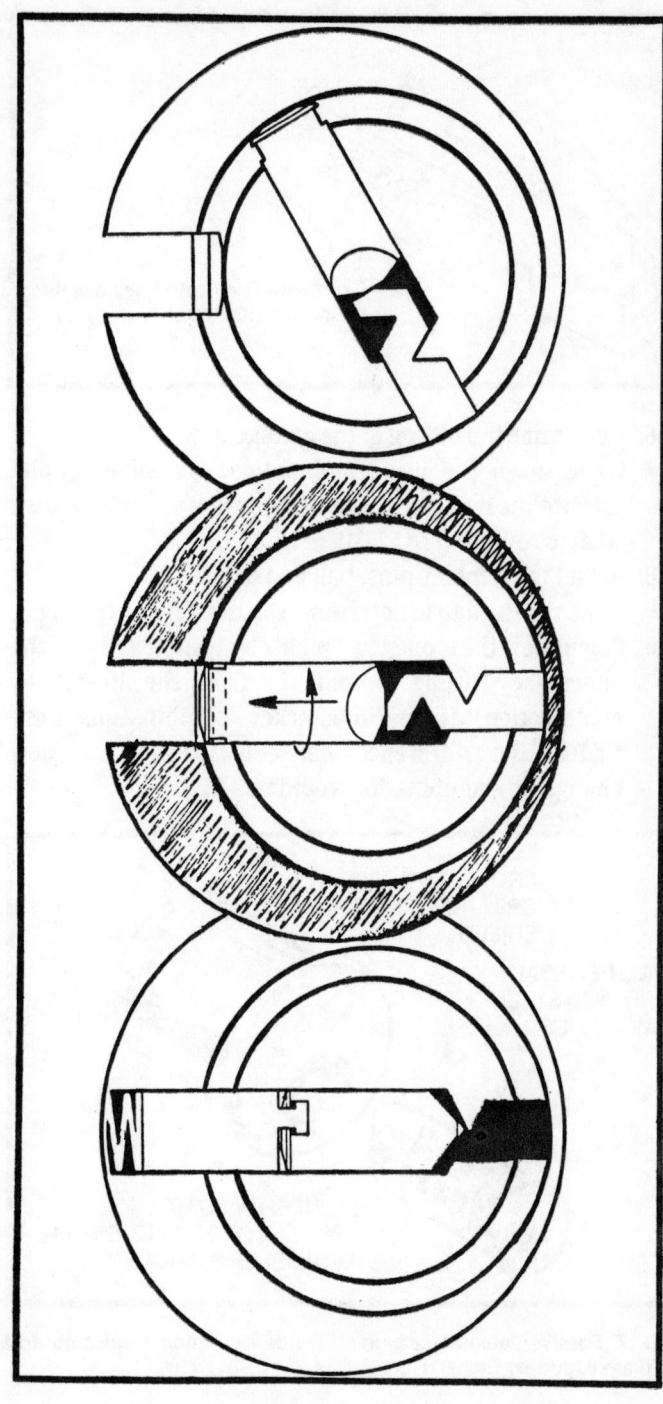

Fig. 13-15. Pins must be rotated and brought to the shear line for the lock to open.

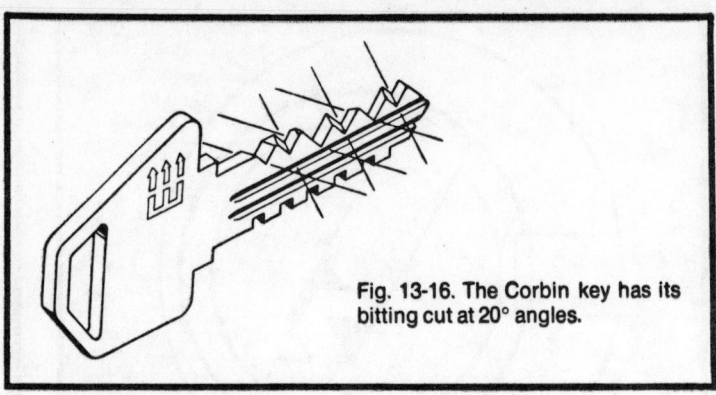

Fig. 13-16. The Corbin key has its bitting cut at 20° angles.

6. Determine the bitting of the change key.
7. Write down the new combination. As an example, suppose the original change-key bitting is 513525 and we wish to reverse it to 525315 (Fig. 13-20).
8. Install the tumbler pins, ball end down.
9. Use a depth gage to determine the masterkey bitting.
10. Calculate the master pins by subtracting the change-key bitting combination from the masterkey combination. If the masterkey combination were 525763, the difference between it and the new change-key combination would be 448.

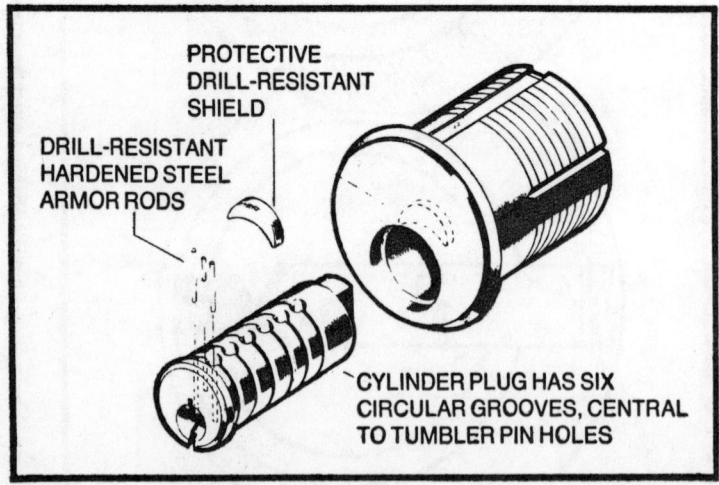

Fig. 13-17. Passive defense measures include hardened steel pins and armor plate. (Courtesy Emhart Corp.)

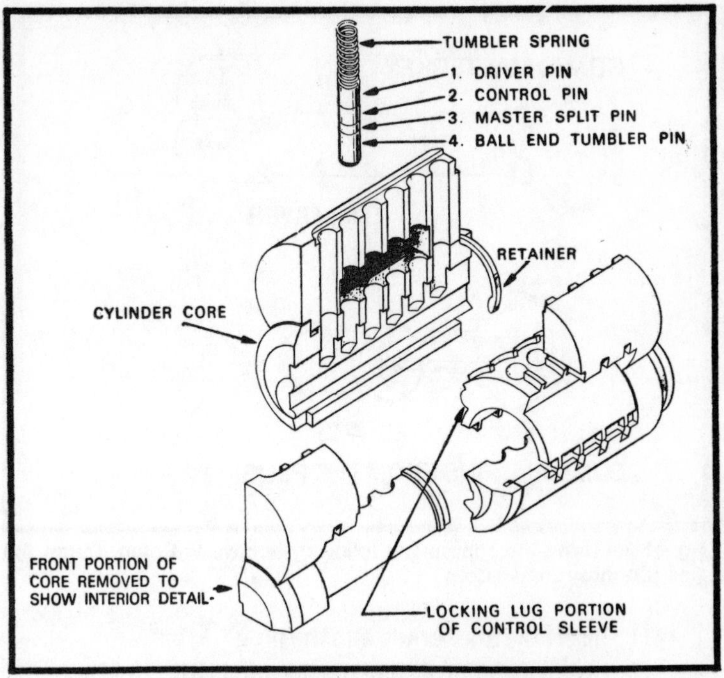

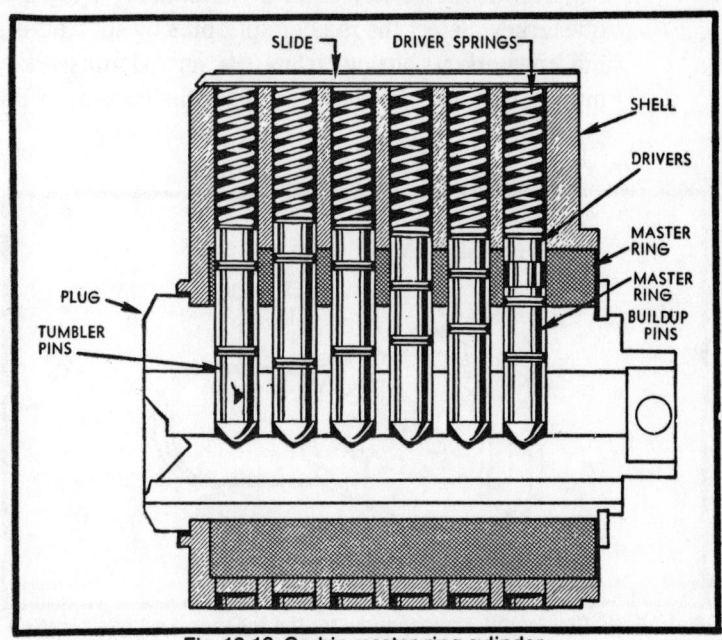

Fig. 13-18. Corbin removable-core cylinder.

TUMBLER SPRING
1. DRIVER PIN
2. CONTROL PIN
3. MASTER SPLIT PIN
4. BALL END TUMBLER PIN

RETAINER

CYLINDER CORE

FRONT PORTION OF
CORE REMOVED TO
SHOW INTERIOR DETAIL

LOCKING LUG PORTION
OF CONTROL SLEEVE

SLIDE   DRIVER SPRINGS

SHELL

DRIVERS

MASTER
RING

MASTER
RING

BUILDUP
PINS

PLUG

TUMBLER
PINS

Fig. 13-19. Corbin master-ring cylinder.

185

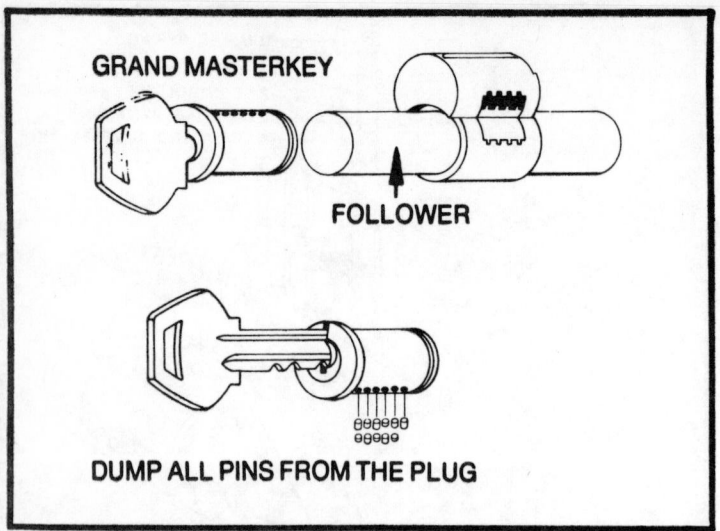

GRAND MASTERKEY

FOLLOWER

DUMP ALL PINS FROM THE PLUG

Fig. 13-20. Using the appropriate follower, remove the plug. Dump the pins. (Courtesy Emhart Corp.)

11. Insert the masterkey into the plug.
12. Install the appropriate master pins (Fig. 13-21).
13. Remove the masterkey carefully and insert the grand masterkey. Select the master split pins by subtracting the masterkey bitting from the grand masterkey bitting. All pins should be flush with the surface of the plug.

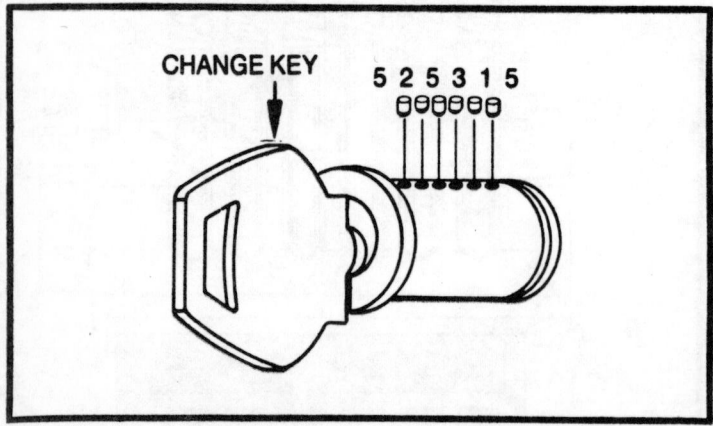

CHANGE KEY 5 2 5 3 1 5

Fig. 13-21. Pin length corresponds to the change-key combination. (Courtesy Emhart Corp.)

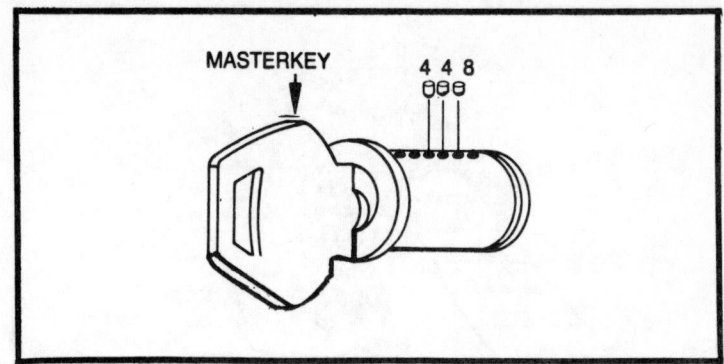

Fig. 13-22. Subtract the change-key combination from the masterkey combination. The difference represents the length of the master pins. (Courtesy Emhart Corp.)

14. Asemble the plug and cylinder.
15. Test all keys.
16. Lubricate the keyway with a pinch of powdered graphite.

Master-Ring Cylinders

The Corbin master-ring cylinder is shown in Fig. 13-22. The change key operates the plug plungers, and the masterkey operates the plunger in the master ring. Sometimes called "two-in-one" cylinders, these cylinders increase the range of key combinations for any given system. To rekey the change key, follow this procedure:

1. Mount the plug in a vise and remove the cylinder slide with a pair of pliers or a small chisel.
2. Remove the springs, drivers, and pins (Fig. 13-23).
3. Ream pin holes through the shell, master ring, and plug (Fig. 13-24).
4. Assuming that the original combination was 414472, reversing the combination gives 274414. This will be the combination of the new change key.
5. Reverse the pins to conform with the new combination. That is, the pin that was first goes into the last chamber; the pin that was second goes in to the fifth chamber, and so on.

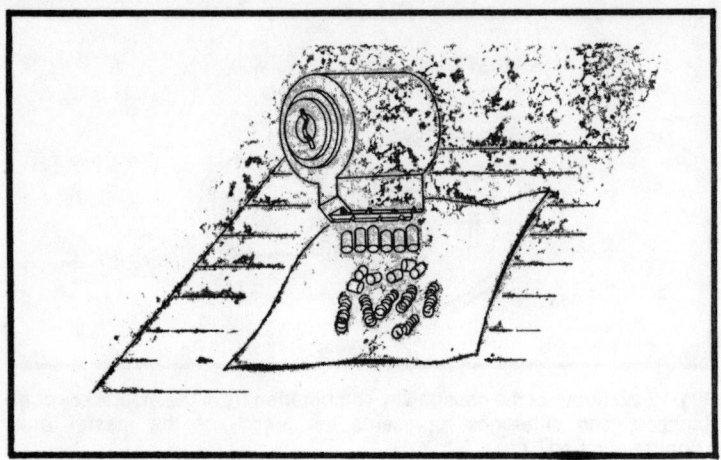

Fig. 13-23. After the slide is withdrawn, remove springs, pins, and drivers. (Courtesy Emhart Corp.)

6. As each pin is installed, tamp it home with a drill bit and turn the key. If the key will not turn, you have confused the pin sequence.
7. Assemble the lock and test.

To rekey the masterkey, follow this sequence:

1. Mount the cylinder in a vise and remove the cylinder slide with a pair of pliers or a small chisel.
2. Remove the springs, drivers, and pins.

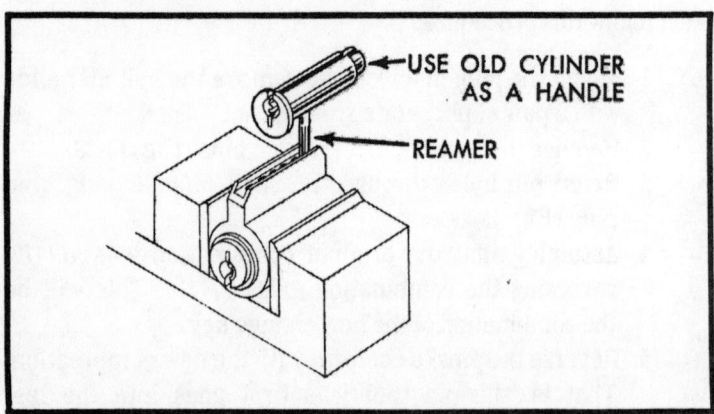

Fig. 13-24. Ream the pin holes through the shell, master ring, and plug. (Courtesy Emhart Corp.)

3. Determine the masterkey bitting with a gage. The combination runs from the shoulder to the tip of the key, the reverse of the usual sequence. As an example, let it be 678572.

4. Write down the change-key combination and reverse it. Suppose the combination is 275414. Reversed. it is 414572.

5. Subtract the reversed change-key combination from the masterkey combination (678572 − 414572). The difference is 264000.

6. Insert the masterkey into the cylinder and select the appropriate buildup pins. In this case the pins are 264.

NOTE

IF ANY NUMBER IN THE CHANGE-KEY COMBINATION IS GREATER THAN THE MASTERKEY NUMBER ABOVE IT. YOU MUST USE A NEGATIVE NUMBER BUILDUP PIN IN THE CHAMBER.

For example. a 678572 masterkey combination with an 814572 change-key combination requires a −2 buildup pin, together with a 6 and 4 pin.

7. As each buildup pin is installed. seat it with a drill bit and turn the key to determine that the correct pin has been installed.

8. Insert the drivers into the cylinder chambers. Either of two drivers are used. depending upon the lock style. Spool drivers (No. J-172) are furnished with mortise cylinders; straight drivers (No. M-099) are used on other cylinders.

9. Insert the springs into their chambers.

10. Holding the springs down with your thumb. try the change key. Do the same for the master and grand masterkey.

11. Mount the cylinder in the slide. hammering the slide down for a secure fit. Be careful not to damage the threads on mortise cylinders.

12. Try all the keys.

Chapter 14
Decoding Locks
by the Numbers

Another skill required by the locksmith is the ability to cut different key types when only the lock manufacturer and key number are known. The locksmith may never see the lock or key; at best he may have only the lock. Certain information may be required of the customer, depending on city or county ordinances. In most cases you will need to ask:

- Who owns the lock?
- Has the owner given authorization to have a key made for the lock? Requests involving high-security locks must be forwarded to the factory together with a written release from the owner.

The code varies with each manufacturer, but includes this information:

- Key blank type
- Cut spacing
- Cut depths

Reading Codes

Direct-digit and indirect codes are most popular. The direct-digit code consists of one or two letters followed by a series of numbers. The letters identify the lock series, and the numbers refer to the depths of the cuts. The series may start at the last tumbler (the one closest to the key tip) or at the first (next to the key shoulder). Indirect codes are safer: the numbers must be decoded before the cuts can be identified. Professional locksmiths have the necessary code books.

The smallest number—either 0 or 1—in both kinds of codes represents the shallowest cut; 9, the last of the series, represents the deepest cut. Once you have the number, you can usually determine the type of lock. For example, key code LL76 is limited to desks. This fact can be useful in that it narrows the range of possible blanks.

The code book says that LL76 is a Yale desk lock and that you can use ILCO 01122A, National Y12, or Yale 9278A blanks. The depth and spacing charts show five depths, each 0.020 in. apart, and five spacings. Counting from the tip to the shoulder, the first four cuts are spaced 0.094 in. apart; the last cut is 0.140 in. from the shoulder. Code number LL76 translates as 24253 and is read from the tip to the shoulder. To cut a key with this code, place a Yale No. 2 depth and spacing key in the vise, together with the appropriate blank. Cut the o, 2 bits. Repeat the operation with Nos. 3, 4, and 5 depth and spacing keys.

Depth and Spacing Keys

Depth and spacing keys simplify the work. These keys are milled to various depths and spacings. Since manufacturers vary these dimensions, you need a complete key set, or at least one that covers the major American locks.

Schlage Wafer-Tumbler Keys

Schlage keys are easy to cut by code since the cuts all have the same depths. Three Schlage keys and one depth key is all that is required.

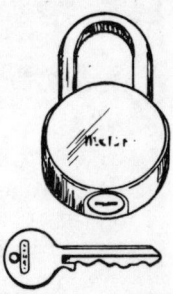

Chapter 15
Double-Bitted
Locks and Keys

The Junkunc Brothers American double-bitted cylinder lock is usually found in padlocks and office and utility locks (Fig. 15-1).

OPERATION

When the double-bitted key is inserted, it passes through the center of the tumblers (as in a disc lock) and aligns them to the shear line, allowing the plug to be rotated. But the key and the tumbler arrangement is different from that of the regular disc lock.

The key cuts are wavy in appearance; thus the tumblers have to align in a wavelike configuration for the lock to open. Further, there are no definitive tumbler cuts on the key. This is because the key holds 10 or more tumblers compressed together and held in a locked position by means of a Z-shaped wire within the tumblers.

All the tumblers are uncoded, meaning they are all of a standard cut. So in order for them to turn within the cylinder, the tumblers have to be cut down. A special keying tool is used for this purpose.

Fig. 15-1. A double-bitted lock and key. (Courtesy American Lock Company.)

CUTTING DOWN THE TUMBLERS

1. Once the tumblers have been inserted into the plug (with the tumblerspring in place), insert the precut key into the plug.
2. Mount the plug in your vise firmly.
3. Attach the keying tool to a ¼ in. drill. Drill the back of the plug. Since the inside diameter of the drill is the same as the outside diameter of the plug, the individual tumblers will cut down to what will be the shear line.
4. Trim the tumblers with a light wire brush to take off any burrs.
5. Insert the plug into the cylinder and test it. Attach the retainer screw and withdraw the key.

The tumblers, since they are uncoded, can be used within any plug. The tumbler spring, because of the shape, holds the various tumblers in the locked position. Only with the insertion

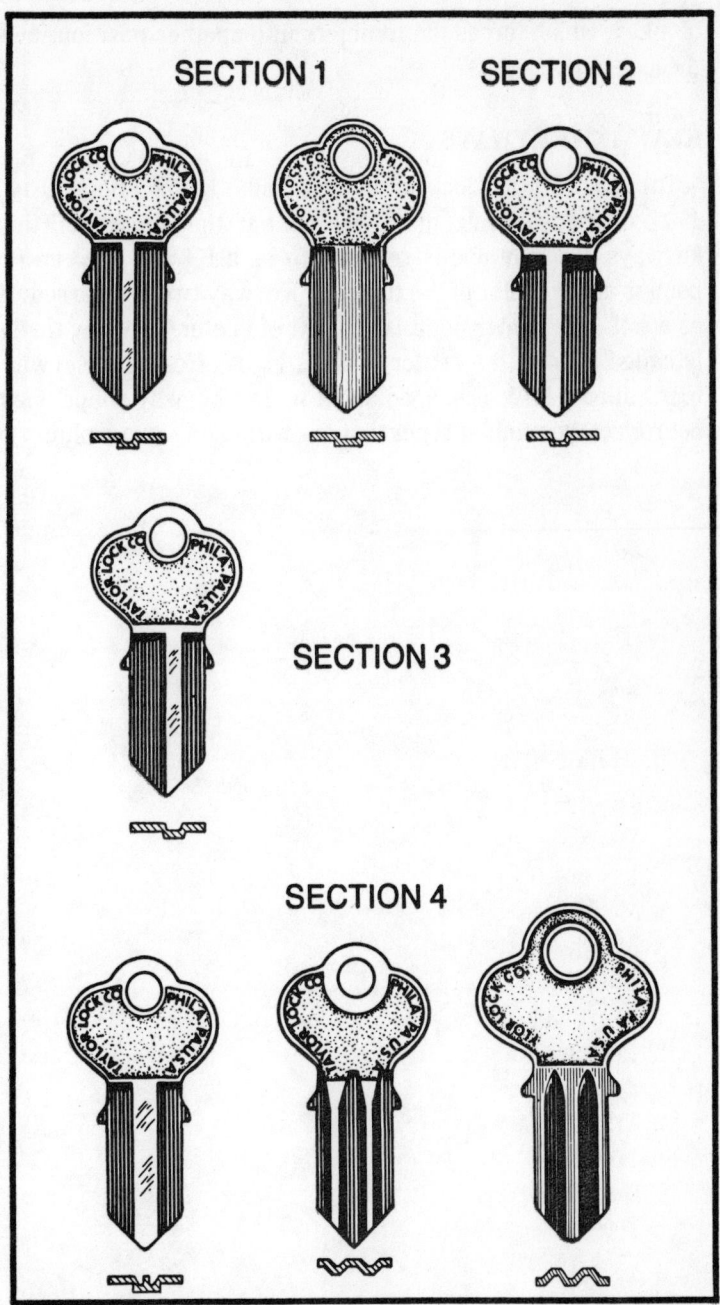

SECTION 1 SECTION 2

SECTION 3

SECTION 4

Fig. 15-2. The four basic double-bitted key sections. (Courtesy Taylor Lock Company.)

of a key, which forces the tumblers into another position, can the lock be opened.

KEYS AND KEYWAYS

The double-bitted lock takes four basic key sections (Fig. 15-2). These sections, of course, match the shapes of the keyways. Keyway one is referred to as a K4 and the center point is at the center of the tumbler. Keyway two is referred to as a K4L; the center point is just left of center. Keyway three is called a K4R; the center point is right of center. Keyway four, called a K4W, is shaped like a *W*. The keyway shape does not reflect the tumbler types that are within any given plug.

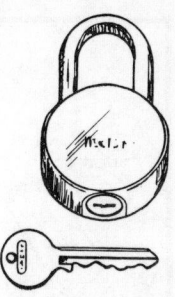

Chapter 16

Vending Machine Locks

The Ace lock, manufactured by the Chicago Lock Company, has become standard on vending machines. The arrangement of the pin tumblers increases security and requires a different type of key—one that is tubular. A typical Ace lock and key is shown in Fig. 16-1. The cam works directly off the end of the plug.

The key has its bitting disposed radially on its end. The depth and spacing of each cut must match the pin arrangement. In addition there are two notches, one on the inside, the other on the outer edge of the key. These notches align the key to the lockface. Otherwise the key could enter at any position.

The key bittings push the pins back, bringing them to the shear line. Once the pins are in alignment, the key is free to turn the plug and attached cam.

The pins within the Ace lock are entirely conventional in construction, with the exception of the bottom one. A ball bearing is sandwiched between the pin and its driver. The bearing reduces friction and increases pin life. The pin in question is not interchangeable with others in the lock. Pin tolerance is extremely critical. There is no room for sloppy key cutting.

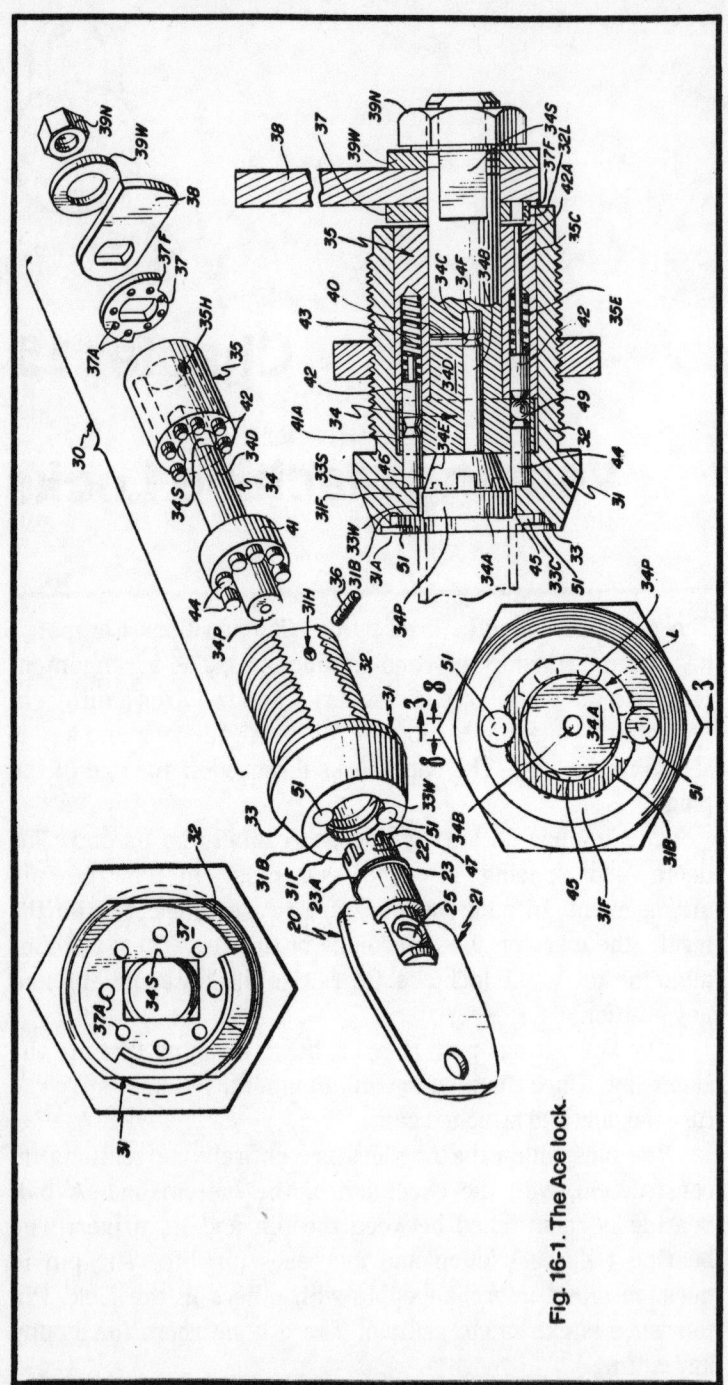

Fig. 16-1. The Ace lock.

DISASSEMBLY

To disassemble the Ace lock, follow this procedure:

1. Place the lock into its holder. Ace makes a special vise for these cylinders.
2. Drill out the retainer pin at the top of the assembly. Use a No. 29 drill bit and stop before the bit bottoms in the hole (Fig. 16-2).
3. Remove what is left of the pin with a screw extractor.
4. Insert the appropriate plug follower into the cylinder. Apply light pressure. Plug follower dimensions are: length 1.50 in.; outside diameter 0.375 in.; inside diameter 0.312 in.
5. Lift off the outer casing from the bushing assembly.
6. Scribe reference marks on the plug sections as an assembly guide.
7. Remove the follower from the cylinder.

NOTE
PERFORM THIS OPERATION CAREFULLY. THE PINS ARE UNDER SPRING TENSION AND MUST BE KEPT IN ORDER. IF NOT, YOU WILL HAVE A MONUMENTAL JOB SORTING THE PINS. THERE ARE 823.543 POSSIBLE COMBINATIONS.

ASSEMBLY

Assembly is the reverse order of disassembly. Replace the retaining pin with an Ace part, available from locksmith supply houses.

REKEYING

Rekeying is not difficult if you approach the job in an orderly manner.

1. Cut the key. There are seven bit depths, ranging from 0.020 to 0.110 in. in 0.015 in. increments.
2. Mount the bushing assembly upright.
3. Remove the pins with tweezers.

NOTE
PIN LENGTHS RANGE FROM 0.025 TO 0.295 INCHES IN IN-CREMENTS OF 0.015 INCHES DRIVERS ARE AVAILABLE IN 0.125, 0.140, AND 0.180 INCH LENGTHS.

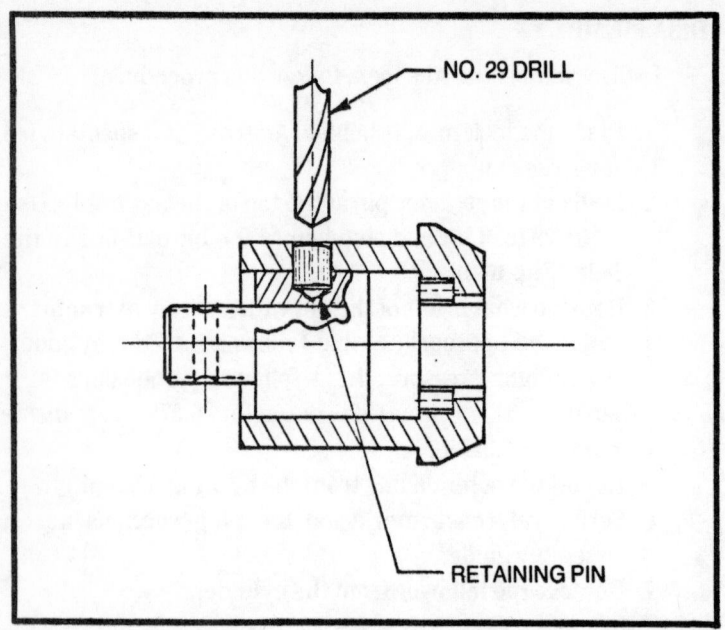

NO. 29 DRILL

RETAINING PIN

Fig.16-2. Use a No. 29 drill bit to remove the retaining pin. Alternately, use a No. 42 bit, thread a small metal screw into the hole and extract the pin and screw by prying upward on the screwhead. (Courtesy Desert Publications.)

4. Select new pins, using the key combination as a guide.
5. Install the pins. The flat ends of the core pins are toward the key (See Fig.16-1). This pattern must be followed:

Core pins 1, 2, and 3 require 0.180 in. drivers.
Core pins 4 and 5 require 0.140 in. drivers.
Core pins 6 and 7 require 0.125 in. drivers.

Pins are numbered clockwise from the top as you face the lock.

6. Insert the key.
7. Install the plug in the cylinder with the scribe marks aligned.
8. Insert a new retaining pin and give it a sharp tap with a punch.
9. Test the key. It may be necessary to rap the cam end of the plug with a mallet.

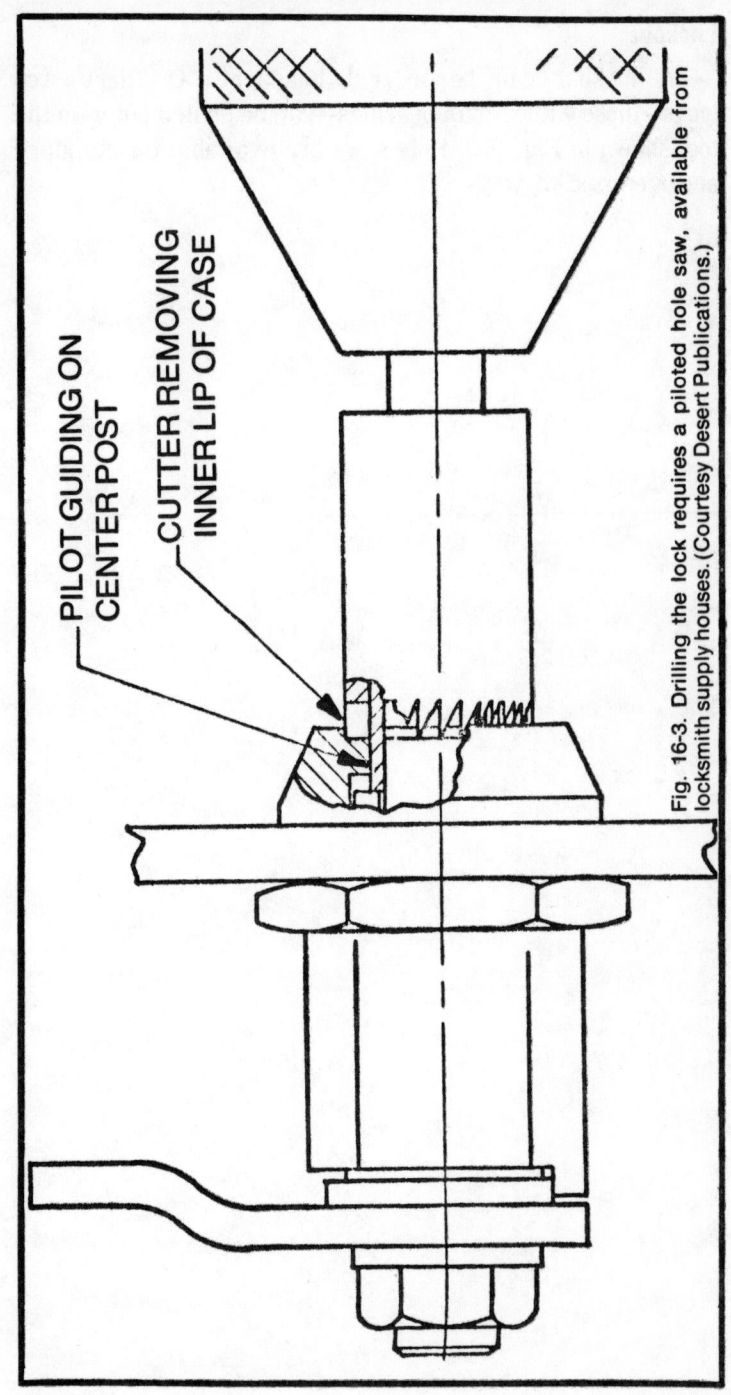

PILOT GUIDING ON
CENTER POST

CUTTER REMOVING
INNER LIP OF CASE

Fig. 16-3. Drilling the lock requires a piloted hole saw, available from locksmith supply houses. (Courtesy Desert Publications.)

Lockout

A lockout can be a real headache. Ordinary Ace locks—those without ball bearings—can be drilled out with the tool shown in Fig. 16-3. Hole saws are available for standard and oversized keyways.

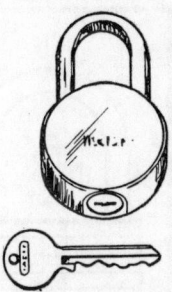

Chapter 17

Keyed Padlocks

Padlocks have many uses. They are used to secure out-buildings, bicycles, buildings under construction, tool boxes, paint lockers, and even automobile hoods. Because of this wide use, it pays a locksmith to have a good knowledge of these locks.

While padlock exteriors vary, the functional and operational differences are few and are similar to other locks. Padlocks may use pin tumblers, wards, wafers, levers, or a spring bar. Some must be shackled closed before the key can be removed; this feature is made possible by a spring-loaded coupling. Key security is improved and there is less likelihood of leaving the lock open.

CHOOSING A PADLOCK

Ask the customer if he has a brand preference and then ask:

- If width, case length, and shackle clearance are critical (Fig. 17-1).
- Where the lock will be used.
- How often the lock will be opened. The price of the lock has a direct relationship to its wearing qualities.

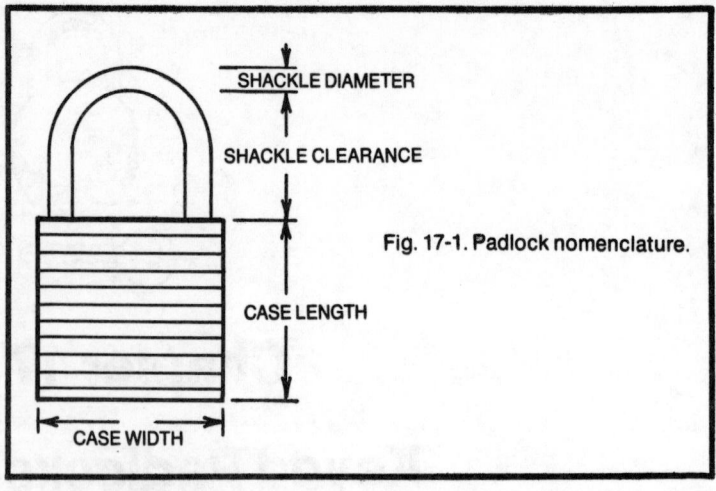

Fig. 17-1. Padlock nomenclature.

Fig. 17-1. Padlock nomenclature.

- If the lock is intended to secure valuable property. An inexpensive lever lock would be adequate to keep children from straying into the backyard, but would be inappropriate for a boat trailer.
- If the lock will be used indoors or outdoors. If an outdoor lock will be protected from the elements.

WARDED LOCKS

Warded locks have limited life spans, particularily when used outdoors. The cheapest locks of this type are only good for

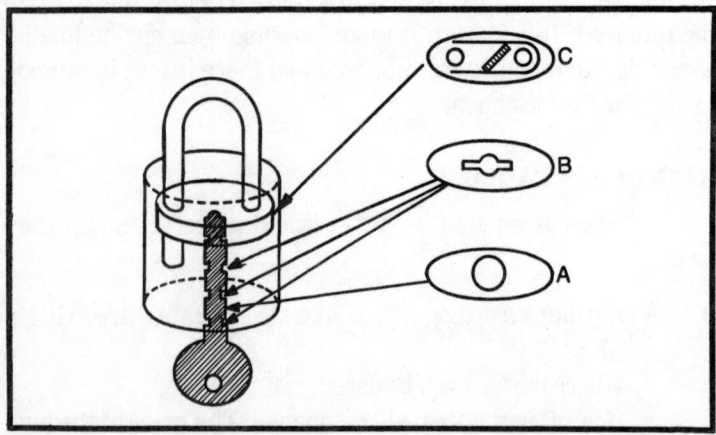

Fig. 17-2. A warded padlock. Section A represents a rotating-disc keyway; section B one of the three wards; section C the spring bar.

a few thousand openings and, when locked, give minimal security.

Most of these locks have three wards, although the cheaper ones have only two. Figure 17-2 illustrates the principle. The key must negotiate the wards before it can disengage the spring bar from the slot in the shackle end. Keys are flat or corrugated. The latter is a mark of the Master Lock Co.

Passkeys

Figure 17-3 ilustrates how a passkey is cut to defeat the wards. The broad tip of the key opens the spring bar; locks with two spring bars require a key cut as shown in Fig. 17-4.

Some corrugated keys can be reversed to fit other locks by filing it on the back of the blades. Alternately, you can file all the unnecessary metal off, converting the key into a passkey.

In many localities passkeys are illegal and, unless you are a locksmithing student from an accredited school, a locksmith trainee, or a licensed locksmith, possession of such a key is a criminal offense. If you have the need for a passkey, keep it in a safe place in your home or office.

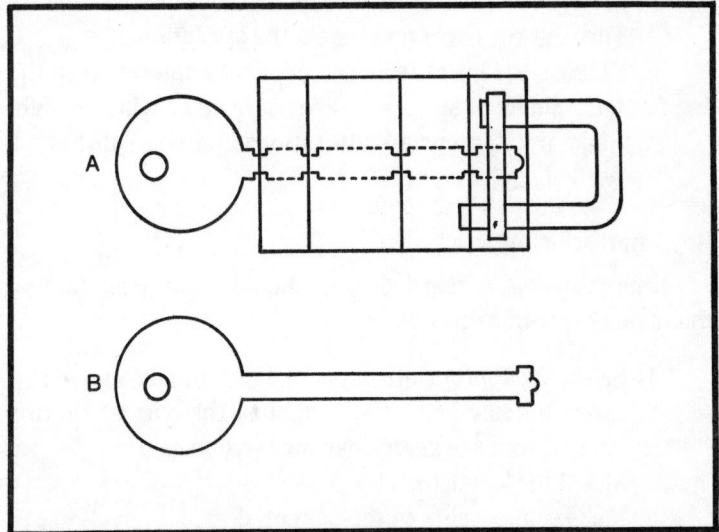

Fig. 17-3. Wards and their limitations. View A shows the ward arrangement and the necessary key bitting; view B illustrates a passkey.

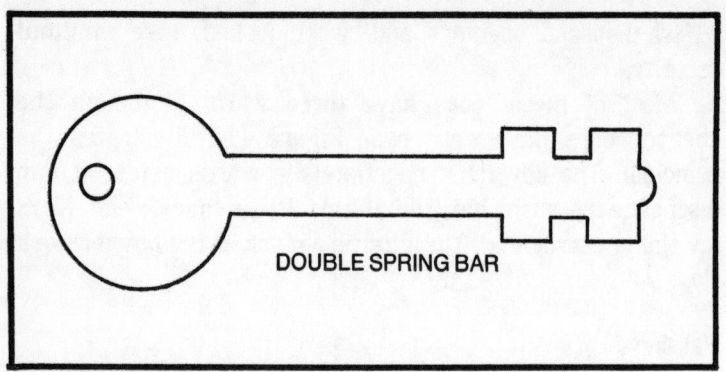

DOUBLE SPRING BAR

Fig. 17-4. A passkey for a lock with two spring bars.

Key Cutting

A warded padlock key is simple to duplicate. Follow this procedure:

1. Using the original key as a guide, select the appropriate key blank.
2. Smoke the original key and mount the original and the blank in your vise.
3. Using a 4 in. warding file, cut away the excess metal until the blank is an exact copy of the original.
4. Turn the keys over and repeat the operation.
5. Remove the burrs from the duplicate and test it in the lock. Should it stick, cuts are not deep or wide enough. Make the appropriate alterations and you will have a perfect duplicate key.

Key Impressioning

Impressioning a warded key should take less than 5 minutes. Follow this procedure:

1. Select the appropriate blank and thoroughly smoke it.
2. Insert the key and twist it against the wards. Do this several times to get a clear impression.
3. Mount the key in the vise.
4. Make shallow cuts where indicated.
5. Smoke the key again and try it in the lock.
6. Remove the key and file the cuts as indicated.

7. Continue to smoke, test, and file until the key turns without protest. Do not go overboard with the file. If the bits are too deep, the key will work but may break off in the lock.

Repairs to the lock itself are out of the question, since it would be cheaper to buy a new one.

AMERICAN (JUNKUNC BROTHERS)

All American (Junkunc Brothers) padlocks share the same patented locking device—two hardened steel balls fitting into grooves in the shackles. This arrangement is the best ever devised. Applying force on the shackle wedges the balls tighter. The H10 model, for example, requires more than 5000 lb to force; test locks have been stressed to 6000 lb and still worked. All these locks use a 10-blade tumbler. The exceptions to these statements are those locks with a deadlock feature. The key must be turned to lock the shackle. These padlocks are made on an entirely different principle and are not discussed here.

American padlocks have a removable cylinder that simplifies servicing. If a customer wants keyed-alike locks, the modification takes only a few minutes. One may also key-alike different models.

These locks have three basic subassemblies: cylinder assembly, locking mechanism assembly, and the shackle assembly.

Cylinder Removal and Installation

Follow this procedure:

1. Open the lock, exposing the retaining screw at the base of the shackle hole.
2. Remove the retaining screw (I in Fig. 17-5). If the screw is stubborn, use penetrating oil, letting it set a few moments before attempting to turn the retaining screw.
3. Strike the side of the lock with a leather mallet. The purpose is to force the retaining pin (C) into the space vacated by the retaining screw. Referring to the

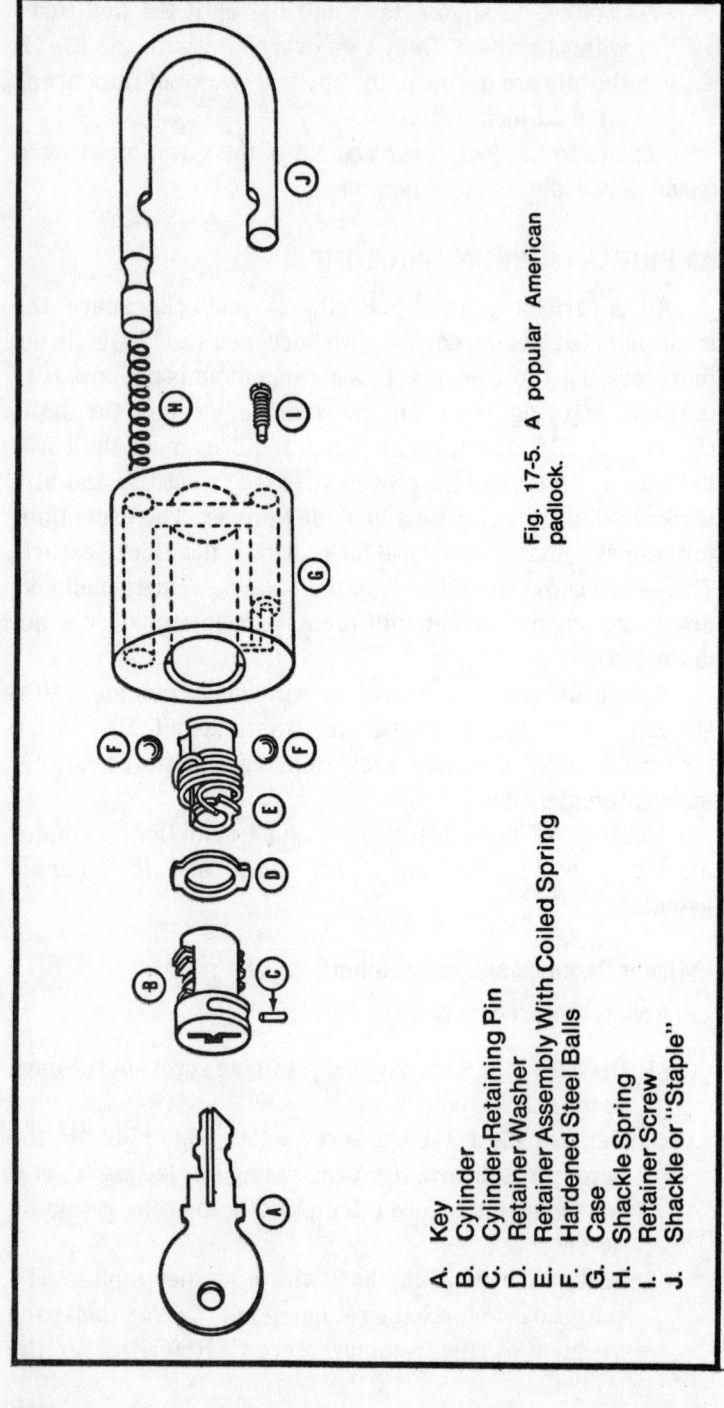

Fig. 17-5. A popular American padlock.

A. Key
B. Cylinder
C. Cylinder-Retaining Pin
D. Retainer Washer
E. Retainer Assembly With Coiled Spring
F. Hardened Steel Balls
G. Case
H. Shackle Spring
I. Retainer Screw
J. Shackle or "Staple"

drawing, note how the pin retains the cylinder in the slot shown.

4. Withdraw the key together with the cylinder.
5. Place a new cylinder in the case.
6. Assemble in the reverse order of disassembly.

NOTE

THE H10 SERIES USES A PLATED SILVER CAP DIMENSIONALLY INDENTICAL TO THE BRASS CAPS ON THE OTHER MODELS.

The Locking Mechanism

The brass retainer is the heart of the assembly (E in Fig. 17-5). Retainers vary in size according to the lock model. A30 and AC20 retainers have a groove milled in the body. L50, K60, and KC40 retainers are plain. The H10 is much larger than the others.

The retainer assembly has three parts—a washer, coil spring, and retainer body. The washer (D) fits over the retainer. One end is hooked to accept the coil spring. The other end of the spring is moored in a hole on the retainer.

To remove the retainer, follow this procedure:

1. Grasp the retainer body (E) with needle-nosed pliers.
2. Rotate the retainer 45° and pull.
3. Remove the steel balls.

To replace, follow these steps:

1. Grease the balls so they will stay put.
2. Replace the balls and spread them apart.
3. Exert pressure against the shackle to hold the balls.
4. Determine that the washer is correctly aligned with the spring. The free end of the spring has to be in line with the retaining-pin hole in the case.
5. Place the retainer assembly about halfway in the case, with the washer riding in the grooves provided. The end of the spring also rides in the groove.
6. Twist the retainer about a quarter turn to the right and down.

Shackle

The various models have different shackle lengths and diameters. The spring (H in Fig. 17-5) must match. For example, an L-shaped spring is used on the L50 lock. An extra pin is used to guide the long spring on the K60.

The shackle is secured in the H10 with a hardened steel pin; other models secure the shackle with the retainer mechanism.

Pin-Tumbler Padlocks

The American five pin-tumbler padlock has been designed so the cylinders may be quickly changed (Fig. 17-6). Late production 100 and 200 series use the same cylinder assembly and keyway, reducing the inventory load.

Assembly and disassembly of the pin-tumbler padlock is only slightly different from that of the other American padlocks. To change the cylinder, follow this procedure:

1. Open the padlock, exposing the retaining screw at the base of the shackle.
2. Remove the retaining screw with a small screwdriver.
3. Pull the cylinder out of the case.

NOTE

LEAVE THE LOCK UNLOCKED: DO NOT DEPRESS THE SHACKLE.

4. Insert a new cylinder in the case. Replace the screws, bringing the cylinder almost flush with the case.

To assemble the locking mechanism:

1. Assemble the retainer (D of Fig. 176) and the retainer washer (C) with the free end of the spring tightly against the left side projection on the washer.
2. Place the shackle spring (H) in the hole on the end of the shackle (I).
3. Insert the shackle spring into the deep well of the lock body.
4. Drop the balls (E) into the case bottom and move them into the pockets with a small screwdriver. There must be room for the retainer assembly.

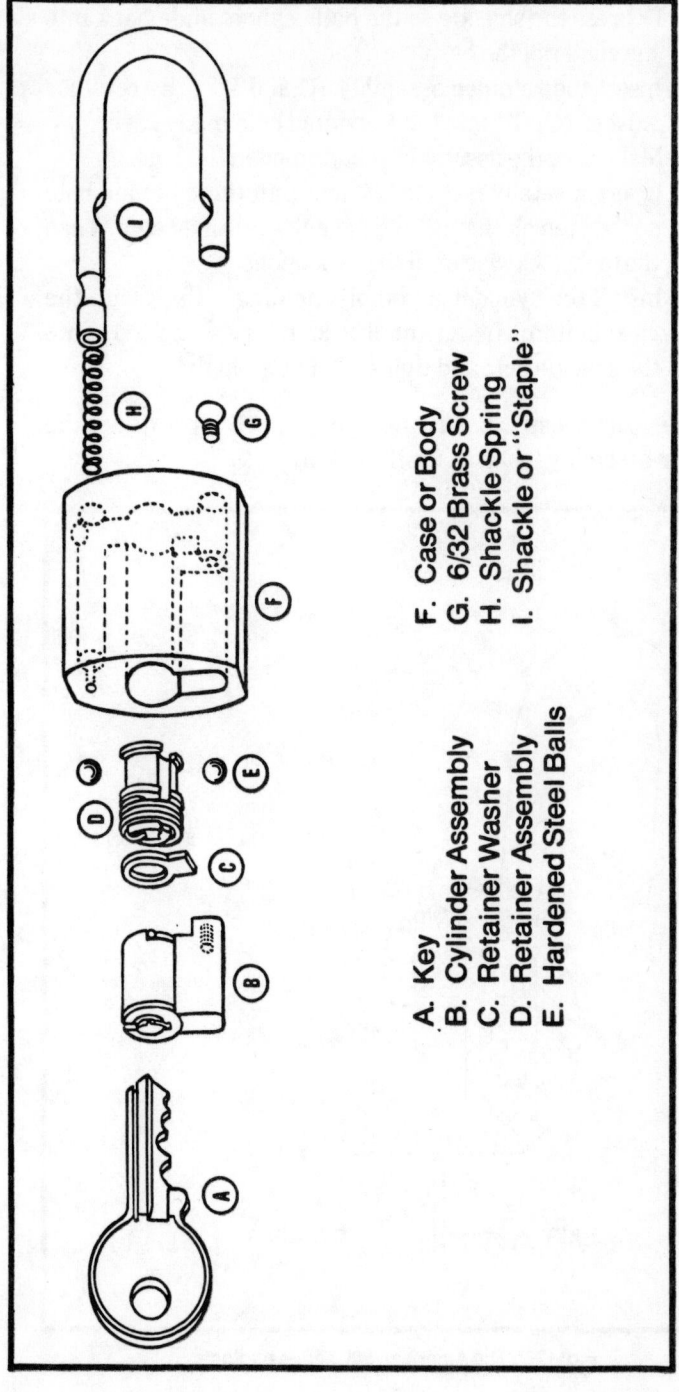

A. Key
B. Cylinder Assembly
C. Retainer Washer
D. Retainer Assembly
E. Hardened Steel Balls

F. Case or Body
G. 6/32 Brass Screw
H. Shackle Spring
I. Shackle or "Staple"

Fig. 17-6. An American pin-tumbler padlock.

5. Depress the shackle so the balls cannot slide back into the center of the case.
6. Insert the retainer assembly (C and D). The retainer washer (C) fits into the elongated hole in the case.
7. Make sure the assembly is bottomed in the hole.
8. Insert a retainer-assembly tool into the cylinder hole so the step on the tool engages the retainer step. Turn the tool clockwise until the lock opens.
9. Install the cylinder assembly, holding it flush with the case bottom. Insert the 6 × 32 brass screw (G) into the shackle hole and tighten it down snugly.

The American Lock Company will provide the retainer-assembly tool free for the asking.

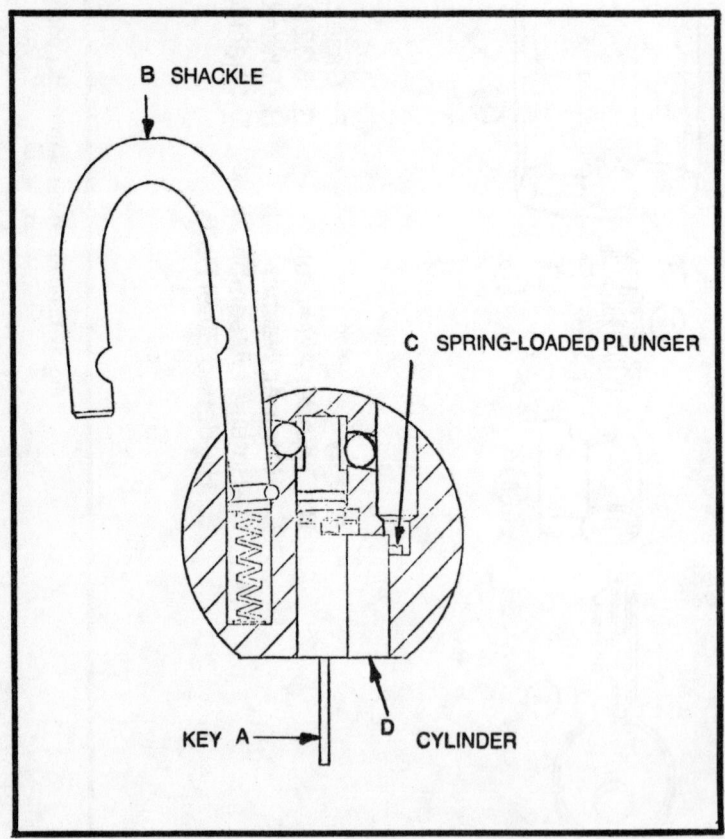

Fig. 17-7. The American 600 series padlock.

Because of the close tolerances of American padlocks, they will not operate properly in extreme cold weather unless Kerns ML3849 lubricant is used. Graphite is acceptable in milder climates.

To remove the cylinder in the 600 series pin-tumbler padlocks, follow this procedure:

1. Unlock the padlock and turn the shackle as shown in Fig. 17-7.
2. Depress the spring-loaded plunger (C) with small screwdriver until the pluger is flush with the wall of the shackle hole. At the same time, pull on the key; withdraw the cylinder.

MASTER

The Master Lock Company makes a variety of lock types, ranging from simple warded locks to sophisticated pin-tumbler types.

An improved version of the familiar Master padlock has recently been introduced—the Master Super Security Padlock (Fig. 17-8). It was designed to give greater security than standard locks and is recommended for warehouses, storage

Fig. 17-8. The Master Super Security Padlock.

Fig. 17-9. A cutaway of the Master Super Security Padlock.

depots, and industrial plants, as well as around the home and yard. Figure 17-9 is a cutaway view of the mechanism. Salient features include:

- A patented dual-lever system to secure the shackle legs. Each lever works independently of the other and is made of hardened steel.
- The long shackle leg is tapered to align the locking levers.
- The case laminations are made of hardened steel and are chrome-plated for weather protection.
- The case is larger than that of standard locks for better protection.
- A rubberoid bumper prevents the lock body from scratching adjacent surfaces. This feature is not found in other locks.

Tension tests show that this high security lock can tolerate a force of 6000 lb on the shackle without damage.

Servicing

Follow this procedure:

1. Drill out the bottom rivets.
2. Remove the plates, one at a time and keep them in order.
3. Remove the lock plug.
4. Assemble the lock using new rivets.

ILCO

ILCO secures the cylinder with what the trade calls the "loose rivet" method. A single brass rivet extends through the case and into the cylinder. The rivet is headless (hence the term "loose") and is secure by virtue of its near invisibility (Fig. 17-10).

To remove the pin, follow these steps:

1. Hold the case up to the light and look for the shadow created by the pin. The pin is located ⅛ in. from the bottom of the case and is centered on the side.
2. Using a No. 48 bit, drill a hole in the end of the pin.
3. Thread a small screw into the hole.
4. Gently pry up on the screw and remove it together with the pin.

To insert a new pin:

1. Select a piece of brass for stock that exactly matches the diameter of the pin hole.

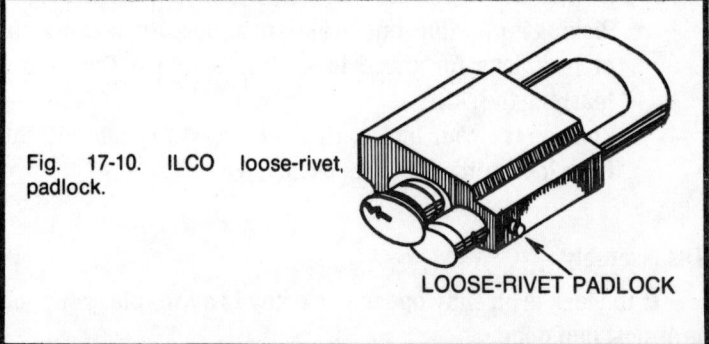

Fig. 17-10. ILCO loose-rivet, padlock.

LOOSE-RIVET PADLOCK

2. Use the next size smaller rod (to make removal easier if it is ever necessary) and cut it so that $^1/_{32}$ in. stands out.
3. Peen over the end of the rod.
4. Burnish the case to camouflage the pin.

Another ILCO variant is a cylinder retaining plate on the bottom of the case. The rivet that holds the assembly together is cunningly disguised in the maker's name stamped on the plate.

To remove the retaining plate, follow this procedure:

1. Draw a line under the ILCO name.
2. At a right angle to the first line, draw a second line (on the outside edge of the O in ILCO).
3. Drill a shallow hole to accept the tip of a 6×32 machine screw.
4. Fit the screw with a washer and thread it into the hole.
5. Using the screwhead for purchase, pry the retainer out of the case.
6. Withdraw the cylinder.

When the cylinder has been serviced and replaced into the case, mount the retainer in its original position, tapping it home with a flat-ended punch. Fill the hole with brass stock.

WAFER-DISC PADLOCK

Wafer-disc padlocks are recognized by their double-bitted keys and can be a headache to service. If you do not have a key, the lock can be opened by either of two methods:

- Picking is possible, but it takes practice. Purchase a set of lockpicks for these locks and spend a few hours learning the skill.
- You can try your luck with a set of test keys, available from locksmith supply houses.

Disassembly

If the lock is already opened or a key is available, your job is almost half done.

1. Release the retaining ring clip with a length of stiff wire inserted into the toe shackle hole.
2. Remove the cylinder plug and lay out the parts.

Keys

Duplicate keys can be cut by hand or with a machine designed for this purpose. These machines are expensive. Locksmiths get around the problem by stocking cylinder inserts and precut keys. The insert replaces the original wafer mechanism. A precut key requires that the discs be realigned to fit. You will need some blank discs and should have access to the cutting tool described in the chapter on double-bitted locks and keys.

In extreme cases you can insert the key and file the disc ends for shell clearance. This is not the sort of thing that a real professional would do, but it works.

Do not attempt to cut double-bitted keys by the impressioning method. A new plug and key is the better choice.

PIN-TUMBLER PADLOCKS

There are hundreds of pin-tumbler locks on the market, but most fall into two categories based on the case construction—laminated or extruded.

Laminated Padlocks

Slaymaker pin-tumbler padlocks are made up of a series of steel plates held together by four rivets at the casing corners. Follow these service procedures:

1. Use a hollow mill drill to shear off the rivet heads on the case bottom; this technique allows the rivets to be reused.
2. Remove the bottom plate.
3. Remove the entire cylinder section in one piece.
4. Two cylinder-housing types are used. The most popular requires a follower tool to keep the pins intact.
5. Make the necessary repairs.
6. Insert the cylinder into the casing and replace the bottom plate over the four rivet ends.

7. After checking the action, use a ball peen hammer and repeen the rivet ends.

Extruded Padlocks

Extruded locks are made from a single piece of metal—usually brass.

You will need these tools to service these locks:

- A small nail with the point cut off
- Key blanks
- Pliers
- A small punch
- Hammer
- A small light (overhead lights are too bright; a flashlight is acceptable)

To disassemble the extruded lock, first locate the plug pins or cover on the bottom of the case. Since the case is highly polished, this is sometimes hard to do. Tilting the lock under the light will show a faint outline of the cover cap or plug pins.

To assemble extruded locks held together by exposed pins, follow this procedure.

1. Determine the diameter of the plug pins.
2. Select a drill bit smaller than the diameter of the pin.
3. Drill slowly and only a fraction of an inch deep—the pin does not extend into the lock very far. If you drill through it you can damage the spring.
4. Extract the plug.
5. Remove each spring and pin.
6. Remove the cylinder. The cylinder is held by a pin that may be covered by a small pin plug. Rapping the lock against a hard surface may shock the plug loose. If not, dislodge the pin plug with a small screwdriver.
7. Make the necessary repairs.
8. Assemble the lock.
9. Fill the pin holes with brass wire.
10. File the wire flush with the case and polish.

Extruded locks with a retaining plate over the pins require a slightly different procedure. If the lock is open, drive the plate out with a punch inserted at the shackle hole. If locked, follow this procedure:

1. Drill a small hole off-center in the plate.
2. Using an icepick or other sharp instrument, drive the plate down, toward the shackle. This will buckle the plate and cause the edges to rise.
3. Pry the plate loose by working a sharp instrument around the edges.
4. If the plate is not severely damaged, it can be reused. Bow the plate slightly so that the center bulges outward when the plate is installed.
5. Peen the edges of the plate to form a tight seam.

Other locks mount the plug and cylinder by means of horizontal pins running across the width of the lock. These pins can be seen under strong light.

Key Fitting for Pin-Tumbler Padlocks

Follow this procedure:

1. Insert the blank into the plug.
2. Place No. 1 pin into its chamber.
3. Using a blunted nail as a punch, drive the pin into the blank. A single light hammer tap is enough to impression the blank.
4. Remove the blank.
5. File the blank at the impression.
6. Insert the key and try to turn the cylinder. File as necessary.
7. Move to the next pin and repeat the process.

HELPFUL HINTS

It's easier to work with padlocks that are open. If the lock is not open, pick it open. Sometimes the shackle bolt can be disengaged with the help of a hat pin inserted through the keyway. However, locks are getting better and this technique doesn't work as well as it used to.

- Polish the lock before returning it to the customer. Use emery paper to remove the deep scratches, then burnish with a wire wheel. Finish by buffing.
- The best security in a padlock comes when the shackle is locked at both the heel and toe. The double bolt action (or balls) is the ultimate in padlock security, making it nearly impossible to force the shackle.
- When picking fails (and even the best of locksmiths may occasionally have this problem), use penetrating oil. Yale, Corbin, and ILCO pin-tumbler locks are especially responsive to penetrating oil.

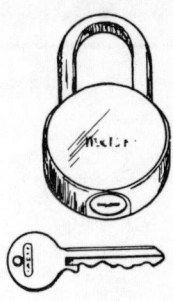

Chapter 18
Home Services

Over the years a locksmith acquires a great deal of knowledge; much of this information is learned from correcting his own mistakes. This is a hard school. The best way to solve problems is to never let them arise in the first place.

This chapter includes some pretty basic techniques. Some of this material may seem obvious. You may wonder why it's in here. The answer is that these techniques have been tested and approved by the experts. They will save you time, money, and the embarrassment of callbacks.

The following few pages are supplied courtesy of Corbin (Emhart Corporation). The first section applies to all locksets, regardless of make. The remaining sections detail service and troubleshooting procedures for specific Corbin locksets, as well as other makes of locks.

CHECKLIST OF COMMON PROBLEMS

Should you encounter difficulties in the operation of a lockset, I recommend that you first review this checklist of common problems and solutions to see if you can clear up the difficulty.

- Is the door locked?

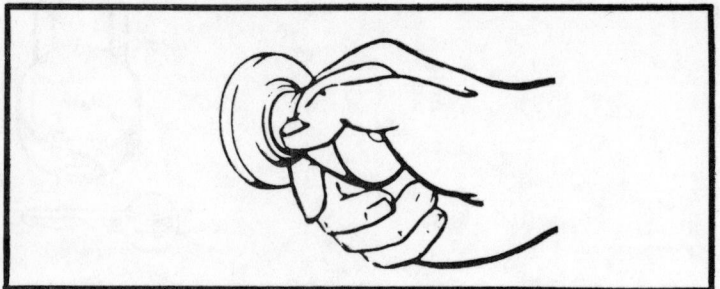

- Are you using the right key?

- Latchbolt or dead bolt does not engage or disengage the strike. or binds in the strike. Usually due to bolt-strike misalignment.
- Has the door warped?

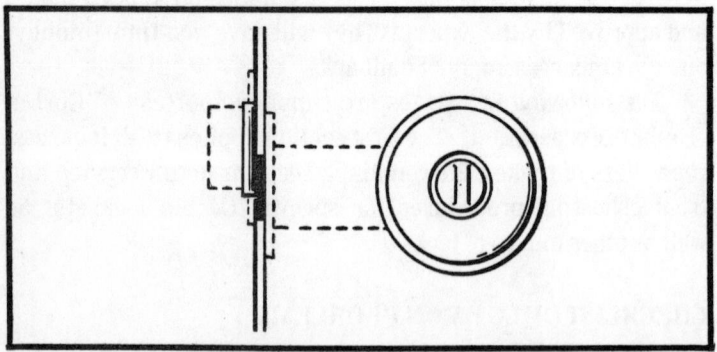

- Is the door binding? Frames which are out of plumb are frequently the cause of faulty operation of locksets and binding of bolts in the strike.

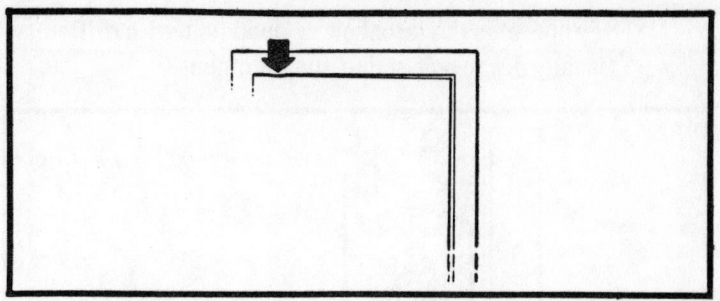

- Are the hinges loose? Tighten the screws, filling holes if necessary, or rehang the door if the screws will not hold.
- Are the hinges worn? If excessive wear has occurred on the hinge knuckles, the door will not be held tightly. Replace the hinges.

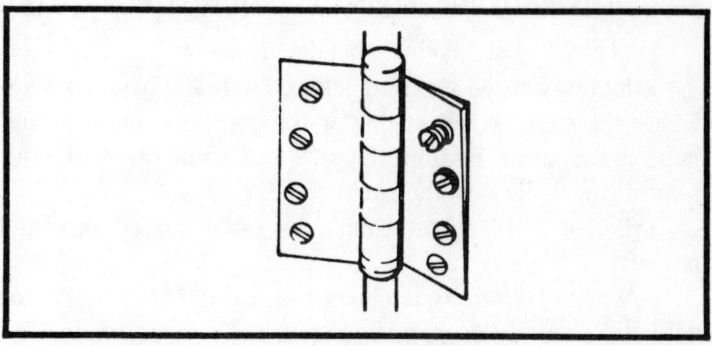

- Is the frame sagging? If sag cannot be corrected and the door and frame returned to plumb relationship, planing or shaving of the door and repositioning or shimming the strike may relieve this condition.

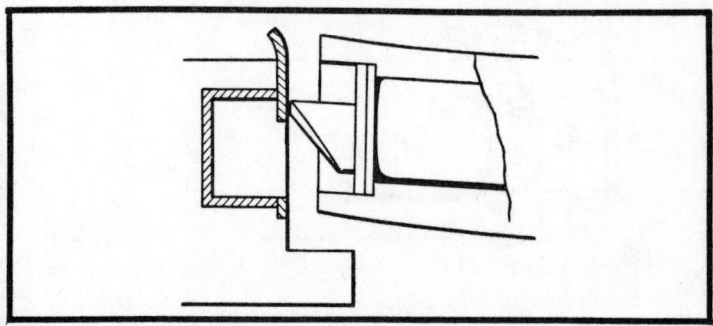

- Key operates the latchbolt or dead bolt with difficulty. Usually due to bolt-strike misalignment.

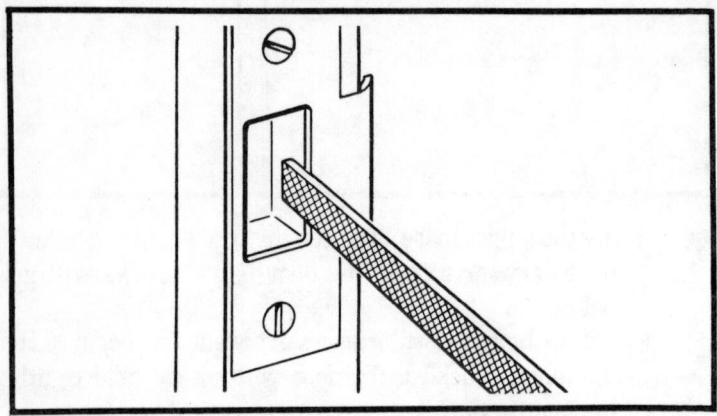

TROUBLESHOOTING CORBIN CYLINDRICAL LOCKSETS

Problem 1—Latchbolt will not deadlock.

Solution—Caused by deadlocking latch going into strike. Either the strike is out of line or the gap between door and jamb is too great. Realign the strike or shim the strike out towards the flat area of the latchbolt (Fig. 18-1).

Problem 2—Latchbolt cannot be retracted or extended properly.

Solution—Caused by latchbolt tail and latchbolt retractor not being properly positioned (Fig. 18-2). Remove the lockset from the door. Reinsert the latchbolt in the door. Looking through the hole in the door, the tail should be centered between the top and bottom of the hole. Remove the latchbolt

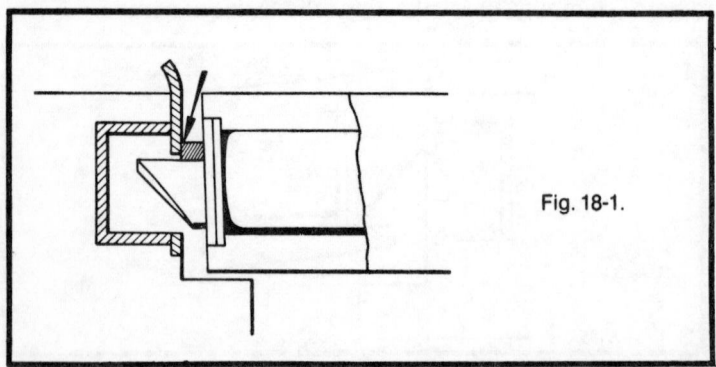

Fig. 18-1.

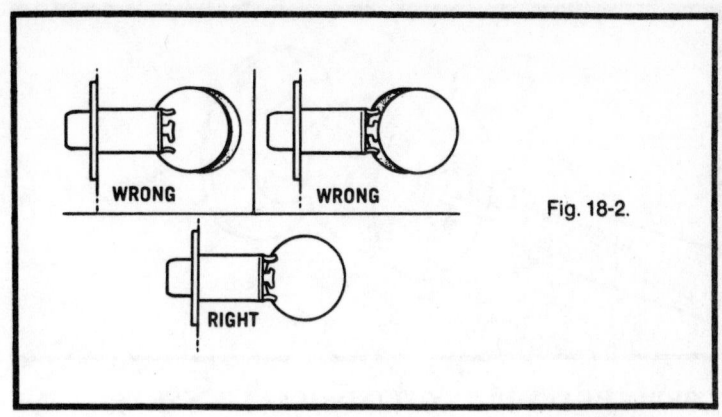

WRONG WRONG

RIGHT

Fig. 18-2.

and insert the lockcase. Looking through the latchbolt hole in the lock face of the door, the latchbolt retractor should be centered in the hole. Adjust the outside rose for proper position. Rebore the holes, if necessary, to line up the retractor and tail.

Problem 3—Latchbolt will not project from the lock face (Fig. 18-3).

Solution—Latchbolt tail and retractor may be misaligned. See Problem 2. If this is not the cause, the spring is probably broken.

Problem 4—Key works with difficulty.

Solution—Lubricate the keyway (Fig. 18-4). Do *not* use petroleum products. Spray powdered graphite into the cylinder or place powdered graphite or lead pencil shavings on the key. Move the key slowly back and forth in the keyway. Bitting (notches) on the key may be worn.

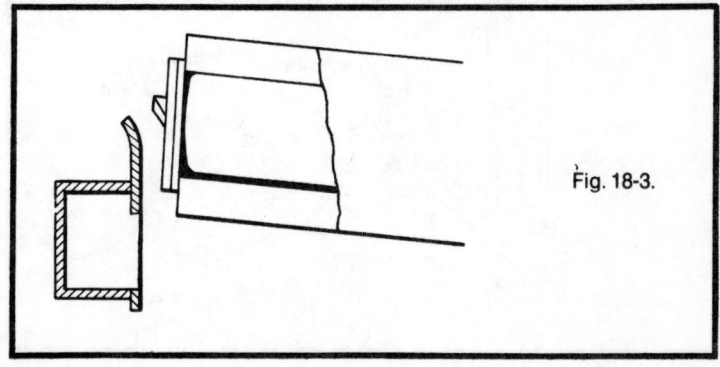

Fig. 18-3.

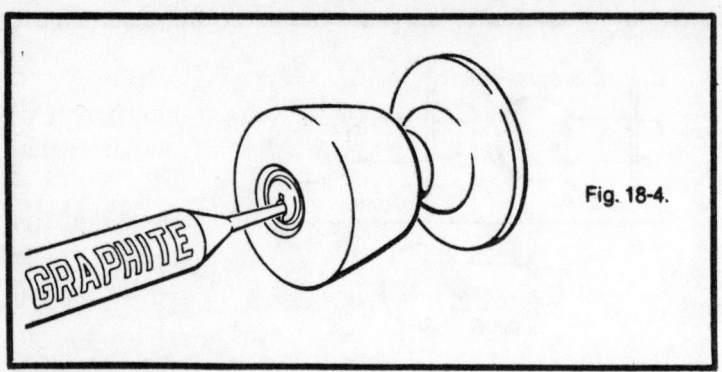

Fig. 18-4.

CORBIN HEAVY-DUTY CYLINDRICAL LOCKSETS: SERVICE PROCEDURES

Servicing these locks is not difficult if the task is approached methodically.

To Tighten the Locksets

1. Tighten the inside rose thimble with a wrench. If the thimble needs to be taken up a great deal, tighten the *outside* rose at the same time to prevent possible misalignment and binding of both (Fig. 18-5).
2. If the lockset is still not tight, back off the thimble, using a screwdriver push down the knob retainer and remove the knob and rose. If the spurs on the back of the rose are bent, straighten and reposition the rose so the spurs are imbedded in the door. If the spurs are broken off, insert a rubber band or nonmetallic washer under the rose. Tighten the thimble.

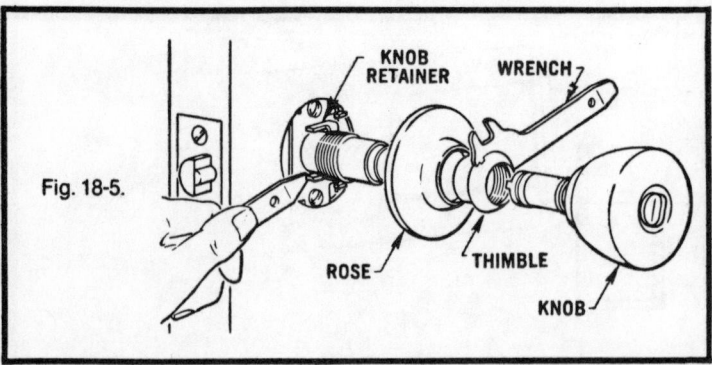

Fig. 18-5.

KNOB RETAINER
WRENCH
ROSE
THIMBLE
KNOB

To Remove and Install the Locksets

1. Remove the key from the knob.
2. Loosen the inside rose thimble with a thimble wrench. Pull the lock slightly to release the rose spurs from the door.
3. Disengage the inside knob retainer with a thimble wrench and pull out the inside knob.
4. Slide the rest of the lockset from the outside of the door.
5. Remove the bolt faceplate screws and slide the latch unit from the lockface of door.
6. To reinstall, reverse procedures above. When inserting case and keyed knob from the outside of the door, be sure the bolt retractor in its case properly engages the latchbolt tail.

To Remove and Replace the Cylinder in the Locksets

1. Follow the procedure in the preceding section.
2. Remove one case screw and slightly loosen the other.
3. Swing out the outside knob retainer (it pivots on the case screw slightly loosened above).
4. Using a screwdriver, pry the knob filler cover off the outside knob.
5. Using special Waldes ring pliers No. 3, remove the large Waldes ring from the groove and withdraw the shank and cylinder (Fig. 18-6).
6. To reinstall the cylinder, reverse the above procedure. Be sure the knurled side of the Waldes ring is face up.

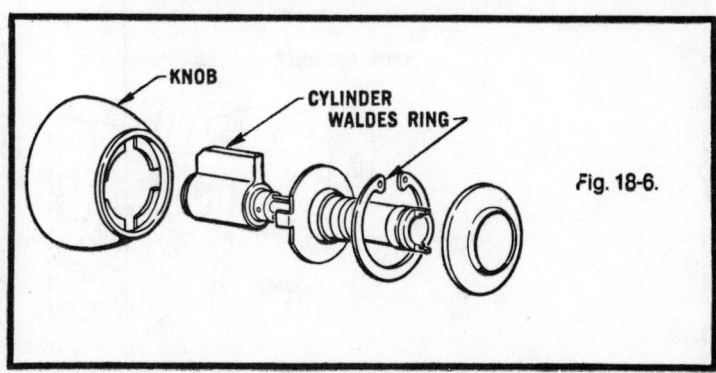

Fig. 18-6.

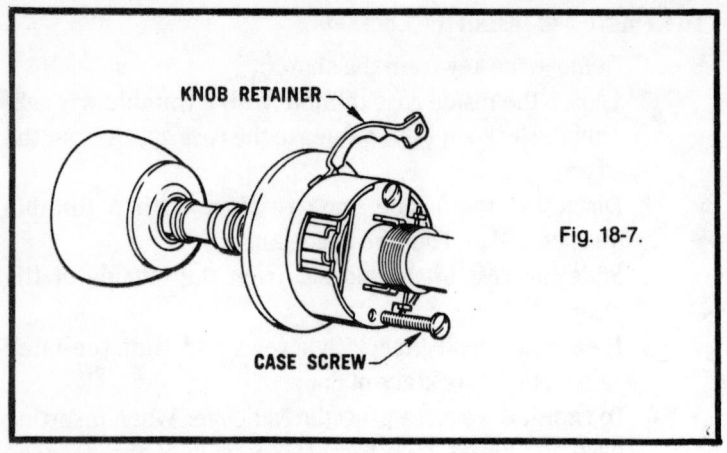

KNOB RETAINER

CASE SCREW

Fig. 18-7.

To Change the Hand of the Locksets

Remember that the lock cylinder should always be in position to receive the key with the bitting (notches) facing upward.

1. Follow the procedure in the preceding section.
2. Slightly loosen *one* case screw and back off the other (Fig. 18-7).
3. Swing out the outside knob retainer (it pivots on case screw slightly loosened above).
4. Lift the knob out and rotate it 180°. Replace it in case. Swing the knob retainer back into place.
5. Insert the case screw and tighten both screws.
6. Reinstall.

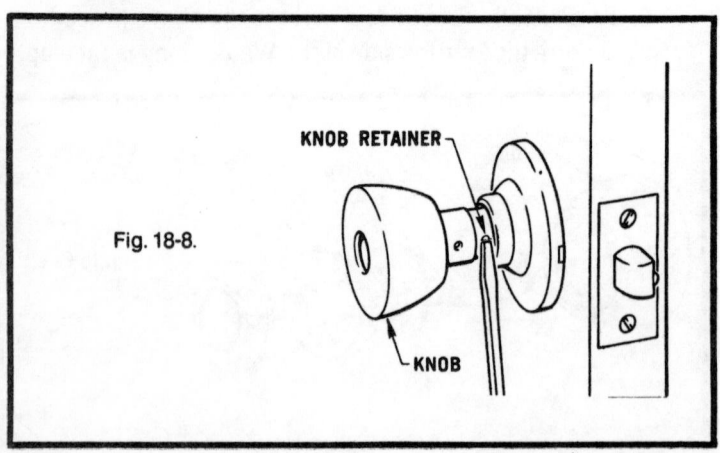

KNOB RETAINER

Fig. 18-8.

KNOB

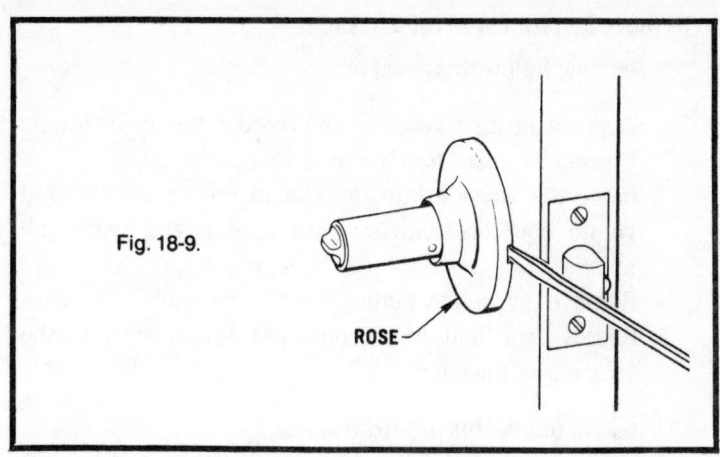

Fig. 18-9.

ROSE

CORBIN STANDARD-DUTY CYLINDRICAL LOCKSETS: SERVICE PROCEDURES

To Tighten the Locksets

1. Depress the knob retainer and remove the inside knob.
2. Unsnap the rose from the rose liner (Fig. 18-9).
3. Place the wrench into the slot in the rose liner and rotate clockwise until tight on the door. If the lockset is extremely loose, tighten the outside rose and inside liner *equally* (Fig. 18-10).
4. Replaced the inside rose; depress the knob retainer and slide the knob onto the spindle until the retainer engages the hole in the knob shank.

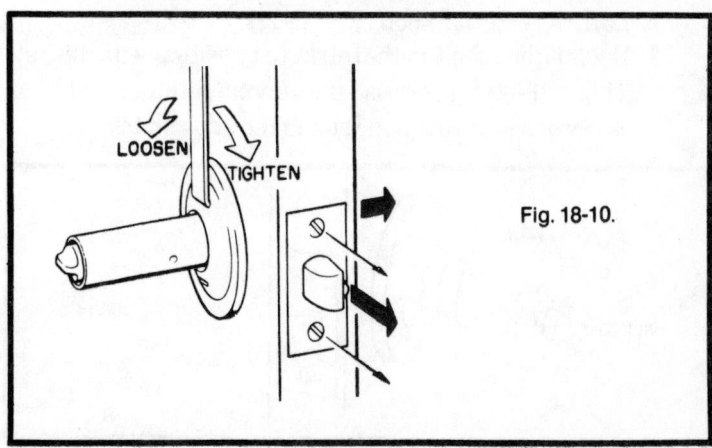

LOOSEN
TIGHTEN

Fig. 18-10.

To Remove and Reinstall the Locksets

To remove, follow this procedure:

1. Depress the knob retainer and remove the inside knob.
2. Unsnap the rose from the rose liner.
3. Place the wrench into the slot in the rose liner and rotate counterclockwise until disengaged from the spindle.
4. Remove the lock by pulling the outside knob.
5. Remove the bolt after removing the screws in the lockface of the door.

To install, follow this procedure:

1. Adjust the lock for the door thickness by turning the outside rose until the edge of the rose matches one of the lines marked on the shank of the knob. (First line is for 1⅜ in. door, second for 1¾ in. door). When using trim rosettes, increase the adjustment to compensate for the thickness of the metal.
2. Install the latchbolt in the face of the door.
3. Install the lock from the outside of the door so the case engages the slot in the latchbolt and the tail interlocks the retractor.
4. Replace the rose liner, rose, and knob on the inside of the door.

To Reverse the Knobs

1. Lock the exterior knob (Fig. 18-11).
2. Hold the latchbolt in the retracted position with the key (Fig. 18-12). Depress the knob retainer with a screwdriver through the slot in the knob shank.

LOCKED

Fig. 18-11.

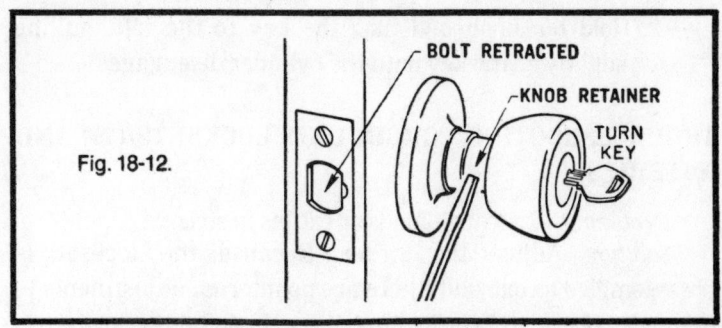

Fig. 18-12.

BOLT RETRACTED
KNOB RETAINER
TURN KEY

3. Pull the knob off the spindle (Fig. 18-13). Rotate the knob so the bitting on the key will be *up*.
4. Replace the knob by lining up the lance in neck of knob with the slot in the spindle.
5. Push the knob on the spindle until it hits the retainer button. Depress the retainer button, and push the knob until it snaps into position.

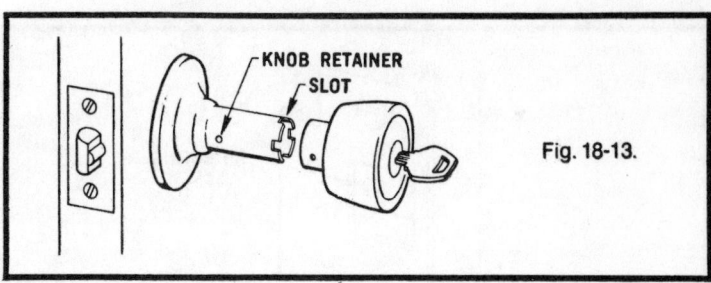

KNOB RETAINER
SLOT

Fig. 18-13.

To Remove the Cylinders

1. Remove the outside knob. Turn the key in either direction until it can be partially extended from the plug (Fig. 18-14).

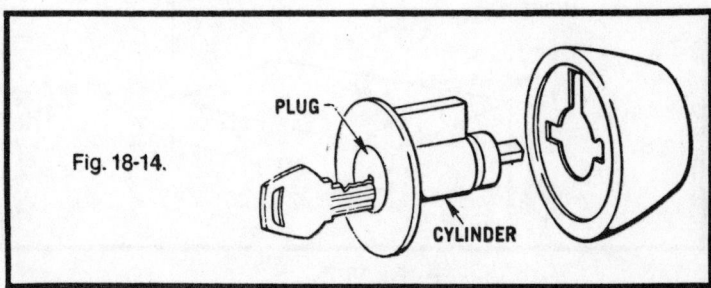

Fig. 18-14.

PLUG
CYLINDER

2. Hold the knob and turn the key to the left, pulling slightly on the key until the cylinder disengages.

TROUBLESHOOTING CORBIN UNIT LOCKSETS (300 AND 900 SERIES)

Problem 1—Latchbolt binds or rattles in strike.

Solution—Adjust the strike. Because the lockset is preassembled as one unit, there are no internal adjustments to be made. Check to be sure that the cutout for the lockset is square and at the right depth so the face of the lockset is flush with the face of the door. Once this has been done, adjust the nylon adjusting screw in the strike (Fig. 18-15).

Problem 2—Key does not activate the knob or latchbolt.

Solution—Check for worn key. If bitting (notches) in the key is worn down or the key is bent, the locking mechanism will not operate properly. If the key is not worn, spray powdered

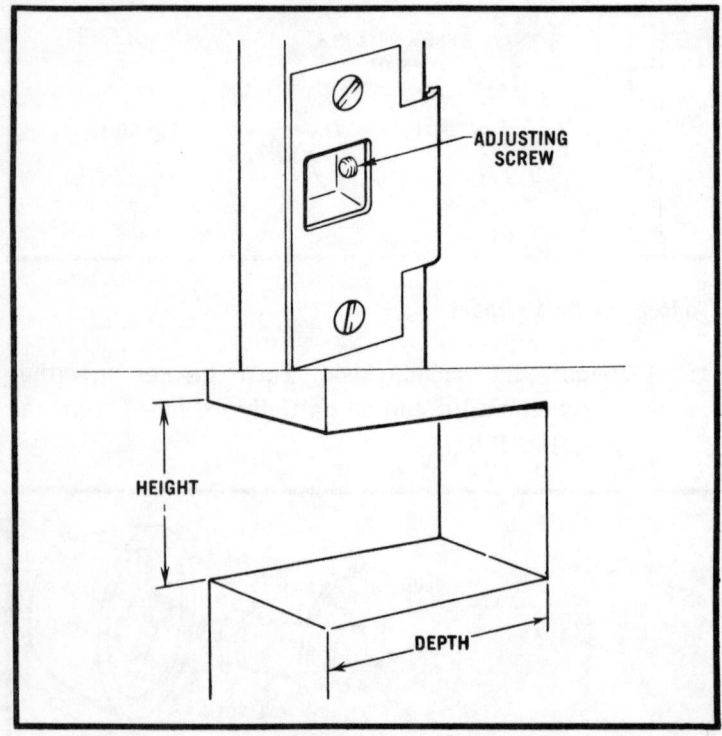

Fig. 18-15.

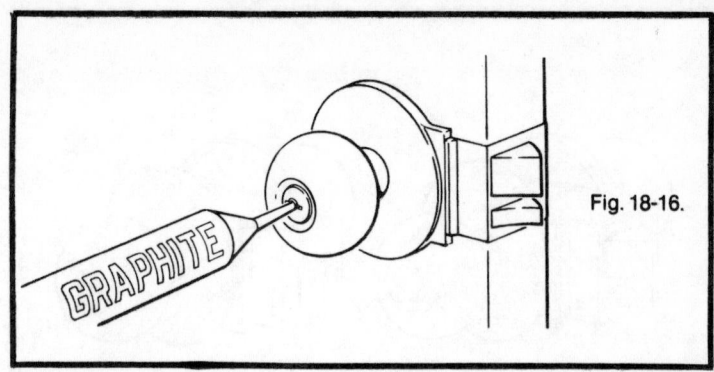

Fig. 18-16.

graphite into the keyway or put graphite or lead pencil filings on the key and move it back and forth slowly in the keyway (Fig. 18-16). Never use petroleum products. Check to be sure that a binding latch is not the cause.

Problem 3—Lockset is loose in the door.

Solution—Tighten the escutcheon screws (Fig. 18-17). Be sure to tighten evenly.

Problem 4—Lockset has the wrong bevel for the door.

Solution—Reverse the lockset. Unit locksets with horizontal keyways may be changed from right to left hand regular bevel, or vice-versa—or from right hand reverse bevel to left hand reverse bevel, or vice-versa—by merely turning the lock upside down (Fig. 18-18).

CORBIN UNIT LOCKSETS (300 AND 900 SERIES): SERVICE PROCEDURES

Usually the unit locksets are easier to service than the cylindrical locksets. To service this series of locksets proceed as follows:

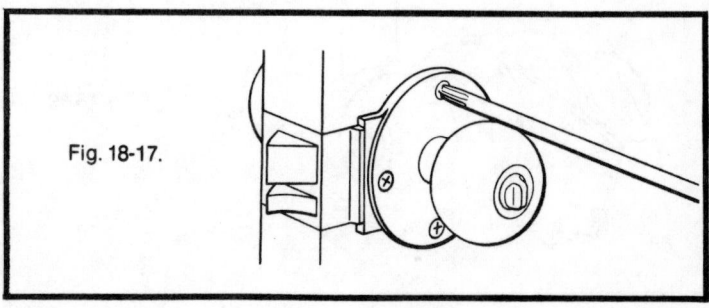

Fig. 18-17.

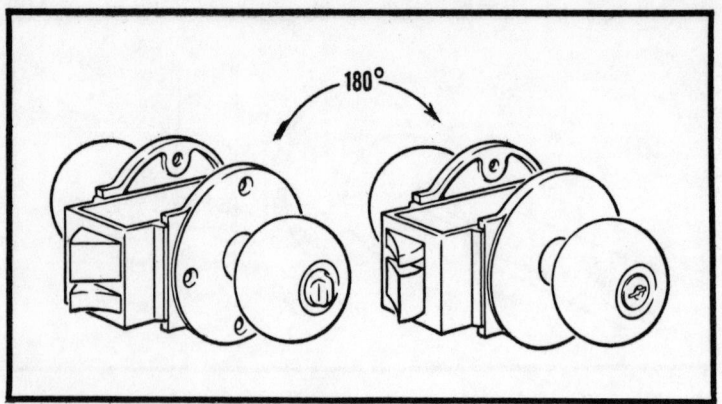

Fig. 18-18.

To Remove the Locksets From a Door

1. Remove the through-bolts on the inside of the door (Fig. 18-19).
2. Push the outside escutcheon away from the door so the lugs clear the holes.
3. Slide the assembly out of the door.

To Remove 900 Series Knobs

1. Remove the attaching screws (Fig. 18-20).
2. Snap off the dust cover.
3. Pry the wire retaining ring from the knob retaining key located in a slot in the frame tube. Remove the retaining key. Remove the inside knob which is fastened to the knob shank.

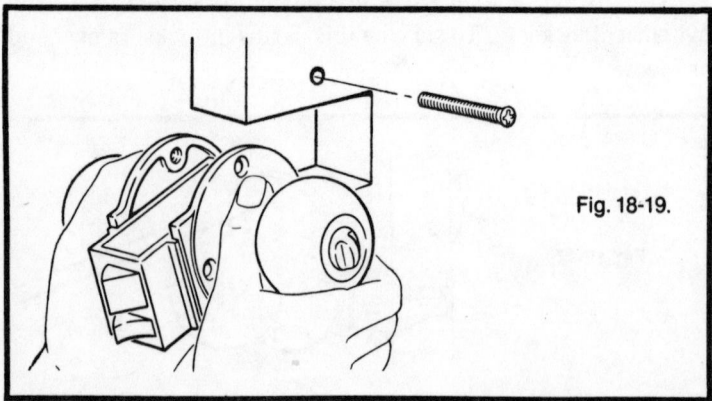

Fig. 18-19.

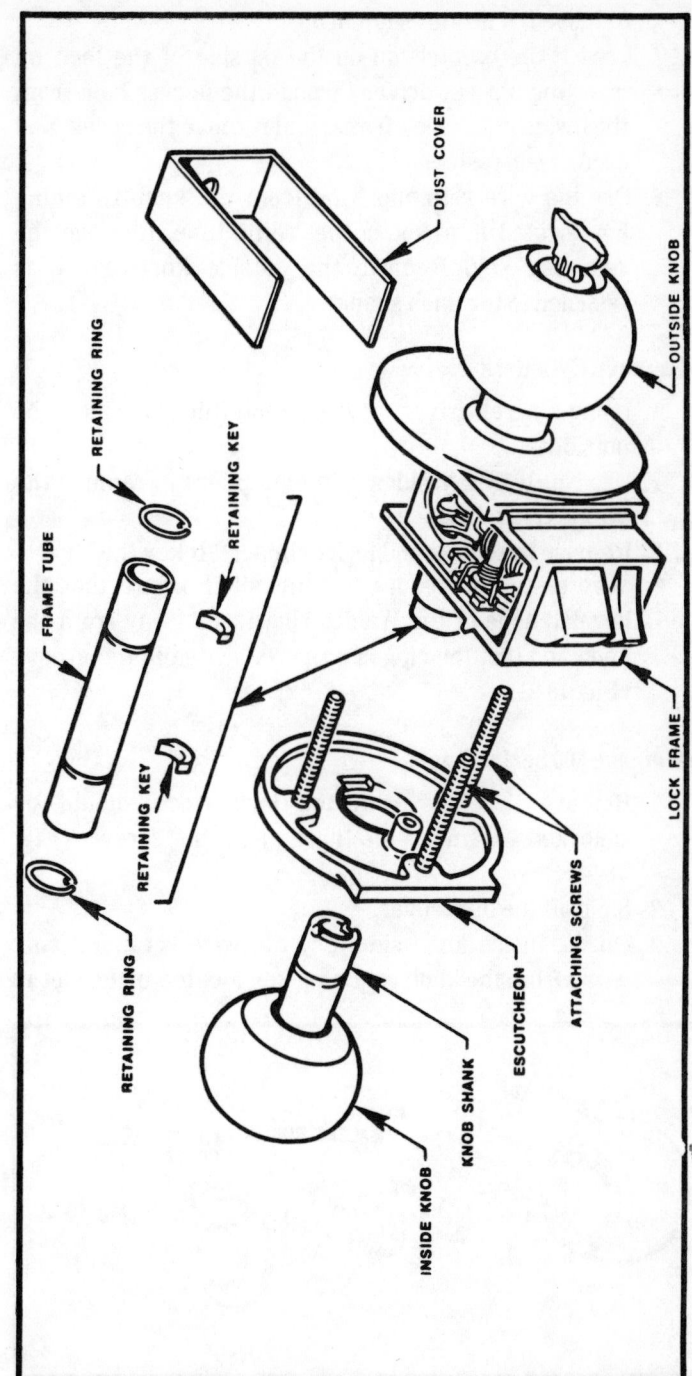

DUST COVER

RETAINING RING

RETAINING KEY

FRAME TUBE

RETAINING KEY

RETAINING RING

OUTSIDE KNOB

LOCK FRAME

ATTACHING SCREWS

ESCUTCHEON

KNOB SHANK

INSIDE KNOB

Fig. 18-20.

4. Remove the inside escutcheon.

5. Loosen the escutcheon on the outside of the lock by inserting a screwdriver through the access hole from the inside of the lock frame and remove the screw and escutcheon fastener.

6. Pry the wire retaining ring from the knob retaining key located in a slot in the frame tube. Remove the retaining key. Remove the outside knob which is fastened to the knob shank.

To Remove Cylinders

1. Using a screwdriver, pry the knob filler cover off the outside knob.

2. Use No. 103F92 Waldes retaining pliers to remove the Waldes ring.

3. Remove the shank and lock cylinder. Rekey the lock.

4. Reverse the procedure to reinstall. Be sure that the beveled edge of the Waldes ring faces away from the knob and that the ring is properly seated in the groove (Fig. 18-21).

To Remove 300 Series Knobs

1. Remove the key from the lock (lock should be unlocked). Remove the three attaching screws (Fig. 18-20).

2. Snap off the dust cover.

3. On the inside knob side, pry the wire retaining ring away from the knob retaining key located in the slot in

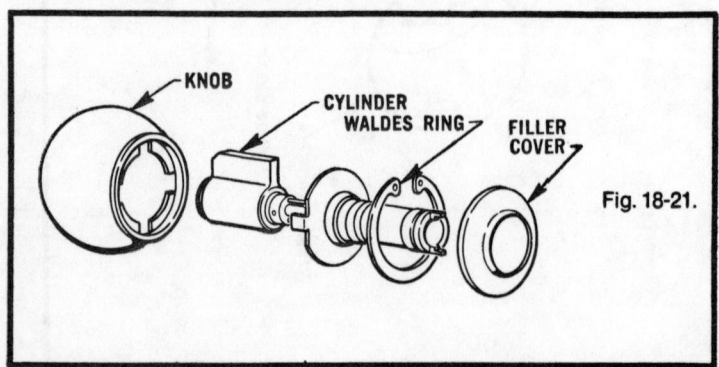

KNOB

CYLINDER
WALDES RING

FILLER
COVER

Fig. 18-21.

the frame tube. Remove the retaining key. Remove the knob that has the shank assembled to the knob.

4. Remove the inside escutcheon.
5. Loosen the escutcheon on the outside of the lock by inserting a screwdriver through the access hole from the inside of the lock chassis and remove the screw and washer.
6. Repeat step 3 on the outside knob and shank.

TROUBLESHOOTING CORBIN MORTISE LOCKSETS (7000, 7500, 8500 SERIES)

Problem 1—With door open, latchbolt doesn't extend or retract freely.

Solution—Check for binding against the rose. Adjust the knob. Loosen the roses or trim on the door. If the bolt now operates freely, the roses or trim must be realigned. A knob aligning tool is recommended. Check installation templates for proper position (Fig. 18-22). If the bolt does not operate properly with trim and roses loosened, remove the lockset from the door. If the lockset operates properly when removed from the door, use a chisel to make the mortise larger so that the lockset enters freely.

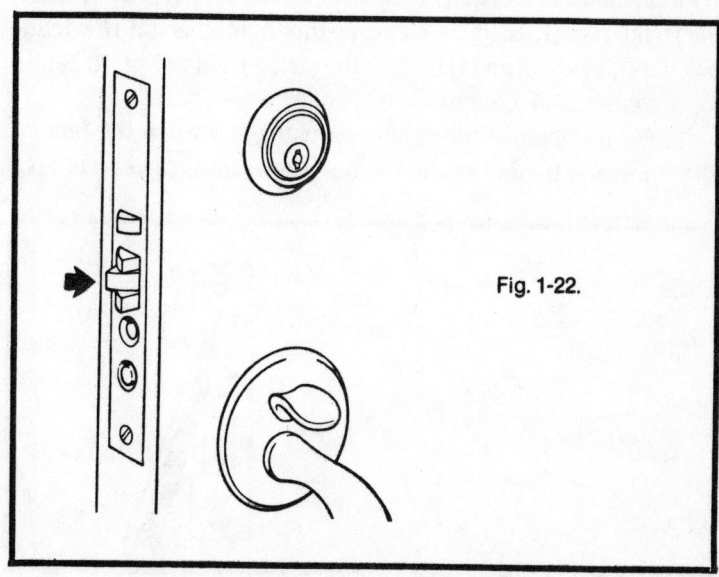

Fig. 1-22.

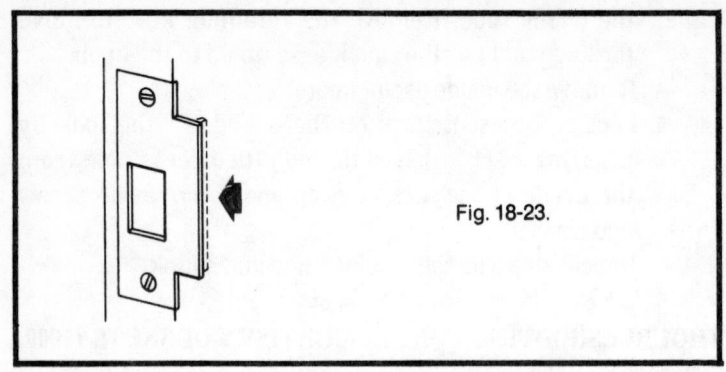

Fig. 18-23.

Problem 2—With door closed, latchbolt doesn't extend or retract freely, or door won't latch at all.

Solution—Open the door. If the latchbolt still doesn't operate properly, see *Problem 1*.

Problem 3—Latchbolt ''stubs'' on the strike lip.

Solution—Bend the strike slightly back toward the jamb. Wax or paraffin makes an excellent lubricant, as does silicone spray (Fig. 18-23).

Problem 4—Deadbolt doesn't enter the strike.

Solution—This is probably due to misalignment of strike and bolt, particularly in cases where door sag has taken place. Both latchbolt and deadbolt holes in the strike must be filed or the strike repositioned. Do *not* force the thumbpiece if the deadbolt doesn't extend and retract in the strike freely (Fig. 18-24).

Problem 5—To remove the cylinder.

Solution—Loosen the cylinder locking screws in the face of the lockset. If the scalp covers the set screws, remove

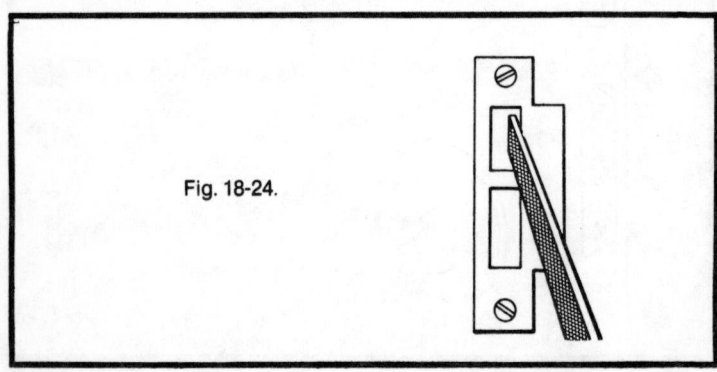

Fig. 18-24.

the scalp. Unscrew the cylinder. When replacing, be sure the locking screws are firmly seated (Fig. 18-25).

Problem 6—Key does not operate the latchbolt or dead bolt.

Solution—Loosen the cylinder locking screws. The cylinder is in the wrong position in the door so the cylinder cam does not engage the locking mechanism properly. Turn the cylinder a whole turn to the left or right until it works properly. The keyway must always be in position to receive the key with bitting (notches) up. Tighten the cylinder locking screws.

Problem 7—Key turns hard when retracting the dead bolt.

Solution—If the bolt operates freely with the door open, check the bolt-strike alignment. Check the scalp to be sure it is not binding the bolt, or that paint over the bolt is not causing the bind.

Problem 8—Key works hard in the cylinder.

Solution—Lubricate the cylinder with powdered graphite or place graphite on the key and move it back and forth slowly in the keyway. Never use any petroleum lubricants (Fig. 18-26).

Problem 9—Key breaks in the lock.

Solution—Remove the cylinder from the lock. Insert a long pin or wire into the back end (cam end of the cylinder). Move it

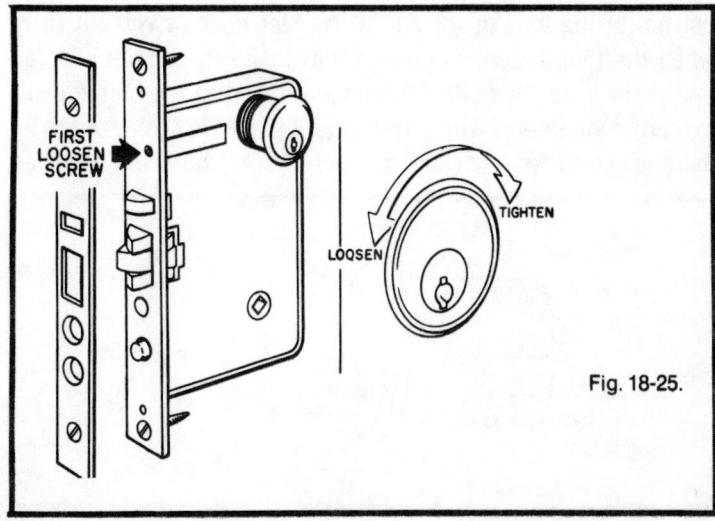

Fig. 18-25.

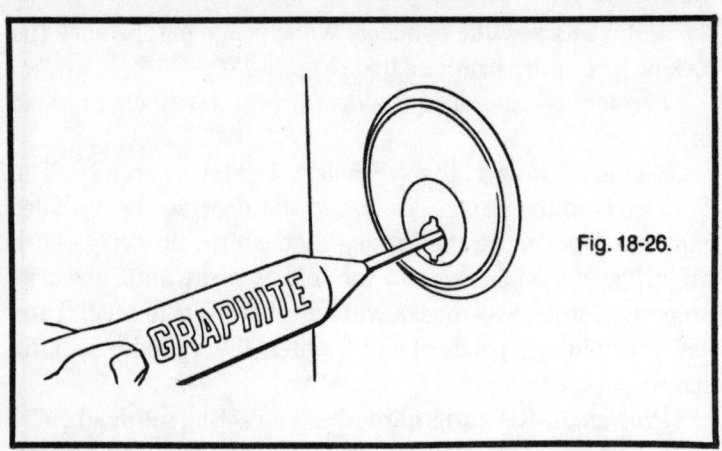

Fig. 18-26.

back and forth until the broken key stub is forced out through the front of the cylinder. Clean the cylinder with ethyl acetate and lubricate it with graphite before reinstalling (Fig. 18-27).

Problem 10—Thumbpiece trim doesn't retract latchbolt completely or doesn't extend bolt completely.

Solution—Check for binding at inside trim; or if the outside thumbpiece trim is used in conjunction with the inside panic device, check to see it is operating properly and is not dogged down. Remove the thumbpiece and check the position of thumbpiece in relation to the latch trip at the bottom of the mortise lock case. When properly installed, the top of the thumbpiece should be up against the bottom of the latch trip, but not lifting it (Fig. 18-28). If the bolt doesn't retract fully when the thumbpiece is pushed down, the thumbpiece is too low on the door. Move the trim up as needed. If the bolt doesn't extend completely when the thumbpiece is released, the thumbpiece is too high on the door. Move the trim down as

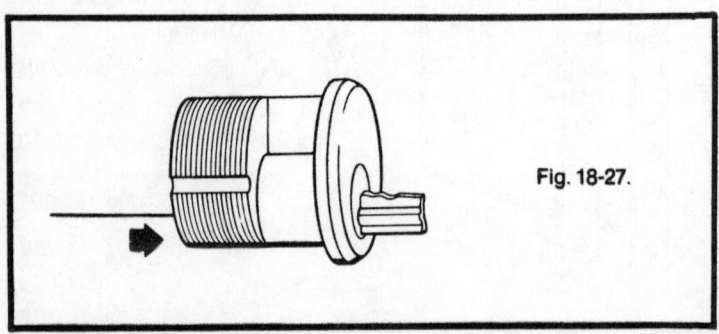

Fig. 18-27.

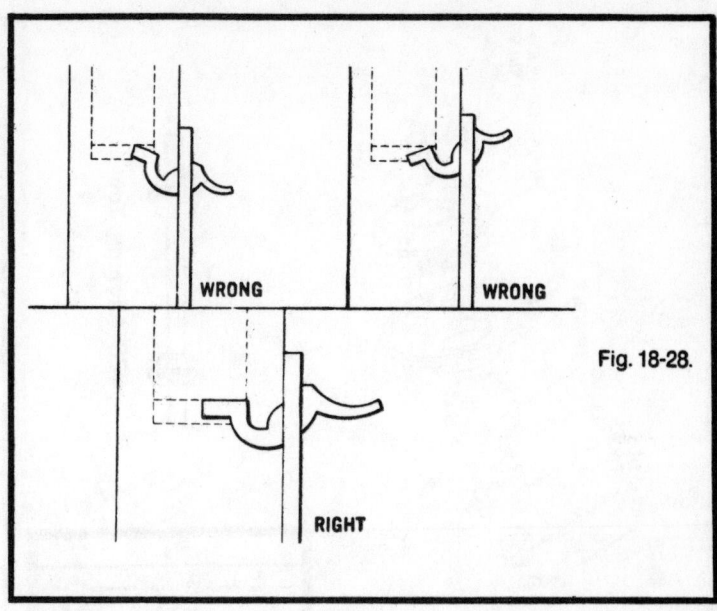

WRONG

WRONG

Fig. 18-28.

RIGHT

needed. If the trim is fixed and cannot be moved, carefully bend the thumbpiece tail up or down.

CORBIN MORTISE LOCKSETS (7000, 7500, 8500 SERIES): SERVICE PROCEDURES

Mortise locksets are not difficult to work with. Servicing them involves only a few basic procedures.

To Install and Adjust the Working Trim

1. Unscrew the spindle. Remove the mounting plates, roses, and thimble from the spindle. Unscrew the sleeve from the inside knob. Remove the spindle from the knob (Fig. 18-29).
2. Install the rose attaching plates. On wood doors install flange washers through the mortise, turning the attaching plates into them. It may be necessary to mortise for the washer to permit the mortise lock to clear it. Install the mortise lock.
3. Align the plates using a No. 028 aligning tool. If not available, assemble the spindles allowing $^1/_{16}$ in. gap between halves of the swivel spindles, with mounting

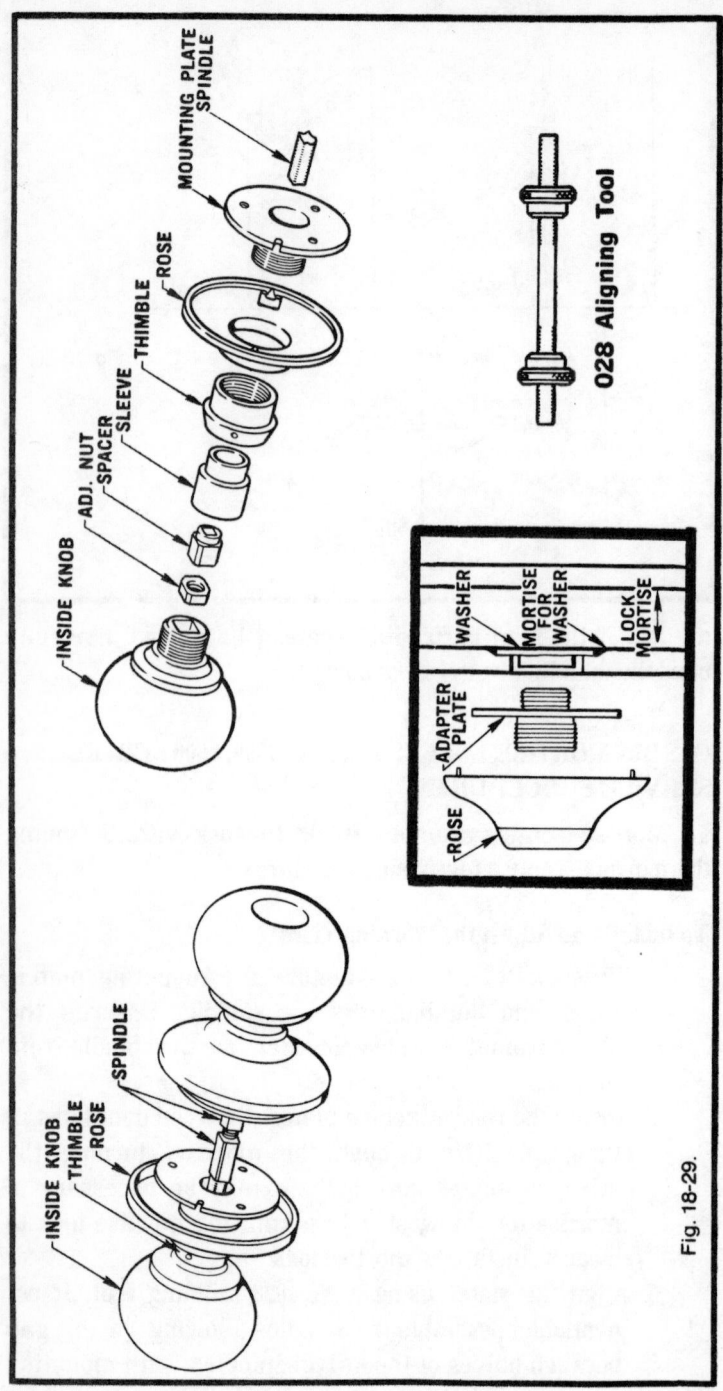

Fig. 18-29.

plates, sleeve spacer, and adjusting nut in position shown in illustration. Tighten the adjusting nut with fingers. Mark screw holes through both mounting plates. Install the screws.

4. Reassemble the spindle, tightening the adjusting nut, and back off ¼ turn to line up with the flats of the spacer. The spindle should have a slight end chuck. Try the knob. If the latch binds, back off the adjusting nut another ¼ turn. Disassemble the spindle.

5. Assemble the roses and thimble. Reassemble the spindle complete with knobs. Tighten the rose thimbles and inside adjusting plate with furnished spanner wrenches. Knobs should turn freely from either side in either direction.

To Install and Adjust the Lever Handle Trim Assembly

1. Check to be sure that the handle hole is 1⅛ in. wide. On wood doors, remove the mortise lockset and make additional mortise for retainer on the inside of the case mortise. Replace the locksets and secure with screws.

2. Place the rose assembly spindle into the lock hub. The label shows top and latch edge. Spot four screw holes through the assembly plate (Fig. 18-30). Make sure that the latchbolt operates freely. Drill holes very carefully.

3. Attach the rose assembly to the door. (With a wood door, attach through door to retainer.)

4. Put on the cover plates. Place the lever handle on the assembly and secure with an Allen set screw.

5. Adjust. If the latchbolt binds, remove the lower plate and slightly change rose assembly position.

To Reverse the Hand

As you will remember from Chapter 11, the hand refers to the position of the hinge on the door and to the direction of swing. Some locksets are universal and fit all four hands. Others must be modified in the field.

1. Refer to Fig. 18-31. Unfasten the two cap screws (A) and remove the cap.

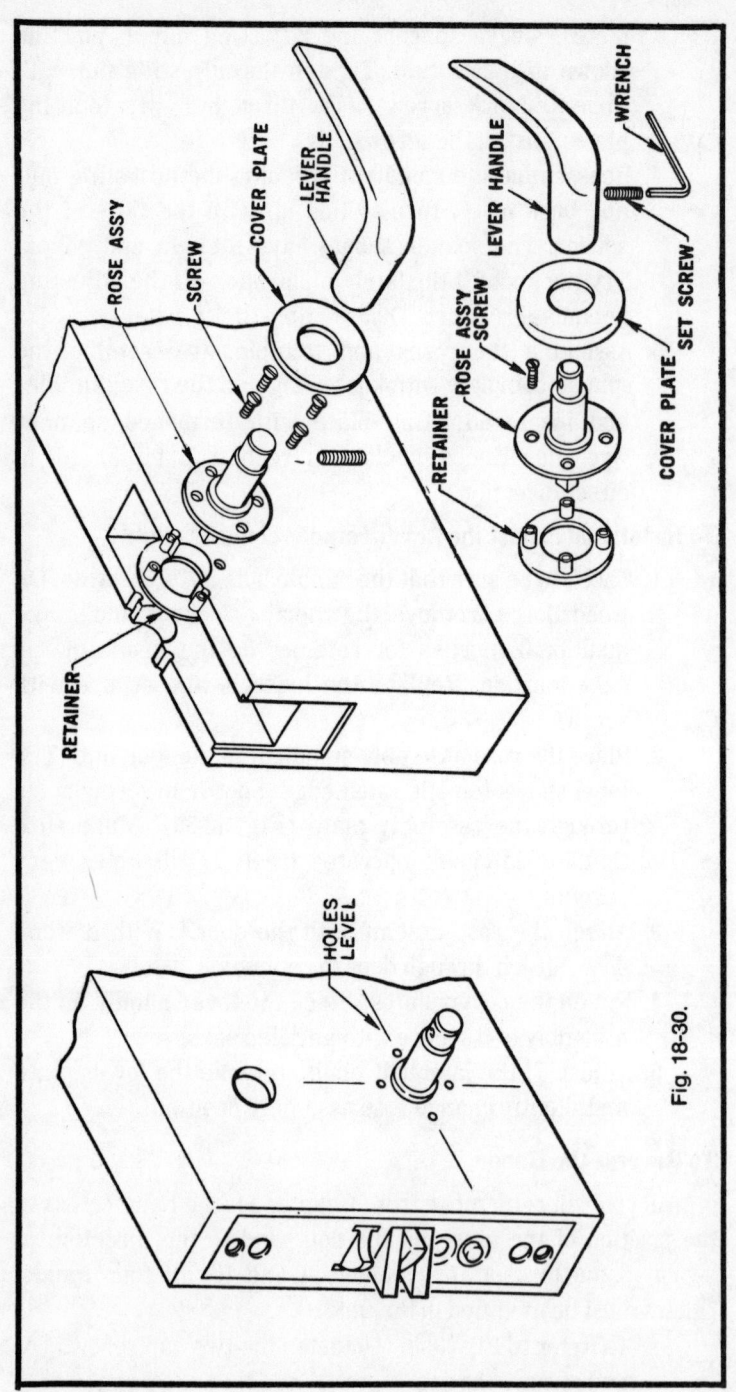

Fig. 18-30.

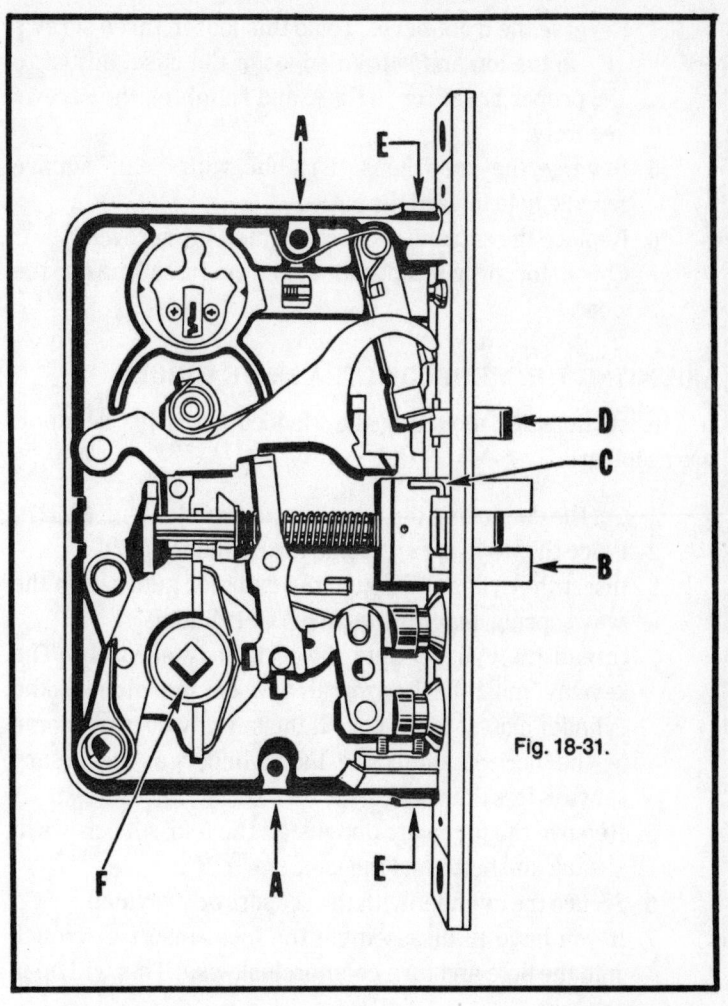

Fig. 18-31.

2. Reverse the latchbolt (B). If the lock has an antifriction latch, first remove the L-shaped pin (C) and reinstall it on the opposite side after reversing the bolt. The short leg should be held against the side of the case with the pressure-sensitive decal.

3. Reverse the auxiliary latch (D) if an auxiliary latch is furnished. To do this first remove the auxiliary latch lever and spring, then invert the latch so the concave surface is toward the stop in the frame, then reassemble the lever and spring.

4. Reverse the front bevel. To do this loosen three screws (E) in the top and bottom edges of the case, adjust to the proper bevel (or to flat), and retighten the screws securely.
5. Reverse the knob hubs (F): hub with ⅜ in. square spindle hole toward the outside.
6. Replace the cap and securely tighten the screws.
7. Check for proper operation before installing in the door.

MOUNTING THE MEDECO ULTRA 700 DEADBOLT

Instructions for mounting the Medeco Ultra 700 deadbolt are as follows:

1. Cut the door using the template as a guide (Fig. 18-32).
2. Place the wave spring inside the cylinder guard.
3. Insert the cylinder through the cylinder guard with the wave spring *under* the head of the cylinder.
4. Thread the cylinder into the lock body securely. The keyway must be horizontal and on the side of the cylinder closest to the bolt. If the keyway does not come to the horizontal, loosen the cylinder as necessary (always less than one turn).
5. Remove the faceplate and install the four set screws at the top and bottom of the lock.
6. Secure the cylinder with the set screws provided.
7. If you have to disassemble the lock, insert a wrench into the hole and turn counterclockwise. This will open the lock.
8. With the lock near or full open, manipulate the lock so the pins slip out of one escutcheon.

Fig. 18-32.

Fig. 18-33.

9. Reverse this procedure to reassemble.

NOTE

THE $^3/_{16}$ IN. ROLL PIN MUST FIT INTO THE HOLE IN THE OPPOSITE ESCUTCHEON FOR ALIGNMENT.

10. With the lock assembled and the cover in place, slide the unit into the door cutout (Fig. 18-33). Position it so that the longest escutcheon is flush with the high edge of the door bevel.
11. Tighten the lock to the door by inserting an Allen wrench in the appropriate holes and turning clockwise. This brings the escutcheons of the lock together so they clamp the door.
12. Turn the mounting screws finger tight, then give them approximately ½ turn more (Fig. 18-34).
13. Install the faceplate with the screws provided (Fig. 18-35).

The Schlage "G" Series Lockset

The mechanics of this lockset have been discussed in an earlier chapter.

Fig. 18-34.

Fig. 18-35.

Installation

These instructions. developed from material supplied by Schlage. apply to 1³⁄₈ to 2 in. wood doors.

1. Mark the height line (the centerline of the deadlatch) on the door (Fig. 18-36). It is suggested that the line be 38 in. from the floor.
2. Mark the centerline of the door thickness on the door edge.
3. Position the template (supplied with the lockset) on the door with the lower deadlatch hole on the height line. Mark center points for one 2⅛ in. hole and two 1½ in. holes through the template (Fig. 18-36). Mark the centers for two 1 in. latch holes on the door edge.
4. Bore one 2⅛ in. hole and two 1½ in. holes in the door panel.
5. Bore two 1 in. holes in the edge of the door. The upper hole should be extended ³⁄₈ in. beyond the far side of

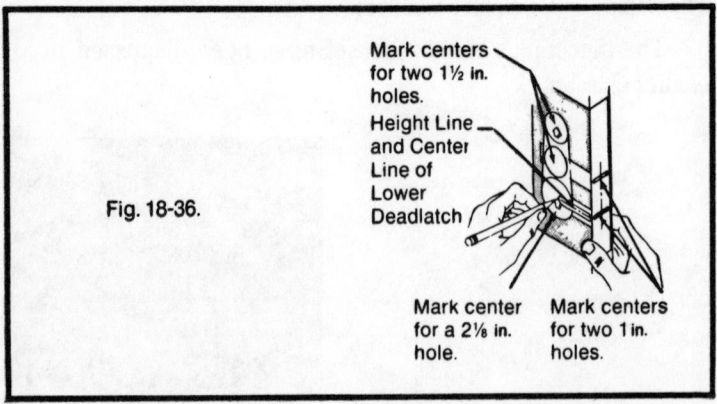

Fig. 18-36.

Mark centers for two 1½ in. holes.

Height Line and Center Line of Lower Deadlatch

Mark center for a 2⅛ in. hole.

Mark centers for two 1 in. holes.

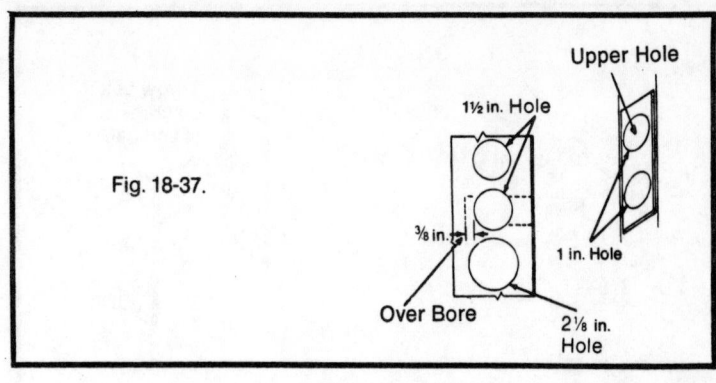

Fig. 18-37.

Upper Hole

1½ in. Hole

1 in. Hole

⅜ in.

Over Bore

2⅛ in. Hole

the middle hole on the door panel (Fig. 18-37). Mortise the edge of the door for the latch front.

6. Mark vertical and height lines on the jamb exactly opposite the center point of the lower latch hole. Mark a second horizontal centerline 2 in. above the height line for the dead bolt hole. Bore a 1 in. diameter hole 1⅛ in. deep ¼ in. below the height line. Bore a second hole to the same dimensions ¼ in. above the second horzontal line. Clear out area between holes for the strike box (Fig. 18-38).

7. Install the strike (it can be reversed for either hand). If you have done the work correctly, the strike screws will be on the same vertical centerline as the latch screws.

8. Disassemble the lock. Remove the inside knob and lift off the outside mechanism (Fig. 18-39).

9. With the dead bolt thrown and the crank slot in the vertical position, insert the latch unit and secure with the screws provided (Fig. 18-40). The deadlatch can be rotated to match the door hand (Fig. 18-41). The

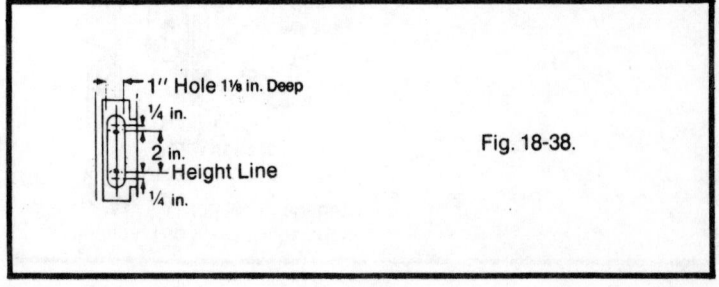

1″ Hole 1⅛ in. Deep

¼ in.

2 in. Height Line

¼ in.

Fig. 18-38.

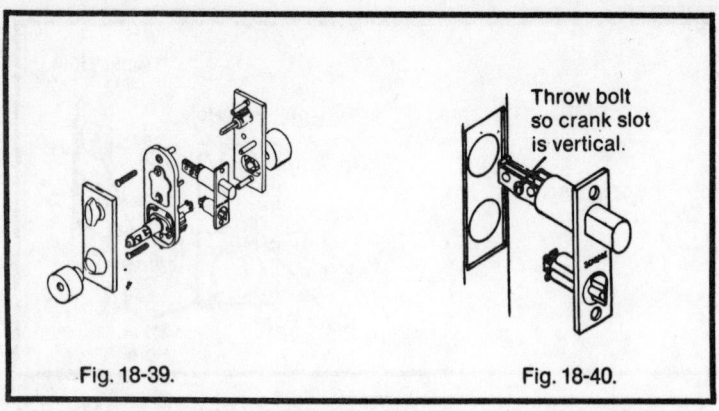

Fig. 18-39.

Throw bolt so crank slot is vertical.

Fig. 18-40.

beveled edge of the deadlatch should contact the striker as the door swings shut.

10. Install the inside mechanism with the knob button released and the crank bar in the vertical position.

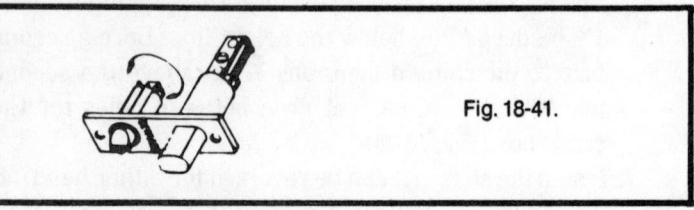

Fig. 18-41.

Insert the end of the crank bar in the slot on the dead bolt (Fig. 18-42). Engage the jaws of the slide with the deadlatch bar, and engage the deadlatch housing with the ears on the inside mechanism. This sounds more confusing than it is; see the insert for Fig. 18-42.

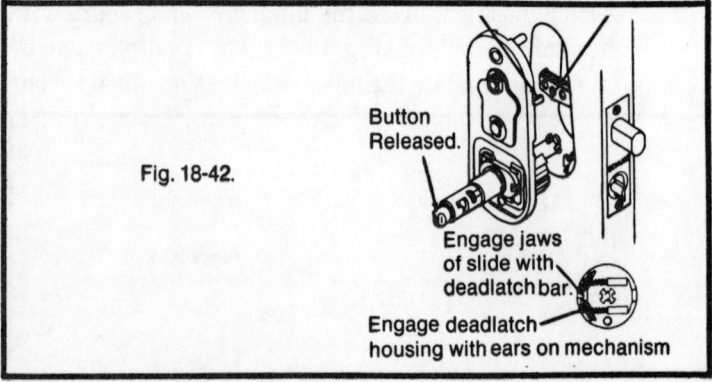

Fig. 18-42.

Button Released.

Engage jaws of slide with deadlatch bar.

Engage deadlatch housing with ears on mechanism

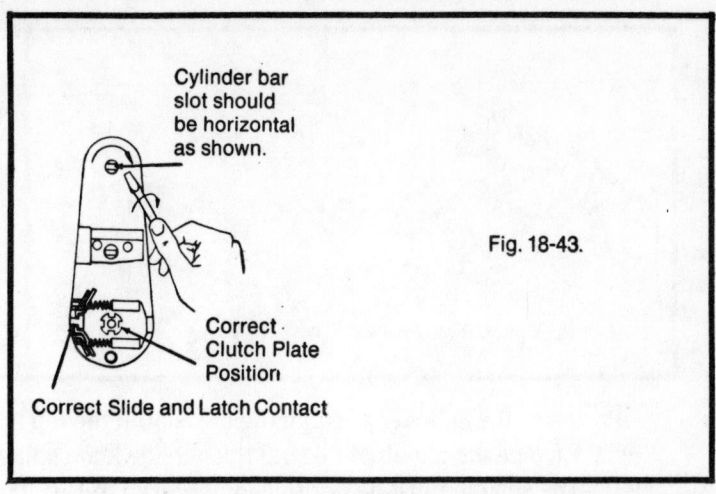

Cylinder bar slot should be horizontal as shown.

Fig. 18-43.

Correct Clutch Plate Position

Correct Slide and Latch Contact

11. From the outside of the door, turn the cylinder bar slot with a screwdriver to check the action of the dead bolt and deadlatch. Once you are satisfied that the lock works properly, retract the dead bolt and release the knob button. Turn the cylinder bar slot to the horizontal position (Fig. 18-43). The clutch plate should be situated as shown.

12. Rotate the cylinder bar to the horizontal position with the knob spindle as shown (Fig. 18-44).

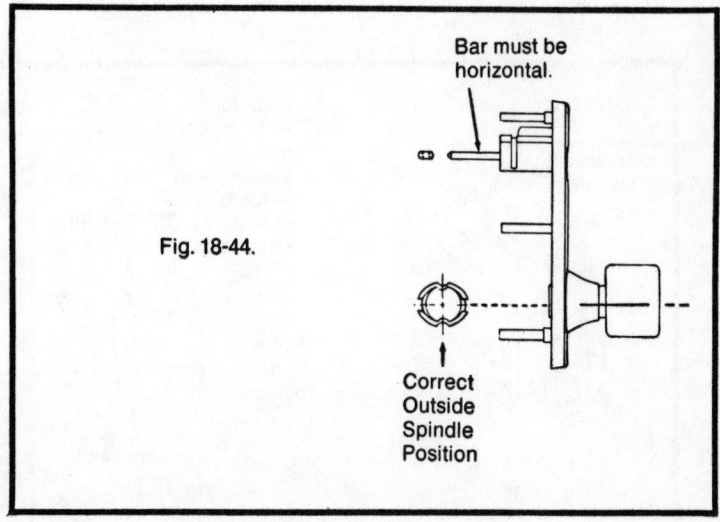

Bar must be horizontal.

Fig. 18-44.

Correct Outside Spindle Position

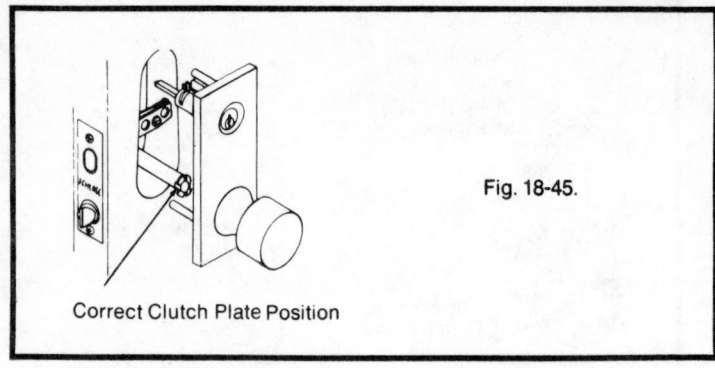

Fig. 18-45.

Correct Clutch Plate Position

13. Insert the cylinder bar into the bar slot in the top hole. Engage the clutch plate with the outside knob spindle. The spindle must be positioned as shown in Fig. 18-44, and the clutch as shown in Figs. 18-43 and 18-45.

14. G85PD locksets require that the cylinder driver bar be flush with the end of the upper bushing (Fig. 18-46).

15. Install the inside rose with the turn unit in the vertical position and the V-notch on the inside of the turn pointing to the edge of the door (Fig. 18-47). Engage the top of the rose with the mounting plate and, depressing the catch, snap the rose into place. Align the lug on the knob with the slot in the spindle and depress the knob catch. Push the knob home.

16. Test the lock.

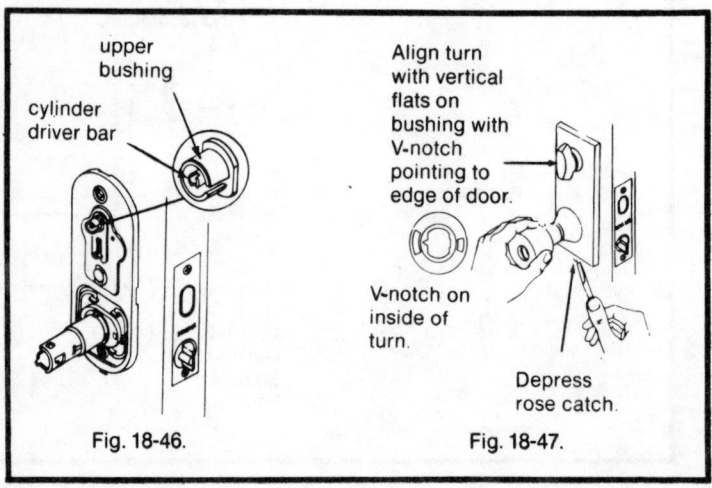

upper bushing

cylinder driver bar

Fig. 18-46.

Align turn with vertical flats on bushing with V-notch pointing to edge of door.

V-notch on inside of turn.

Depress rose catch.

Fig. 18-47.

Changing the Hand

"G" series locksets are available in right- and left-hand versions. However, the hand can be changed in the field.

Figure 18-48 is an assembled view of a right-hand unit. To convert it to left-hand operation, follow these steps:

1. Disengage the retaining rings (Fig. 18-49).
2. Lift off the mounting-plate cover (Fig. 18-50).
3. Remove the linkage-bar plate, linkage bar, bolt bar, driver bar, knob-driver linkagearm, and lower bushing (Fig. 18-51).
4. With a factory-supplied wrench or long-nosed pliers, straighten the tabs (Fig. 18-52).
5. Compress the slide with a screwdriver and pull the assembly out (Fig. 18-53).
6. Rotate the spindle 180° to relocate the knob catch. (Fig. 18-54).
7. Rotate the slide assembly 180°. With the slide compressed, insert the lugs into slots of the mounting plate (Fig. 18-55).

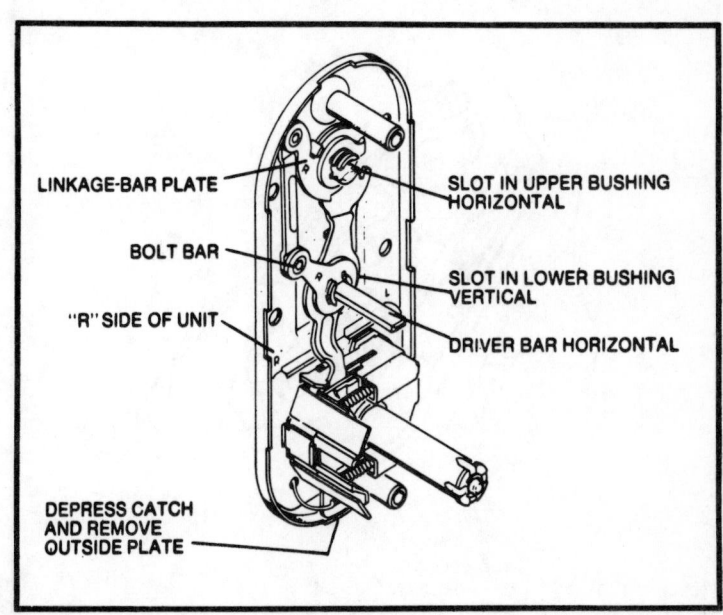

LINKAGE-BAR PLATE

SLOT IN UPPER BUSHING HORIZONTAL

BOLT BAR

SLOT IN LOWER BUSHING VERTICAL

"R" SIDE OF UNIT

DRIVER BAR HORIZONTAL

DEPRESS CATCH AND REMOVE OUTSIDE PLATE

Fig. 18-48.

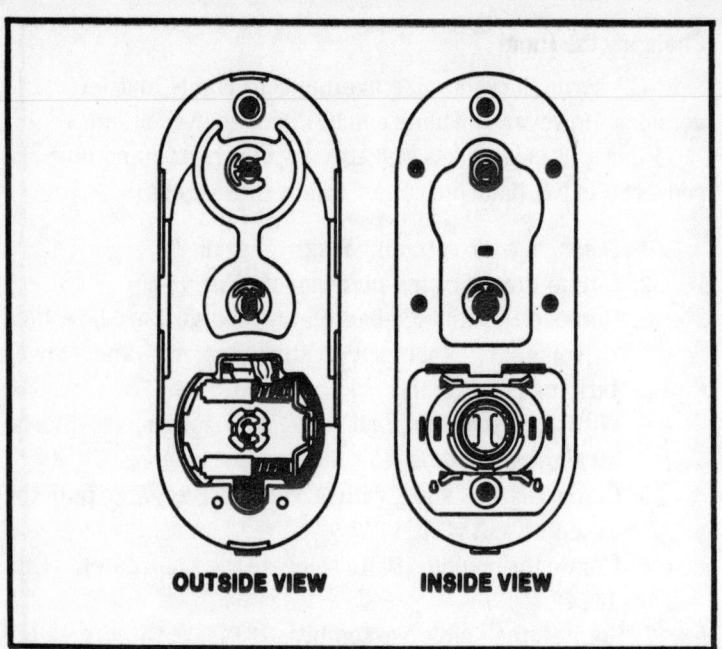

OUTSIDE VIEW   **INSIDE VIEW**

Fig. 18-49.

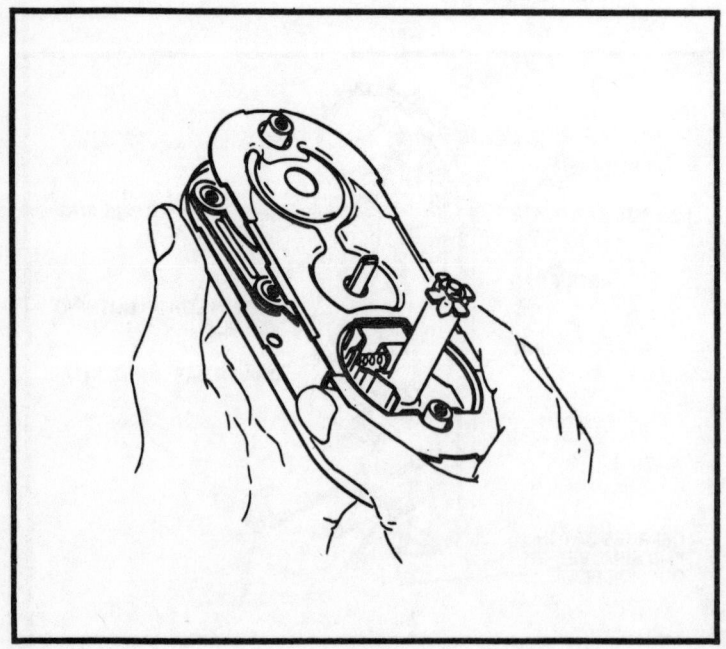

Fig. 18-50.

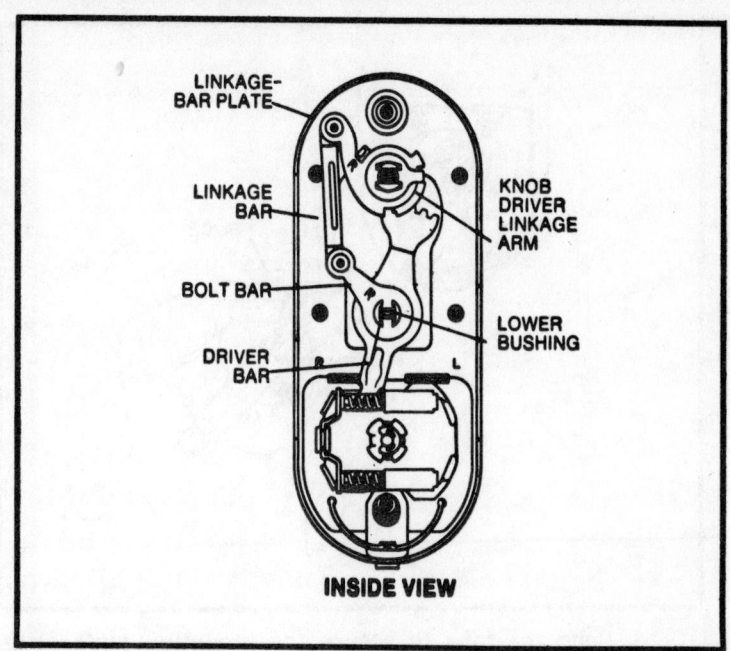

Fig. 18-51.

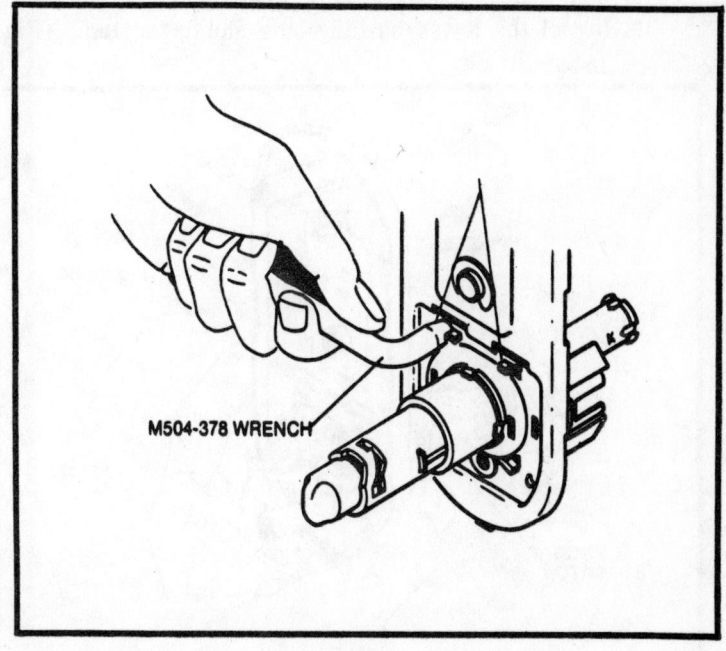

Fig. 18-52.

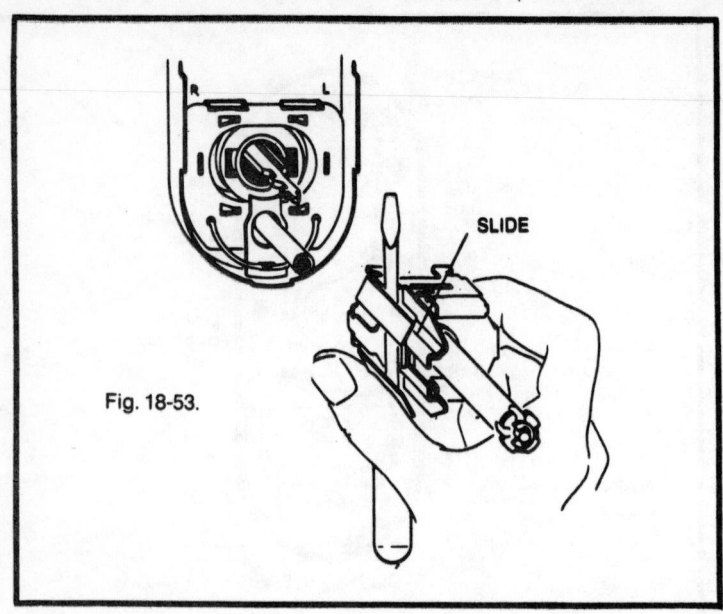

Fig. 18-53.

SLIDE

8. Bend the tabs to secure the mounting plate (Fig. 18-56).
9. Install the linkage arm (Fig. 18-57).
10. Install the lower bushing—the slot is vertical (Fig. 18-58).

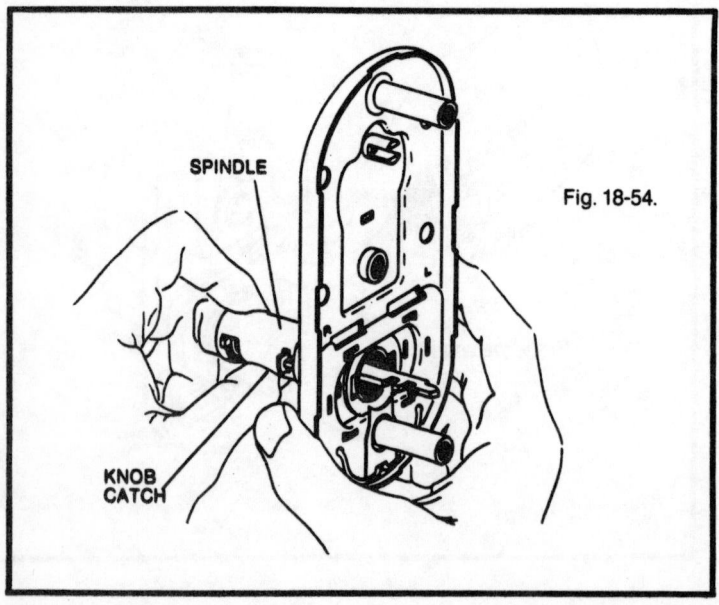

SPINDLE

Fig. 18-54.

KNOB CATCH

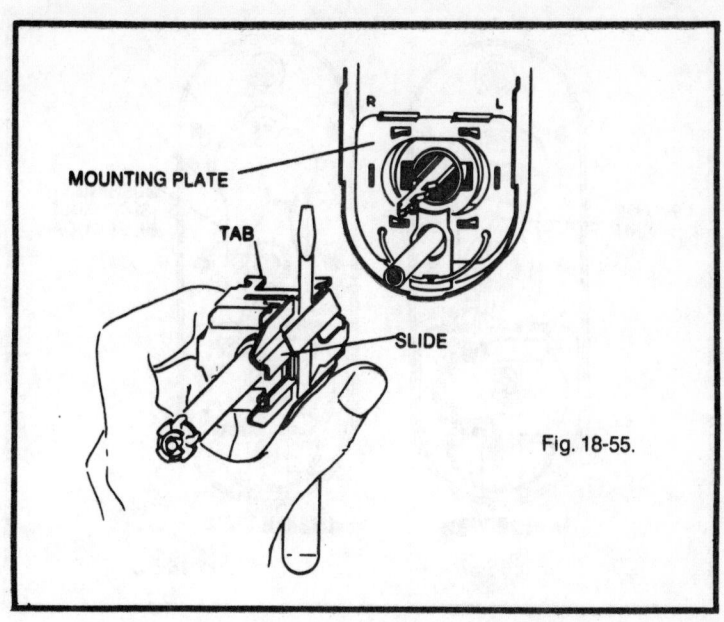

R
L
MOUNTING PLATE
TAB
SLIDE

Fig. 18-55.

Fig. 18-56.

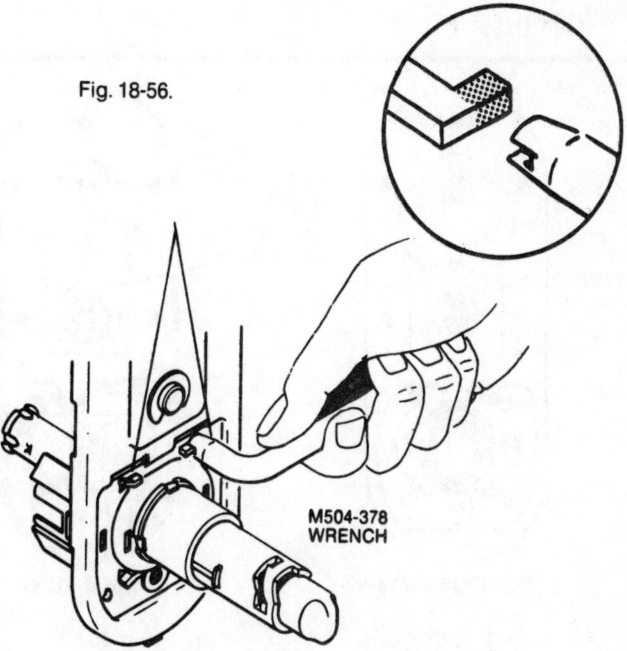

M504-378
WRENCH

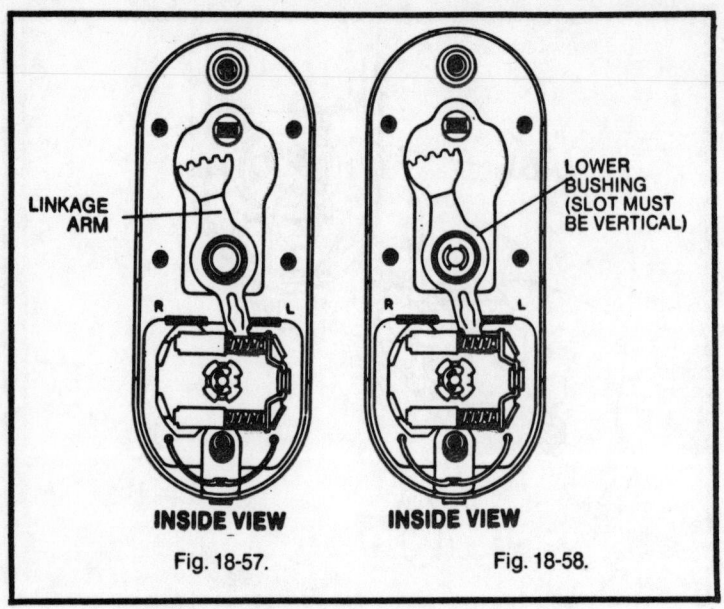

LINKAGE
ARM

LOWER
BUSHING
(SLOT MUST
BE VERTICAL)

R   L

INSIDE VIEW

Fig. 18-57.

R   L

INSIDE VIEW

Fig. 18-58.

11. Install the retaining ring (Fig. 18-59).
12. Install the driver bar—the bar is horizontal (Fig. 18-60).

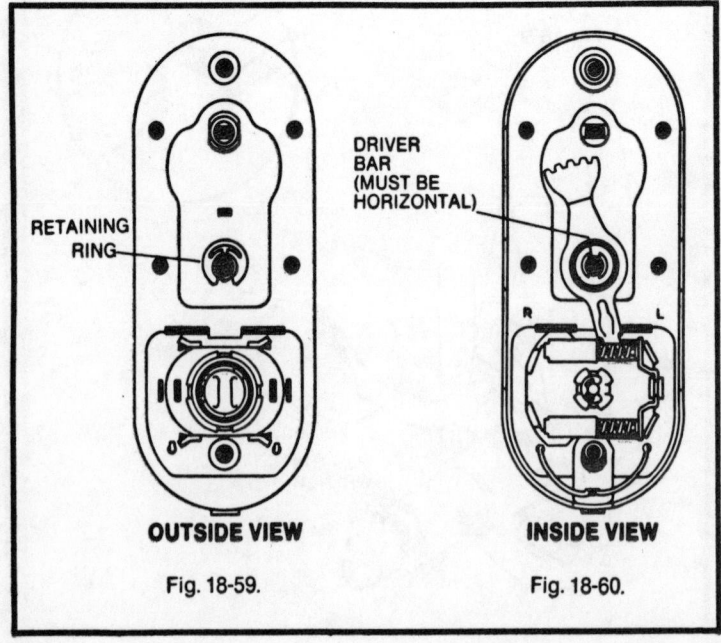

RETAINING
RING

DRIVER
BAR
(MUST BE
HORIZONTAL)

R   L

OUTSIDE VIEW

Fig. 18-59.

INSIDE VIEW

Fig. 18-60.

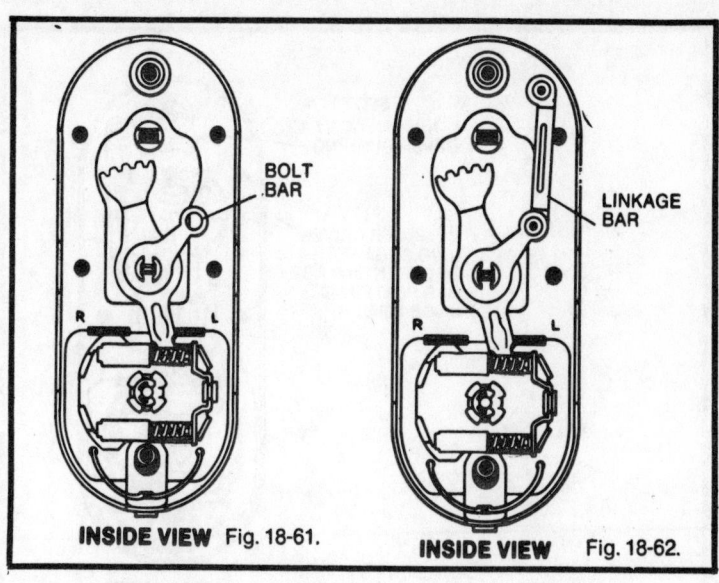

INSIDE VIEW Fig. 18-61.

INSIDE VIEW Fig. 18-62.

13. Install the bolt bar (Fig. 18-61).
14. Install the linkage bar (Fig. 18-62).
15. Install the knob driver (Fig. 18-63).
16. Install the linkage-bar plate (Fig. 18-64).

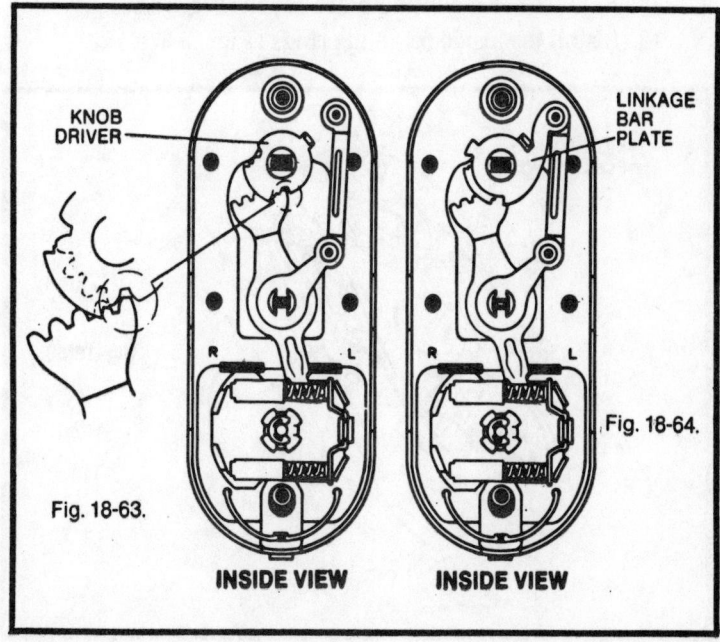

Fig. 18-63.

Fig. 18-64.

INSIDE VIEW

INSIDE VIEW

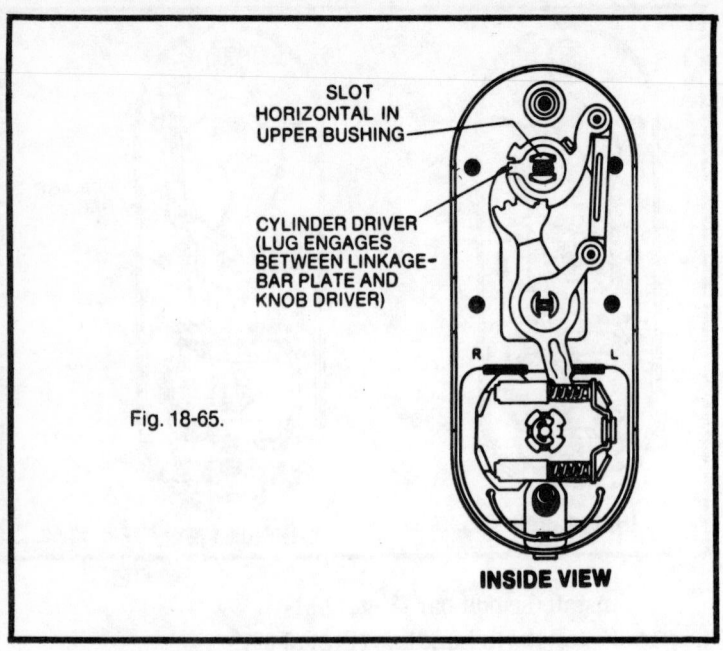

SLOT HORIZONTAL IN UPPER BUSHING

CYLINDER DRIVER (LUG ENGAGES BETWEEN LINKAGE-BAR PLATE AND KNOB DRIVER)

R   L

H

Fig. 18-65.

INSIDE VIEW

17. With the upper bushing slot in the horizontal position, install the cylinder driver (Fig. 18-65).
18. Replace the mounting-plate cover (Fig. 18-66).
19. Install the inside retaining rings (Fig. 18-67).

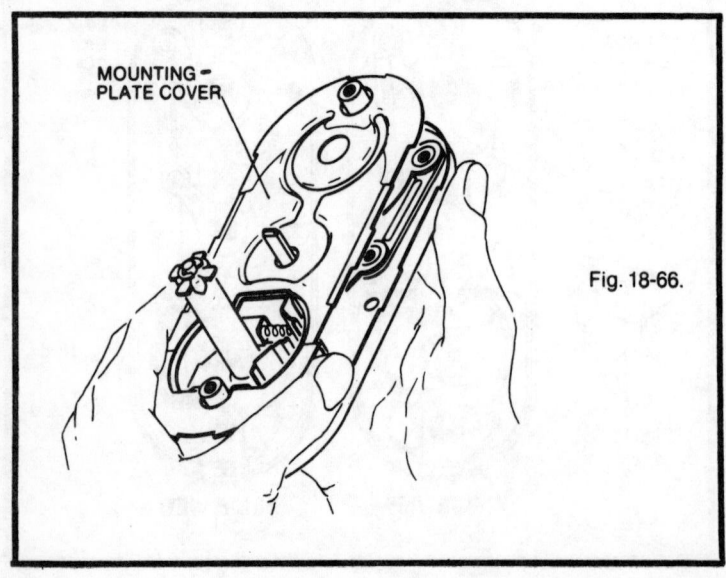

MOUNTING-PLATE COVER

Fig. 18-66.

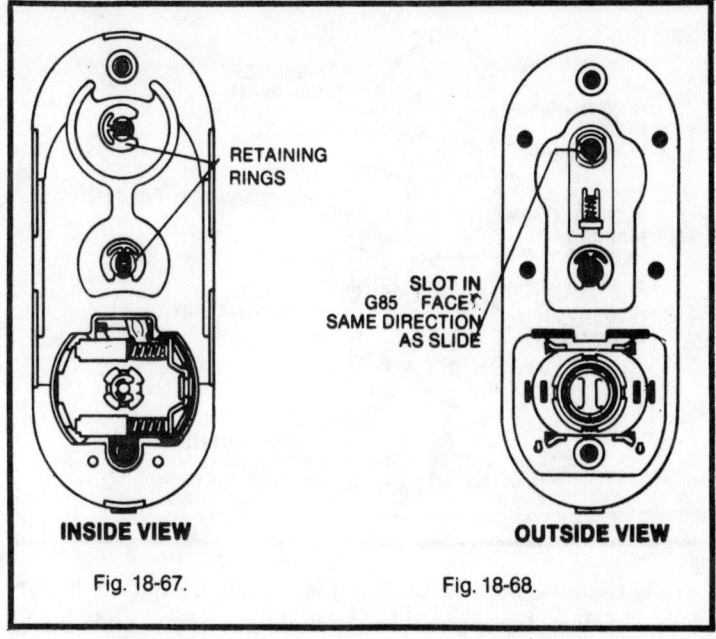

RETAINING RINGS

SLOT IN G85 FACE SAME DIRECTION AS SLIDE

INSIDE VIEW

Fig. 18-67.

OUTSIDE VIEW

Fig. 18-68.

20. Turn the slot in the G85 unit to face the same direction as the slide (Fig. 18-68).
21. Turn the deadlatch 180° (Fig. 18-69).
22. Check your work (Fig. 18-70).

Figure 18-71 is an assembled view of a left-hand unit. To convert it to right-hand operation, follow these steps:

1. Disengage the retaining rings (Fig. 18-72).
2. Lift off the mounting-plate cover (Fig. 18-73).

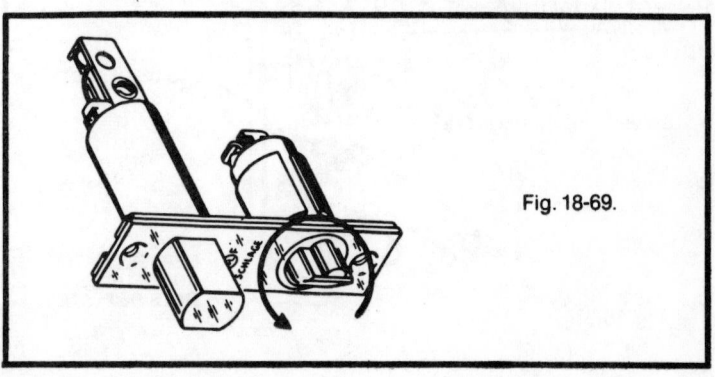

Fig. 18-69.

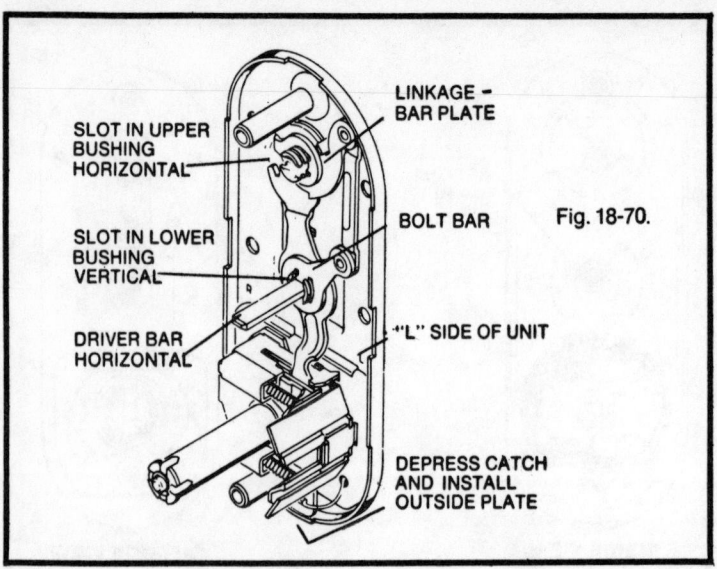

LINKAGE – BAR PLATE

SLOT IN UPPER BUSHING HORIZONTAL

SLOT IN LOWER BUSHING VERTICAL

BOLT BAR

Fig. 18-70.

DRIVER BAR HORIZONTAL

"L" SIDE OF UNIT

DEPRESS CATCH AND INSTALL OUTSIDE PLATE

3. Remove the linkage-bar plate, linkage bar, bolt bar, driver bar, knob-driver linkage arm, and lower bushing (Fig. 18-74).

4. With a factory-supplied wrench or long-nosed pliers, straighten the tabs (Fig. 8-75).

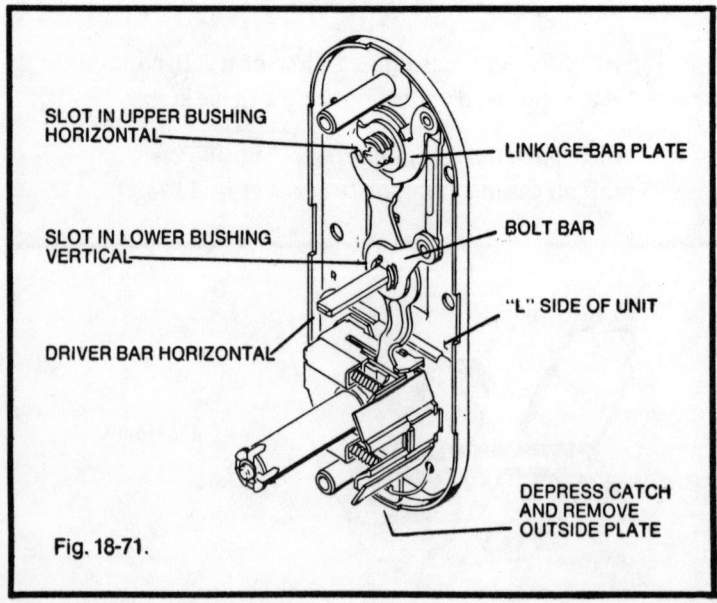

SLOT IN UPPER BUSHING HORIZONTAL

LINKAGE–BAR PLATE

SLOT IN LOWER BUSHING VERTICAL

BOLT BAR

"L" SIDE OF UNIT

DRIVER BAR HORIZONTAL

DEPRESS CATCH AND REMOVE OUTSIDE PLATE

Fig. 18-71.

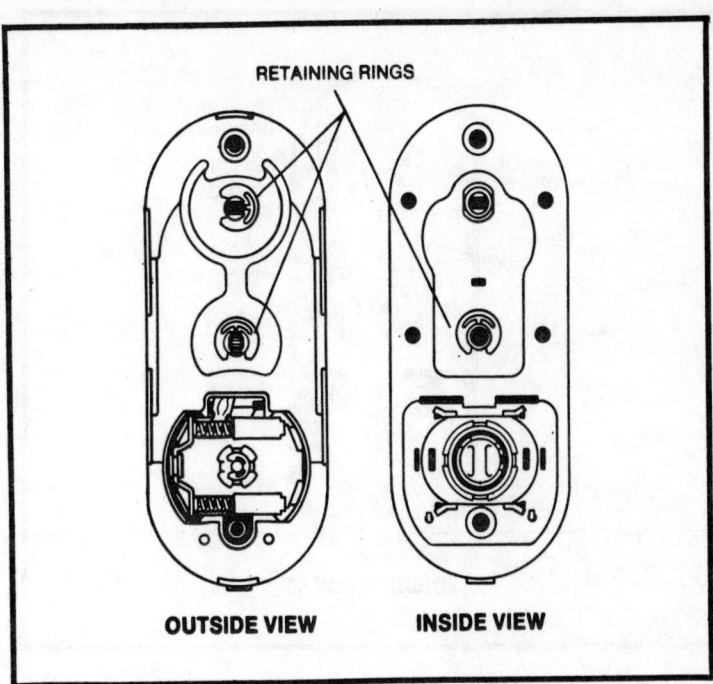

Fig. 18-72.

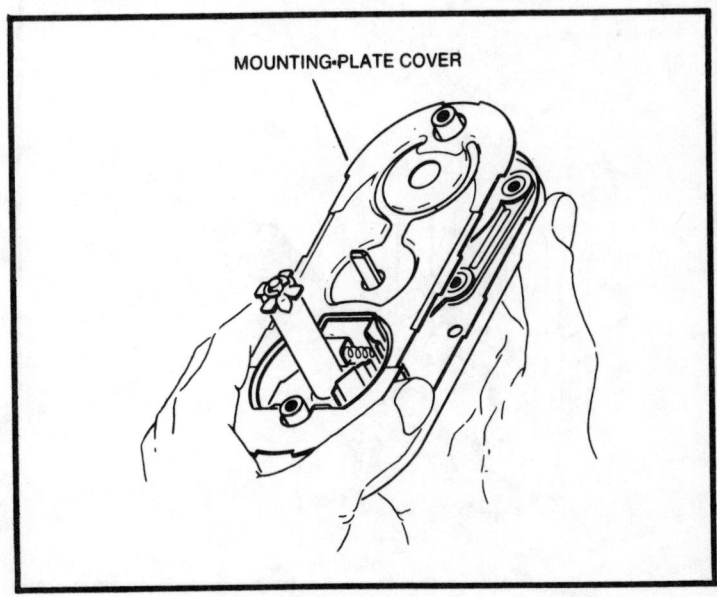

Fig. 18-73.

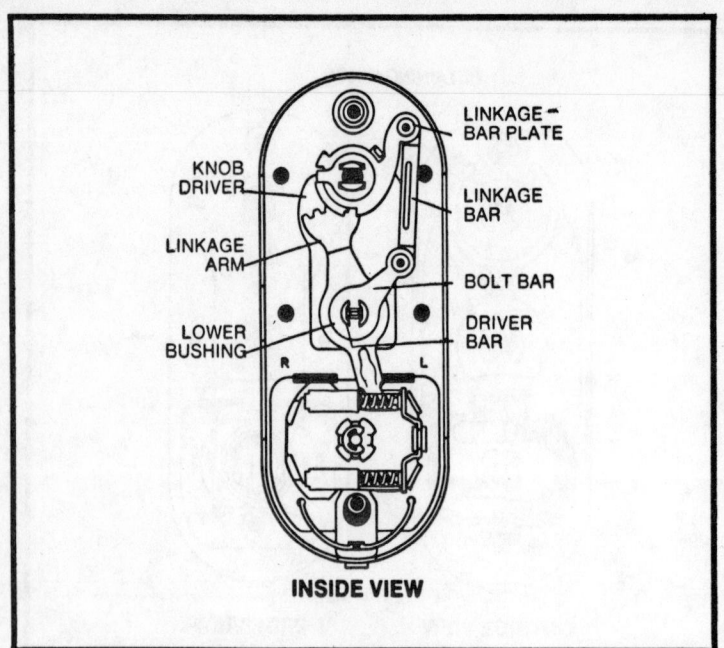

LINKAGE – BAR PLATE

KNOB DRIVER

LINKAGE BAR

LINKAGE ARM

BOLT BAR

LOWER BUSHING

DRIVER BAR

R L

INSIDE VIEW

Fig. 18-74.

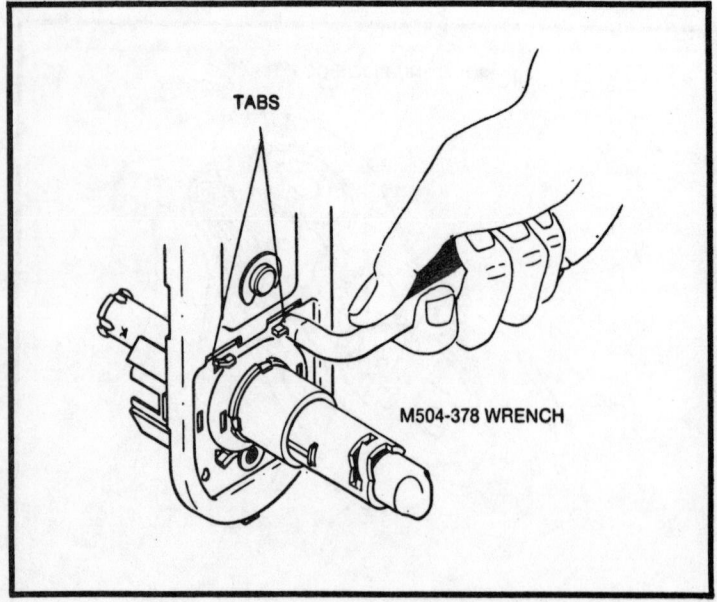

TABS

M504-378 WRENCH

Fig. 18-75.

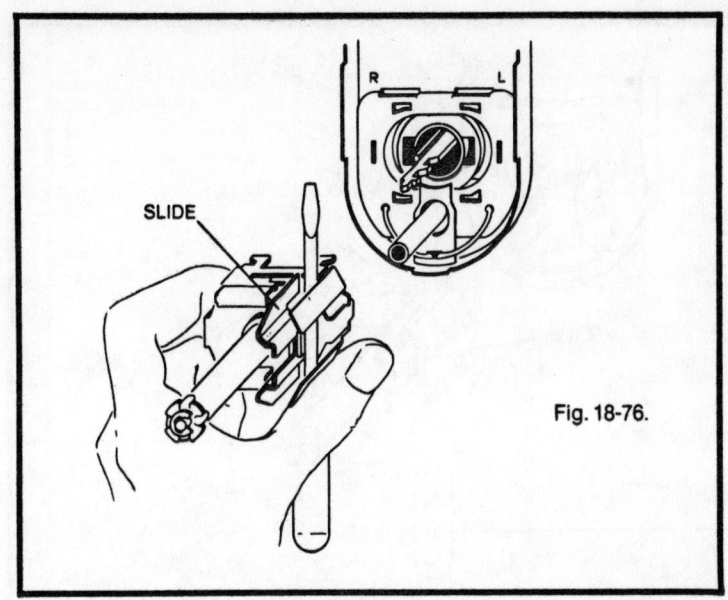

SLIDE

Fig. 18-76.

5. Compress the slide with a screwdriver and pull the assembly out (Fig. 18-76).
6. Rotate the spindle 180° to relocate the knob catch (Fig. 18-77).

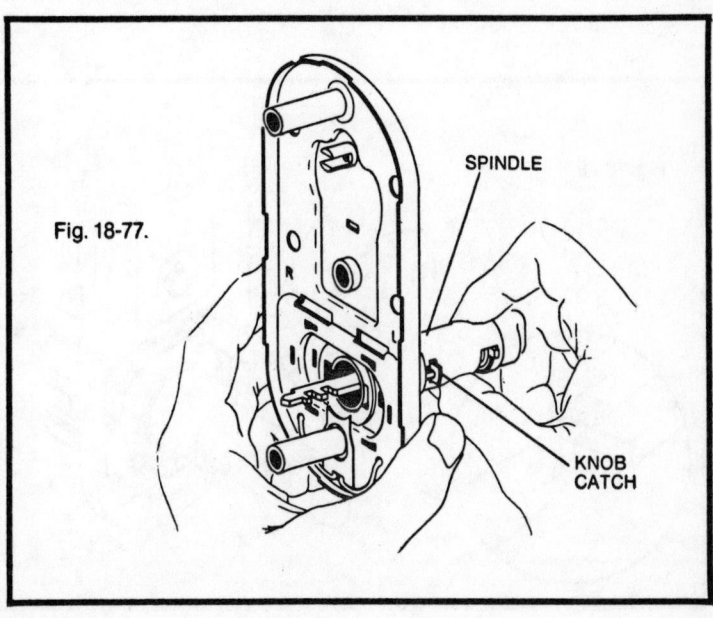

Fig. 18-77.

SPINDLE

KNOB CATCH

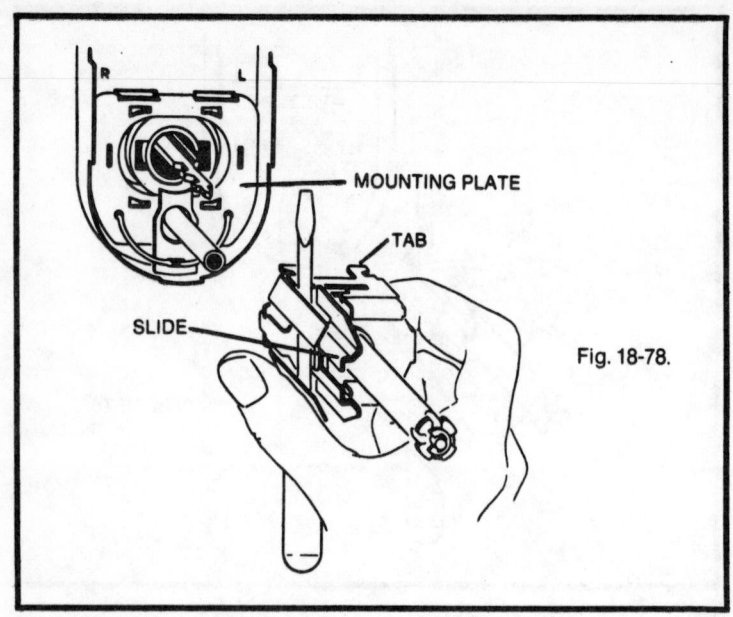

MOUNTING PLATE

TAB

SLIDE

Fig. 18-78.

7. Rotate the slide assembly 180°. With the slide compressed, insert the lugs into slots on the mounting plate (Fig. 18-78).

8. Bend the tabs to secure the mounting plate (Fig. 18-79).

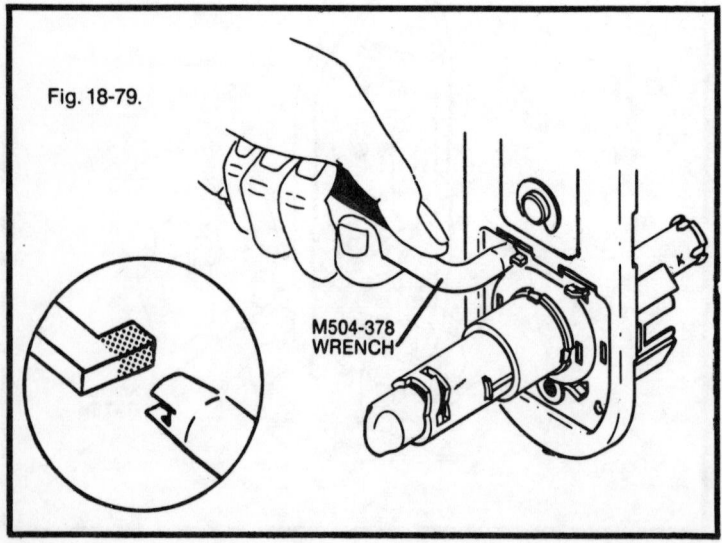

Fig. 18-79.

M504-378 WRENCH

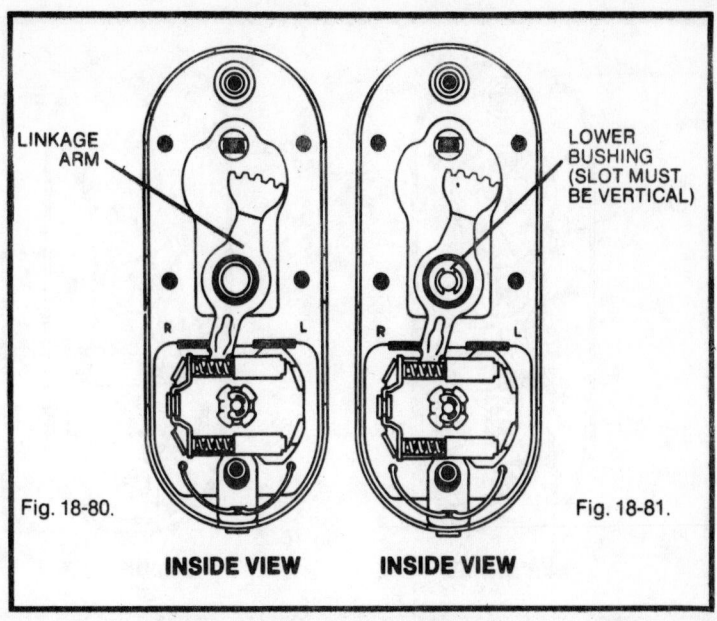

LINKAGE ARM

LOWER BUSHING (SLOT MUST BE VERTICAL)

Fig. 18-80.

Fig. 18-81.

INSIDE VIEW   **INSIDE VIEW**

9. Install the linkage arm. (Fig. 18-80).
10. Install the lower bushing (Fig. 18-81).
11. Install the retaining ring (Fig. 18-82).
12. Install the driver bar (Fig. 18-83).

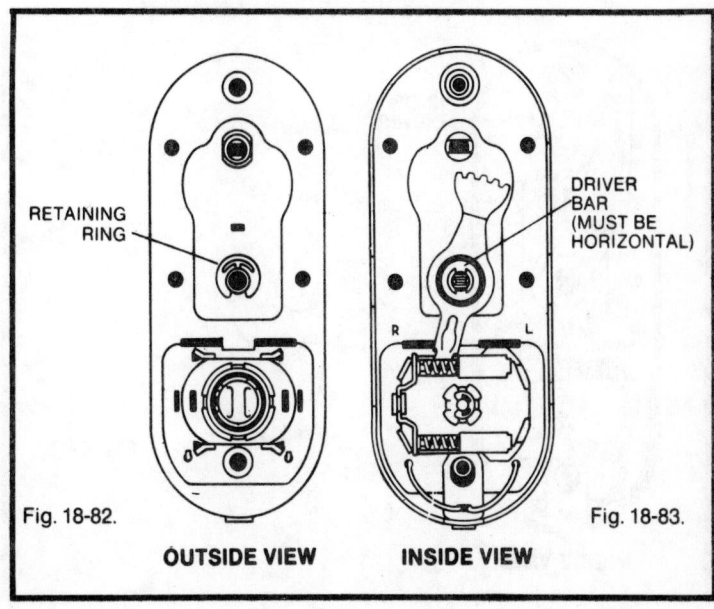

RETAINING RING

DRIVER BAR (MUST BE HORIZONTAL)

Fig. 18-82.

Fig. 18-83.

OUTSIDE VIEW   **INSIDE VIEW**

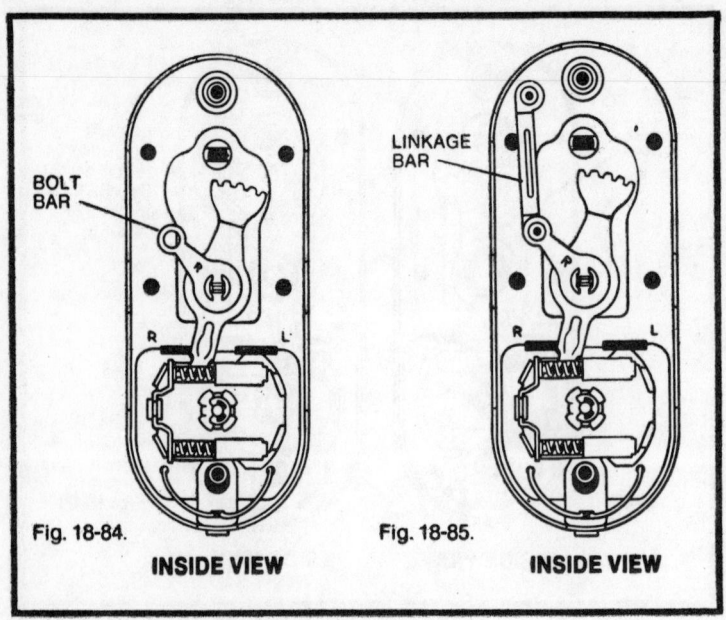

Fig. 18-84.

Fig. 18-85.

13. Install the bolt bar (Fig. 18-84).
14. Install the linkage bar (Fig. 18-85).
15. Install the knob driver (Fig. 18-86).

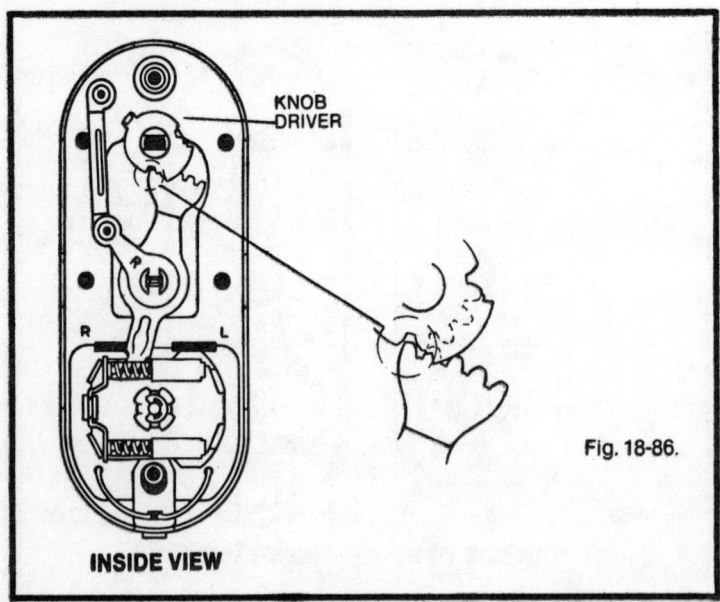

Fig. 18-86.

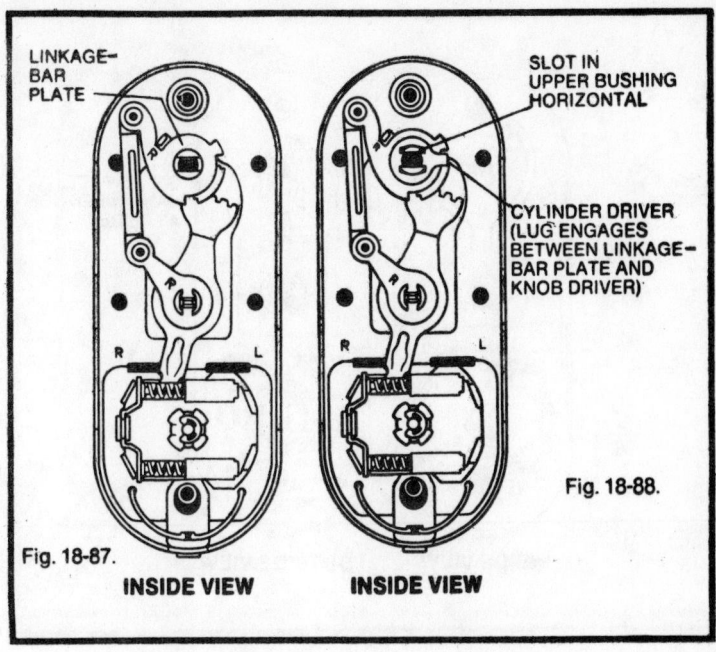

LINKAGE-
BAR
PLATE

SLOT IN
UPPER BUSHING
HORIZONTAL

CYLINDER DRIVER
(LUG ENGAGES
BETWEEN LINKAGE-
BAR PLATE AND
KNOB DRIVER)

Fig. 18-88.

Fig. 18-87.

INSIDE VIEW   **INSIDE VIEW**

16. Install the linkage-bar plate (Fig. 18-87).
17. With the upper bushing slot in the horizontal position, install the cylinder driver (Fig. 18-88).
18. Replace the mounting-plate cover (Fig. 18-89).

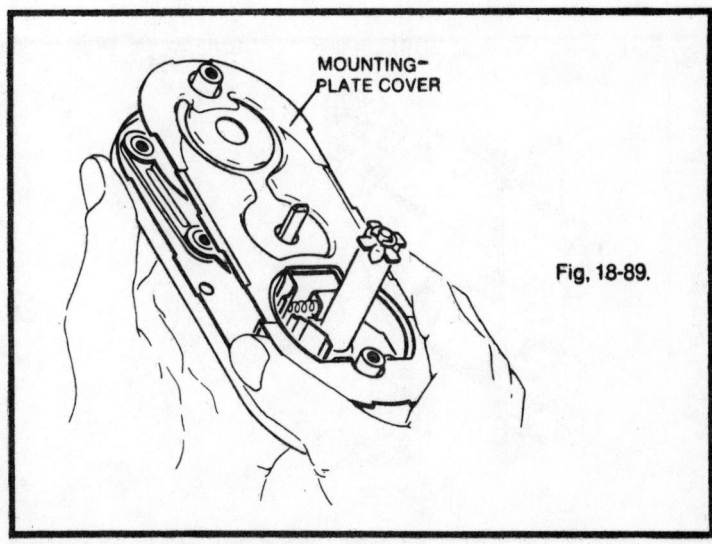

MOUNTING-
PLATE COVER

Fig, 18-89.

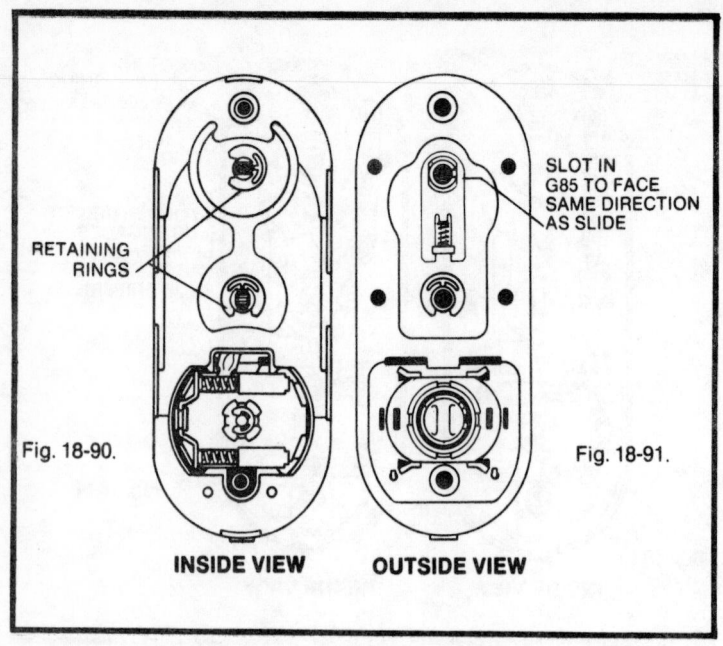

Fig. 18-90.

Fig. 18-91.

RETAINING RINGS

SLOT IN G85 TO FACE SAME DIRECTION AS SLIDE

INSIDE VIEW   **OUTSIDE VIEW**

19. Install the inside retaining ring (Fig. 18-90).
20. Turn the slot in the G85 unit to face the same direction as the slide (Fig. 18-91).
21. Turn the deadlatch 180° (Fig. 18-92).
22. Check your work (Fig. 18-93).

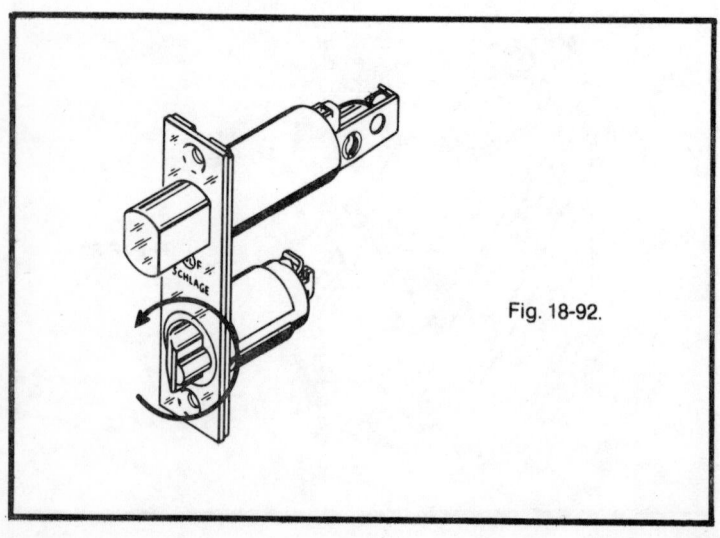

Fig. 18-92.

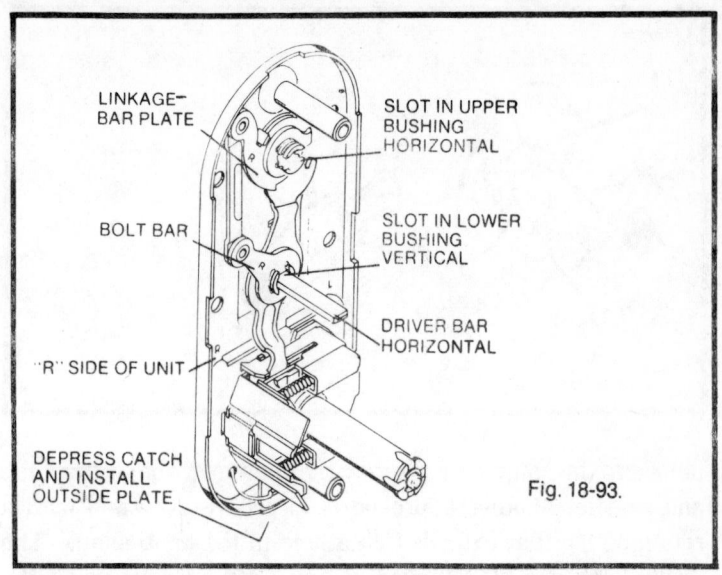

LINKAGE—
BAR PLATE

SLOT IN UPPER
BUSHING
HORIZONTAL

SLOT IN LOWER
BUSHING
VERTICAL

BOLT BAR

DRIVER BAR
HORIZONTAL

"R" SIDE OF UNIT

DEPRESS CATCH
AND INSTALL
OUTSIDE PLATE

Fig. 18-93.

Attaching and Removing Knobs

Figure 18-94 illustrates two of the most frequently encountered methods of securing door knobs. The knob may be threaded over the spindle and held by a set screw, or it may be

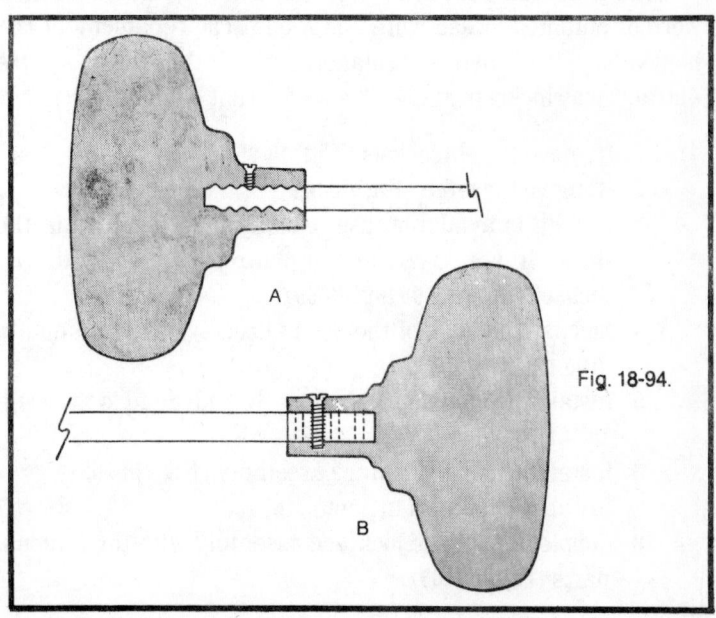

A

B

Fig. 18-94.

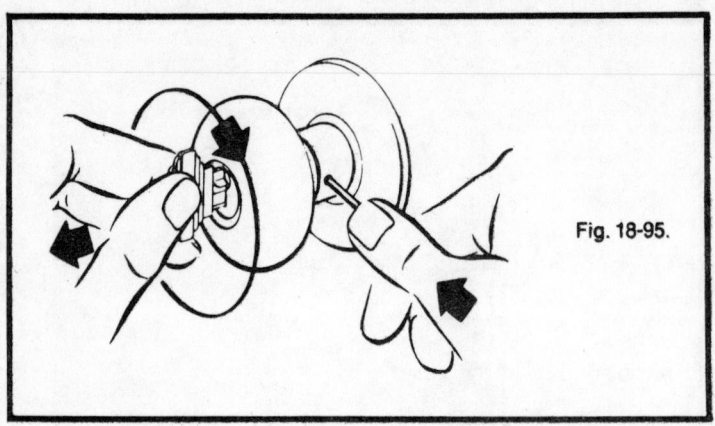

Fig. 18-95.

pinned to the spindle by a screw that passes through the knob and spindle. Another approach is to secure the knob with a retaining lug that extends into a hole in the knob shank. The inside knob can be removed at any time by depressing the retainer; the outside knob can be removed only when the lock is open (Fig. 18-95).

UPDATING A LOCKSET

The following instructions concern the replacement of a worn or outdated lockset with a new one. The replacement is a heavy-duty "G" series Schlage, one of the most secure entranceway locksets made. The work is not difficult.

1. Remove the old lockset (Fig. 18-96).
2. Remove the latch (Fig. 18-97).
3. If a jig is available, use it as a guide for cutting the door; if not, use the template packaged with the lockset (Figs. 18-98 and 18-99).
4. Mortise the edge of the door to receive the strikeplate (Fig. 18-100).
5. Install the double-locking latch and dead bolt (Fig. 18-101).
6. Install the inside lockface assembly (Fig. 18-102).
7. Install the internal mechanism.
8. Couple the outside lockface assembly with the internal parts (Fig. 18-103).

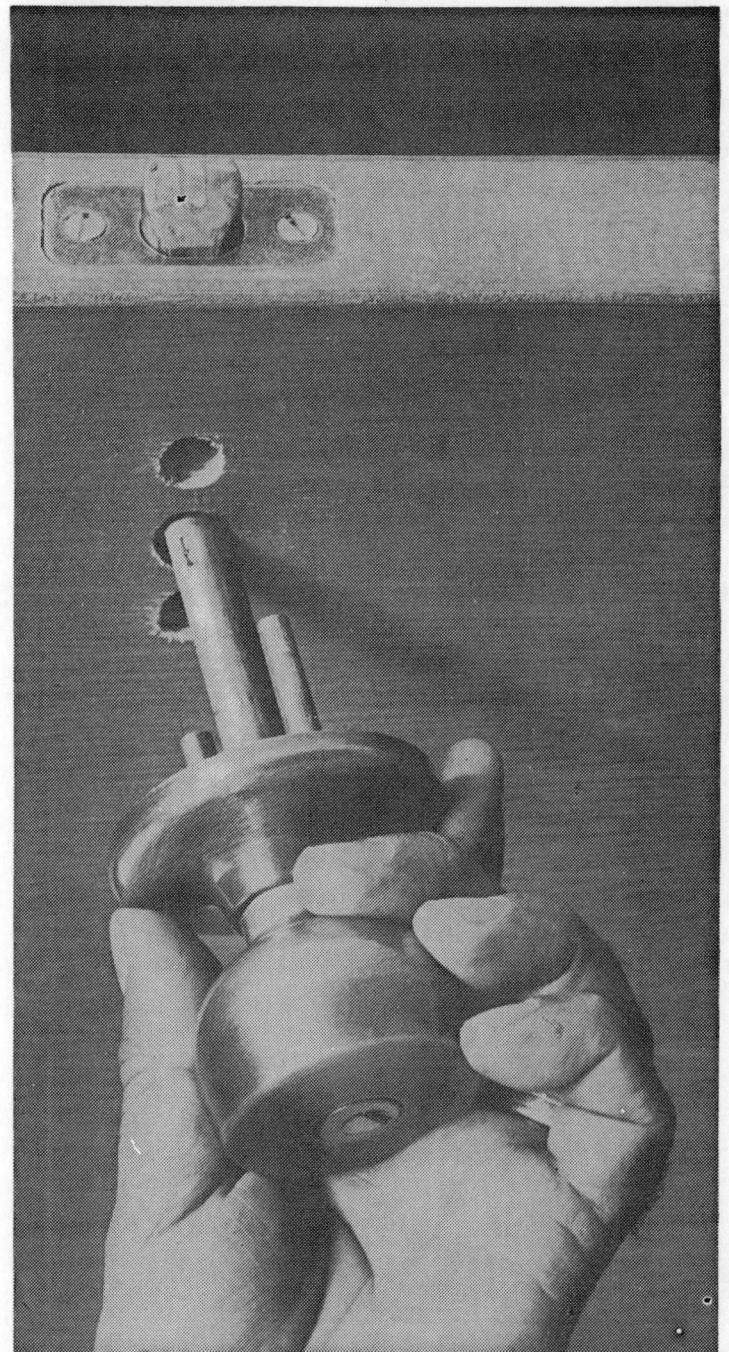

Fig. 18-96.

273

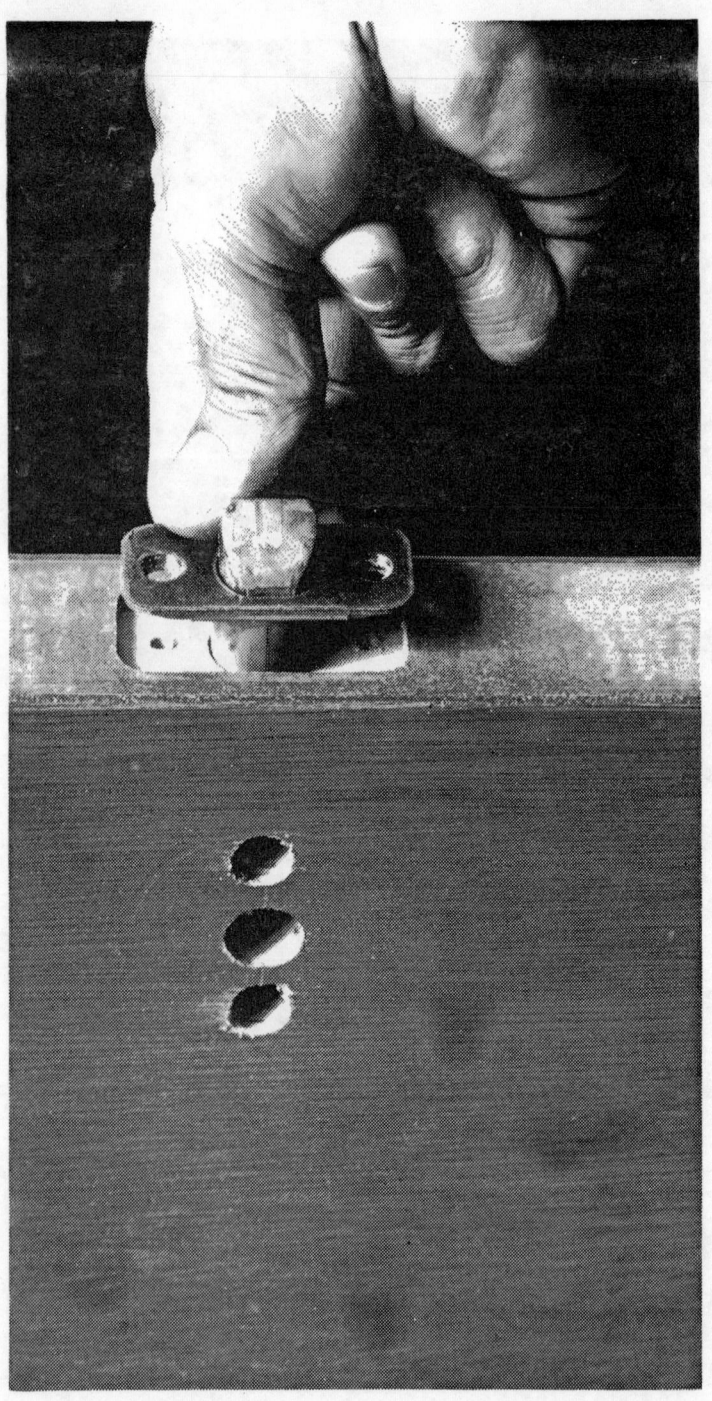

Fig. 18-97.

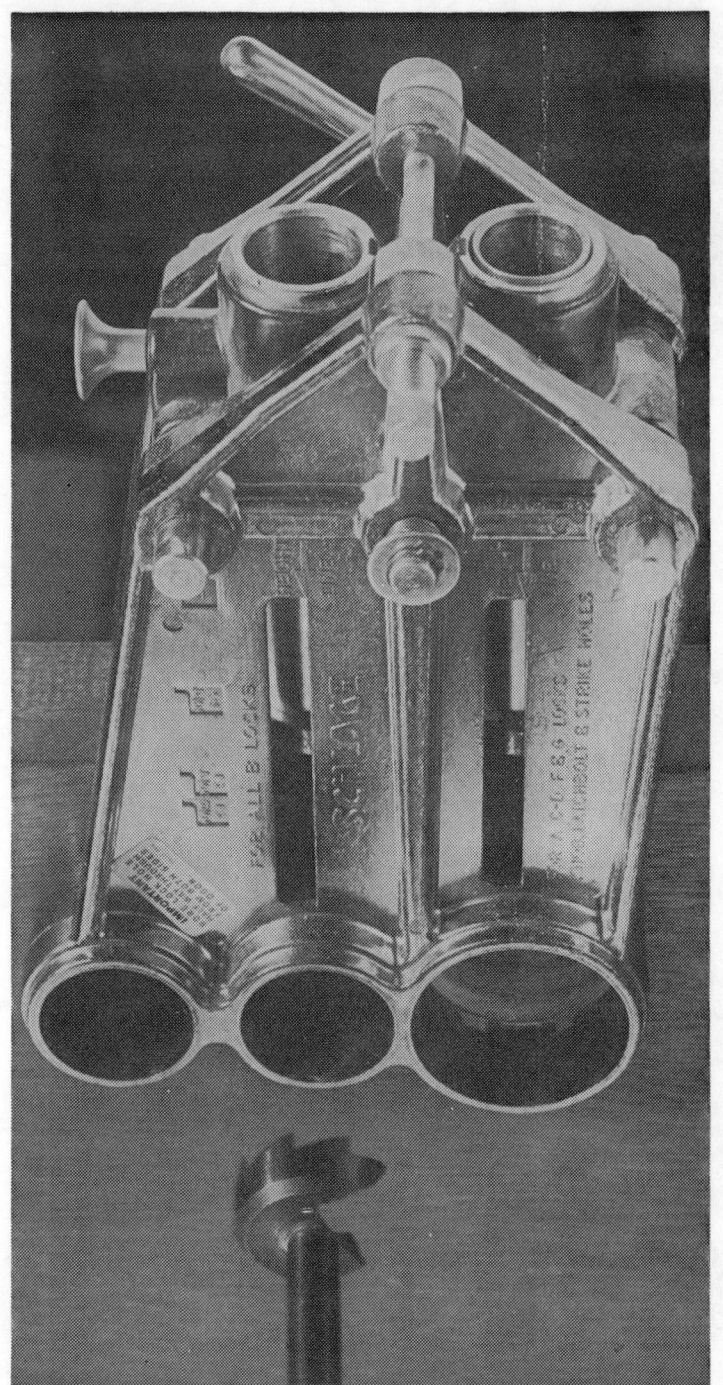

Fig. 18-98.

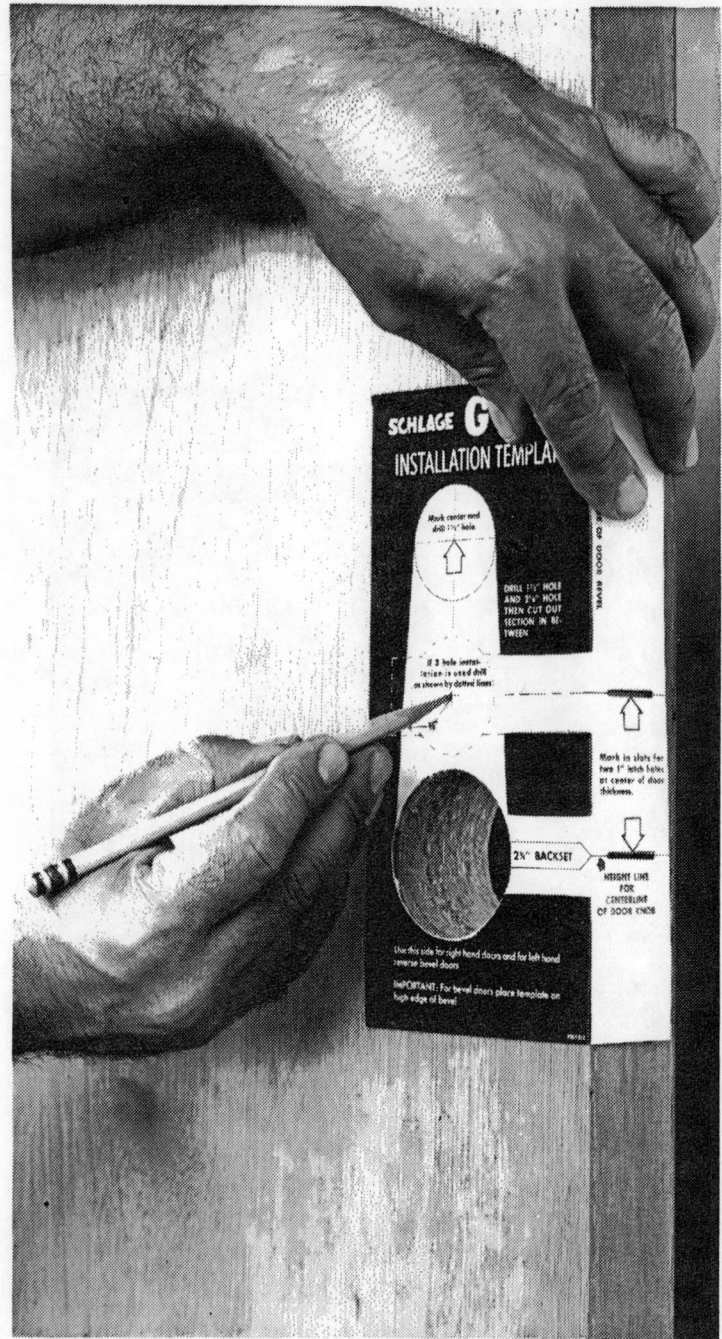

Fig. 18-99.

Fig. 18-100.

Fig. 18-101.

Fig. 18-102.

Fig. 18-103.

9. Secure the inside lockface assembly with the screws provided (Fig. 18-104).
10. Snap the inside cover into place (Fig. 18-105). The lockset is now installed (Fig. 18-106).

STRIKEPLATES

Quality locksets are equipped with a deadlocking plunger on the latchbolt. Otherwise the bolt could be retracted with a knife blade or a strip of celluloid. Nevertheless, it is important to leave very little space between the door edge and the strikeplate. A bolt with a ½ in. throw should have no more than ⅛ in. visible between the door and jamb. If necessary, mount the strikeplate over a steel spacer. This moves the strikeplate closer to the bolt.

The length of the lip is also critical since a short lip will increase wear on the latchbolt and may frustrate the automatic door-close mechanism, leaving the door unlatched. Mortise strikeplates are mounted on the same vertical centerline as the bolt (or should be). Measure the distance from the centerline of the latchbolt to the edge of the jamb and add ⅛ in. for flat strikeplates and ¼ in. for curved types. Figure 18-107 illustrates a collection of Schlage strikeplates for reference.

ENTRANCE-DOOR SECURITY

Security is a function of the door, cylinder, bolt, and strikeplate.
- There is no security with a hollow wooden door.
- Masterkeying should be kept as simple as possible. Each split pin increases the odds in favor of the lock picker.
- Resist demands for extensive cross keying. These demands originate with customers who insist upon single-key performance for executive keys.
- Security begins with proper key-control procedures.
- It is wise to supply extra change-key blanks during the installation phase of the system. Without readily available blanks, the convenient thing to do would be to cut duplicate change keys from masterkey blanks.

Fig. 18-104.

Fig. 18-105.

Fig. 18-106.

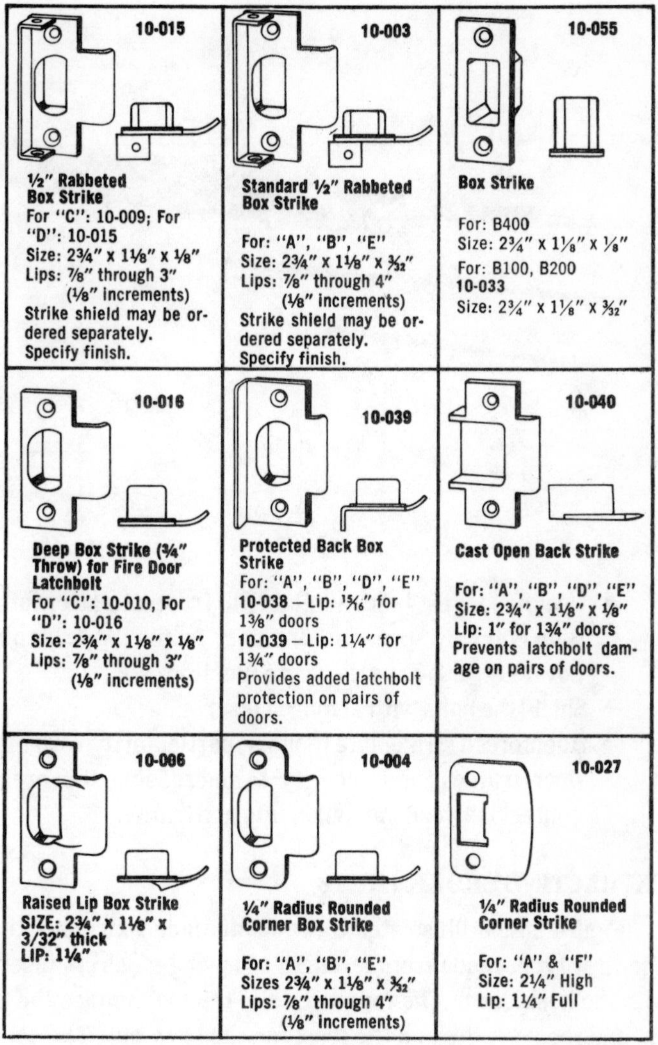
10-015
½" Rabbeted Box Strike
For "C": 10-009; For "D": 10-015
Size: 2¾" x 1⅛" x ⅛"
Lips: ⅞" through 3" (⅛" increments)
Strike shield may be ordered separately.
Specify finish.
10-003
Standard ½" Rabbeted Box Strike
For: "A", "B", "E"
Size: 2¾" x 1⅛" x 3/32"
Lips: ⅞" through 4" (⅛" increments)
Strike shield may be ordered separately.
Specify finish.
10-055
Box Strike
For: B400
Size: 2¾" x 1⅛" x ⅛"
For: B100, B200
10-033
Size: 2¾" x 1⅛" x 3/32"
10-016
Deep Box Strike (¾" Throw) for Fire Door Latchbolt
For "C": 10-010, For "D": 10-016
Size: 2¾" x 1⅛" x ⅛"
Lips: ⅞" through 3" (⅛" increments)
10-039
Protected Back Box Strike
For: "A", "B", "D", "E"
10-038 – Lip: 15/16" for 1⅜" doors
10-039 – Lip: 1¼" for 1¾" doors
Provides added latchbolt protection on pairs of doors.
10-040
Cast Open Back Strike
For: "A", "B", "D", "E"
Size: 2¾" x 1⅛" x ⅛"
Lip: 1" for 1¾" doors
Prevents latchbolt damage on pairs of doors.
10-006
Raised Lip Box Strike
SIZE: 2¾" x 1⅛" x 3/32" thick
LIP: 1¼"
10-004
¼" Radius Rounded Corner Box Strike
For: "A", "B", "E"
Sizes 2¾" x 1⅛" x 3/32"
Lips: ⅞" through 4" (⅛" increments)
10-027
¼" Radius Rounded Corner Strike
For: "A" & "F"
Size: 2¼" High
Lip: 1¼" Full

Fig. 18-107.

Security could be compromised since the change key could fit other locks.

- Double-cylinder locks should be used wherever possible.
- Automatic deadlatches should be specified for all locksets without a dead-bolt function. Otherwise the bolt could be retracted by "loiding," that is, by slipping a flat object between the bolt and the strike.

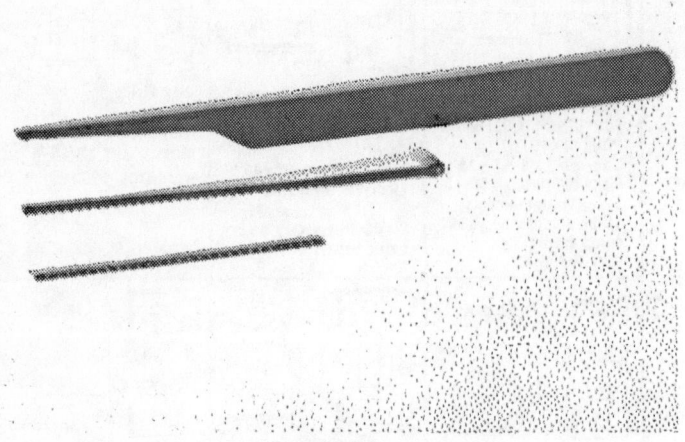

Fig. 18-108.

- Use the longest bolts available to frustrate attempts to gain entry by spreading the door frame. A strikeplate cover can also have the same function.
- Shield the bolts with armored inserts.
- Reenforced strikeplate mounts, particularily with metal door frames, add security by increasing the area of contact between the strikeplate and frame.

EXTRACTING BROKEN KEYS

Figure 18-108 illustrates three homemade extractor tools. The first type, made from a length of spring stock, is inserted with the hook down. The rounded tip of the tool nudges the pins up, and the hook fits into the most forward key cut. The second two are made from coping or fret saw blades. They are inserted so the teeth snag the edges or top of the key blade.

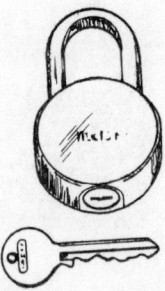

Chapter 19

Auxiliary Door Locks

The term *auxiliary* applied to door locks has two distinct meanings. It can mean a second lock acting in conjunction with a first and, hopefully, increasing the security from intruders. These locks are usually surface-mounted, in which case they are known as rim locks or locking bodies. Locks used on interior doors are also known as auxiliary locks. Most of these are mortise locks, i.e., they mount inside the door.

LOCKING BODIES

Locking bodies come in a variety of styles and shapes. A few of the options are shown in Fig. 19-1. The bolt may lock automatically when the door is closed. The two locking bodies at the top of Fig. 19-1 employ these self-locking latchbolts. Other locking bodies use dead bolts that must be manipulated to lock. Either type may be worked only from the inside or have a keyway on the outside of the door.

Keyless locking bodies with latchbolts have one big disadvantage: they invite lockouts.

Construction

The basic mechanism is a case (or cover), a bolt, and a bolt-actuating mechanism. Latchbolts are spring loaded and

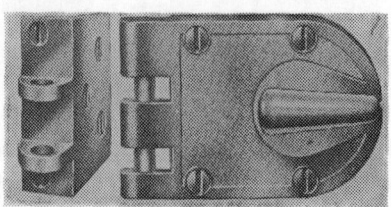

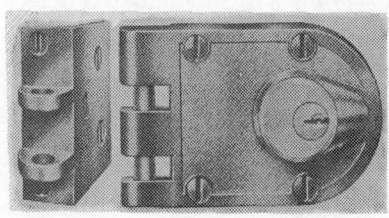

Fig. 19-1. A collection of locking bodies.

are checked by a safety catch. Engaging the safety from the inside puts the bolt and spring in check. The door can be shut without latching.

Locking bodies are adaptable to many different doors. Since they mount on the inside surface of the door, the bevel of the door edge is unimportant. But there are times when you will have to choose a particular locking body because of the configuration of the door frame. If at all possible, use the strike intended for this lock since strikes and locks work best as a matched team. If this is not possible, be certain that the strike you use works properly. Figure 19-2 illustrates two of the many strikes available.

Installation

Every locking body worthy of the name comes with installation instructions, but you will sometimes find yourself working "blind." In the absence of manufacturer's instructions, follow this procedure:

1. Position the locking body on the door and scribe its corners and locking body screw holes.
2. Remove the locking body and align the strike on the door frame.
3. Tentatively mark the strike position.
4. Assuming that an outside cylinder will be fitted, drill a hole through the door at the estimated point of contact of the tailpiece (the extension on the back of the cylinder) and the bolt-activating mechanism.
5. Measure the diameter of the cylinder and chuck up the appropriate bit or hole saw. Most cylinders are 1⅜ in. in diameter.
6. Drill the hole from the outside using the first hole as a guide.

A

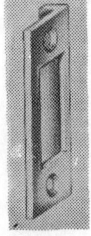
B

Fig. 19-2. Two strikes by Corbin.

7. Break off the splinters and smooth the edges of the hole.

8. Mount the cylinder over the collar and attach it to the retaining plate on the inside of the door.

9. Hold the latch in position and check that the cylinder works properly. You may have to trim the tailpiece.

10. Check the screw holes against the locking body.

11. Start the screw holes with a small nail and drill $1/16$ in. pilot holes $1/3$ in. deep.

12. Mount the locking body and tighten the screws.

13. It may be necessary to chisel out part of the frame behind the strike. Some are surface-mounted.

The customer may want a set of instructions so he can do the work himself. In this case, the following instructions would be helpful:

1. Drill a cylinder hole in the door $2\frac{3}{8}$ in. from the edge for doors opening inward, and $2\frac{1}{2}$ in. for doors opening out.

2. Install the cylinder. Make sure the pins are on top of the keyway.

3. Position the locking body even with the edge of the door and check that the tailpiece operates the latch when the key is used.

4. Attach the locking body.

5. Install the strike.

INSIDE DOOR LOCKSETS

The Corbin 7000 series is typical of the better locksets (Fig. 19-3). Designed for exterior or interior use, it has these features:

- Case—cast iron and lacquered.
- Dead bolt—forged brass with a $5/8$ in. throw. Hardened steel inserts are optional.
- Latchbolt—extruded brass with a $5/8$ or $3/4$ in. throw, depending upon application. Antifriction latchbolts are available.
- Latch holdback (safety catch)—optional.

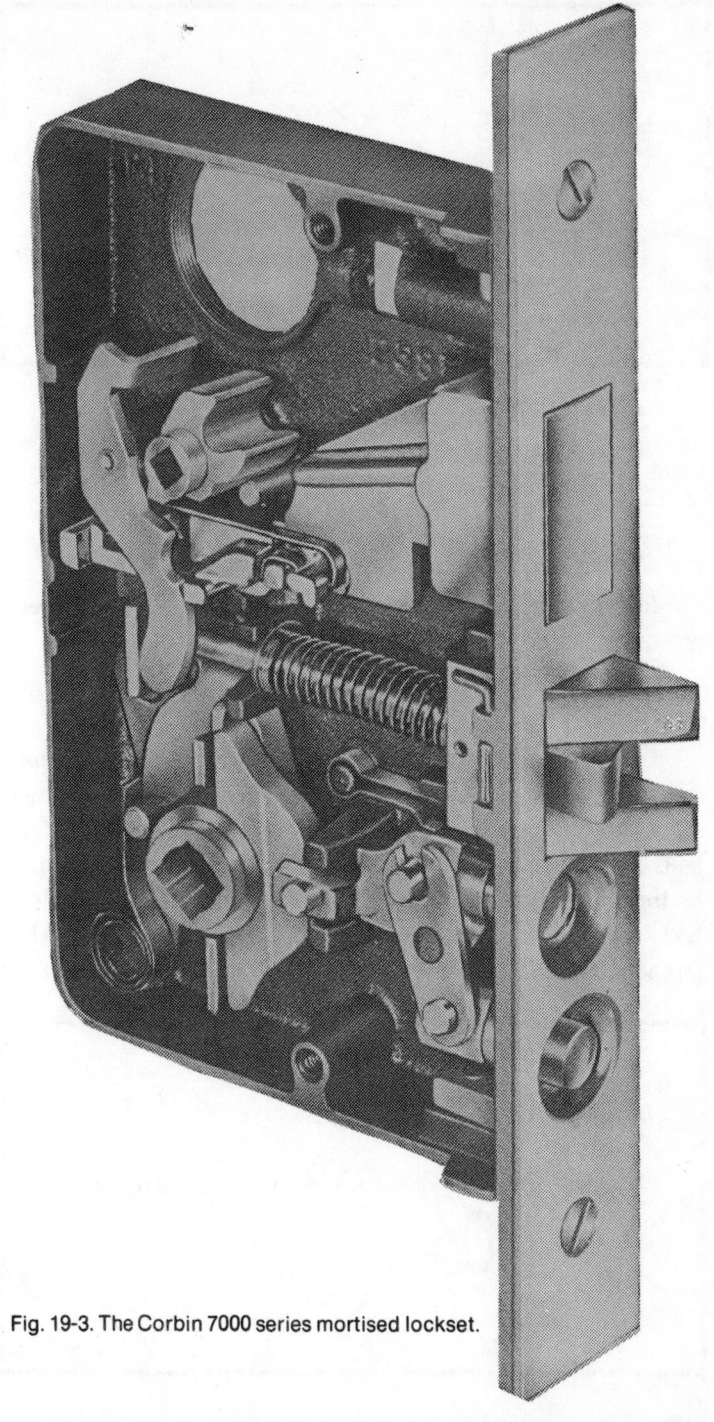

Fig. 19-3. The Corbin 7000 series mortised lockset.

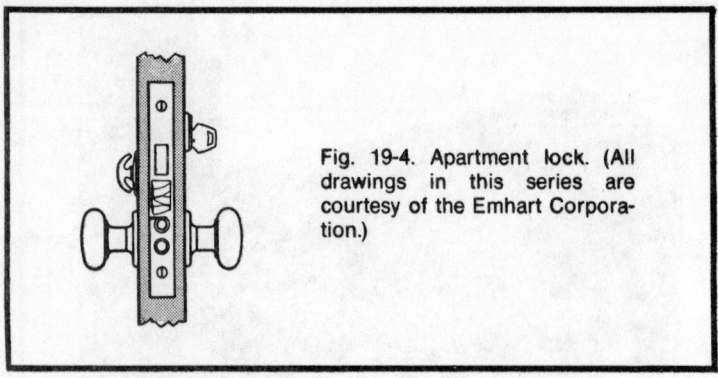

Fig. 19-4. Apartment lock. (All drawings in this series are courtesy of the Emhart Corporation.)

- Hub—cold-forged bronze.
- Strike—wrought brass or bronze with the edge of the lip $1\frac{1}{8}$ in. from the centerline as standard.

Installation

Installation of Corbin and other locksets has been discussed in the previous chapter.

VARIANTS

Indoor locksets have evolved according to function:
Apartment Lock The latchbolt is released by either knob, unless the outside knob is locked by the stop button or dead bolt. The stop button can be released manually at the button or by turning the inside knob. Turning the key retracts the dead bolt; turning the inside knob retracts the dead bolt and the latchbolt (Fig. 19-4).

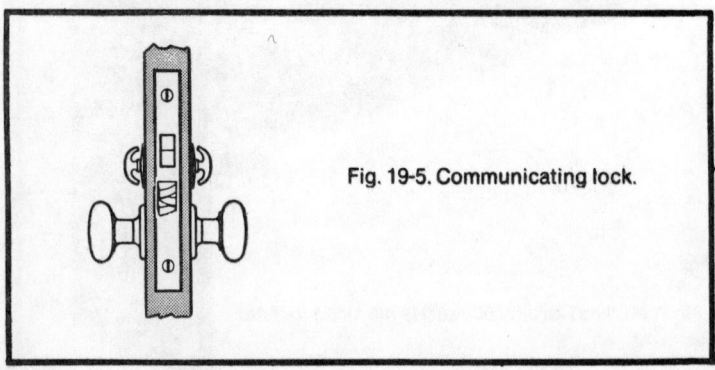

Fig. 19-5. Communicating lock.

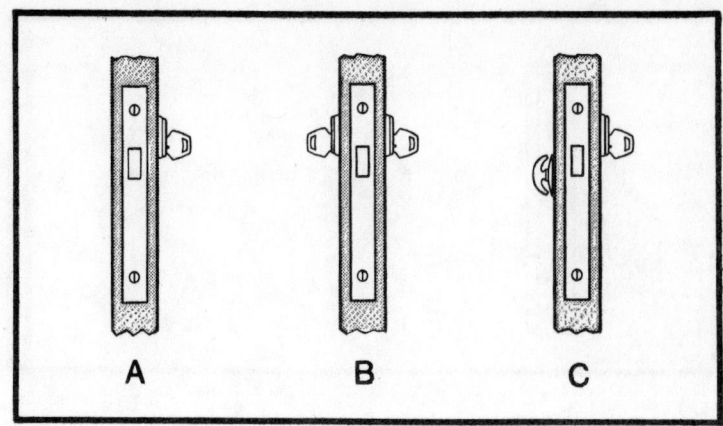

Fig. 19-6. Dead locks. The lock may be controlled by a single key (A), two keys (B), or by a key and a turnpiece (C).

Communicating Lock The latchbolt is released by either knob. A split dead bolt allows operation by either turnpiece (Fig. 19-5).

Dead Lock There are three variations to the familiar dead lock: dead bolt operation by key from one side only (Fig. 19-6A); dead bolt operation by key from both sides (Fig. 19-6B); dead bolt operation by key on one side and turnpiece on the other (Fig. 19-6C).

Dormitory Lock The latchbolt is released by either knob, unless the outside knob is locked by the dead bolt. The dead bolt deadlocks the latchbolt. The dead bolt is activated by the inside knob (which also retracts the latchbolt and unlocks the outside knob) by turning the turnpiece or the key (Fig. 19-7).

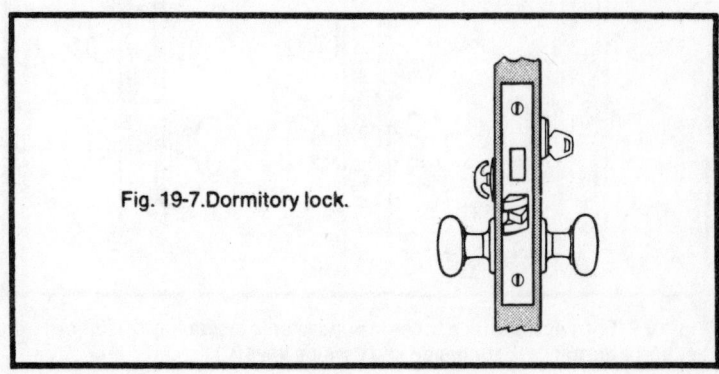

Fig. 19-7. Dormitory lock.

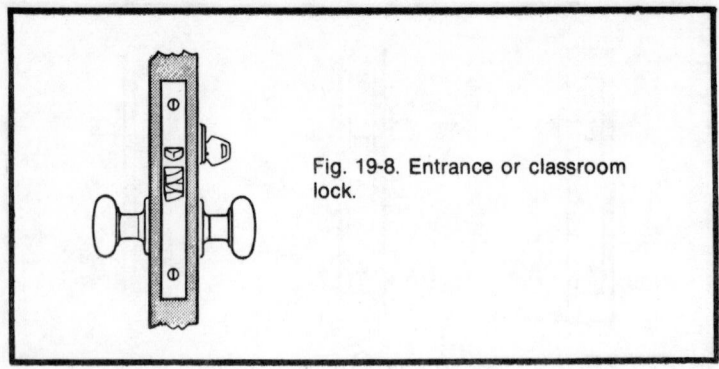

Fig. 19-8. Entrance or classroom lock.

Entrance or Classroom Lock The latchbolt is released by the knob on either side, unless the outside key locks the outside knob. In this mode the inside knob still opens the door (Fig. 19-8).

Entrance or Office Lock There are three variations. The first does not have a turnpiece (Fig. 19-9A). The latchbolt is released by the key from the outside and either knob, unless the outside knob is locked by the key. Alternately, this same type of lockset may be fitted with a turnpiece on the inside (Fig. 19-9B). In this case, the latchbolt is worked by the key on the outside and the knob on either side, unless the outside knob is locked by the key. The dead bolt is released by the inside turnpiece and the outside key. The third variation has a key on both sides (Fig. 19-9C) and works like the lock just

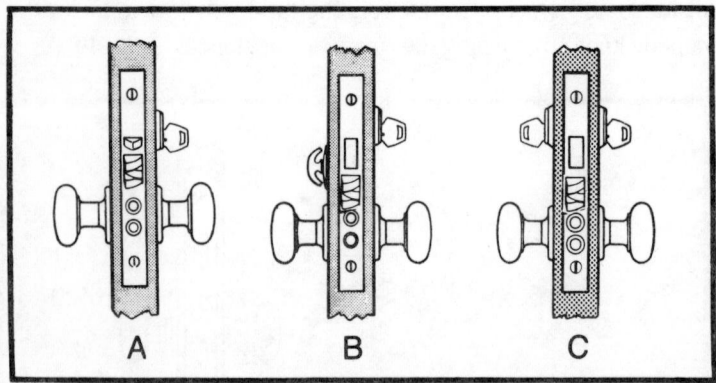

Fig. 19-9. Entrance or office locks may have an outside key (A), an outside key and a turnpiece (B) or inside and outside keys (C).

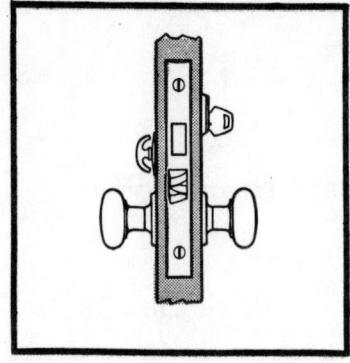

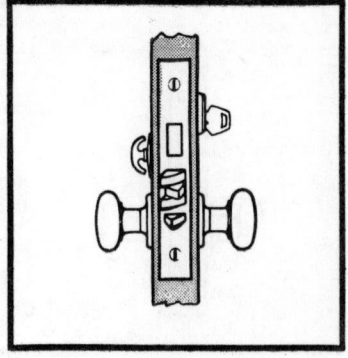

Fig. 19-10. Entrance or storeroom lock.

Fig. 19-11. Hotel room lock.

described, except that the key, rather than the inside turnpiece, works the deadbolt.

Entrance or Storeroom Lock The latchbolt is activated by the knob on either side. The dead bolt operates by the outside key or the inside turnpiece (Fig. 19-10).

Hotel Room Lock The dead bolt is activated by the turnpiece on the inside. The latchbolt is retracted by the inside knob and by the guest, master, and grand masterkeys. The inside knob retracts the dead bolt and latchbolt simultaneously. Only the emergency key can release the dead bolt from the outside (Fig. 19-11).

Passage or Closet Latch Turning either knob releases the latchbolt (Fig. 19-12).

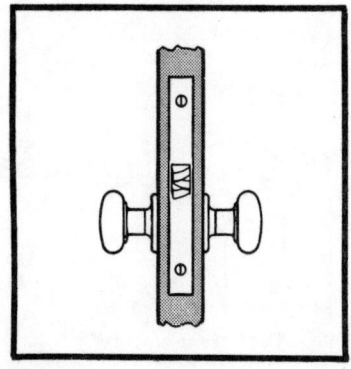

Fig. 19-12. Passage or closet latch.

Fig. 19-13. A privacy lock.

295

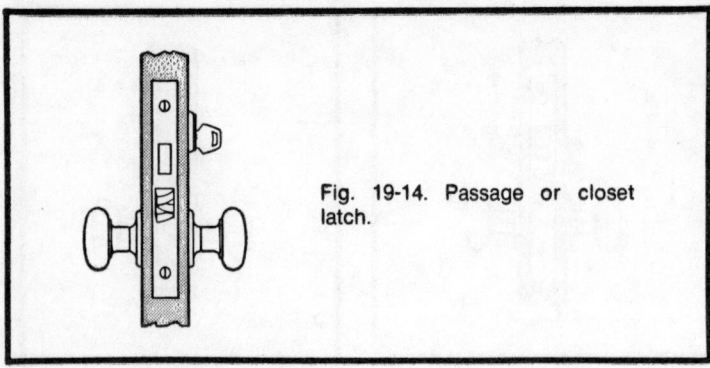

Fig. 19-14. Passage or closet latch.

- *Privacy Lock* The latchbolt is released by either knob. The dead bolt is activated by the inside turnpiece or by the inside knob.
- *Storeroom Lock or Closet Lock* The latchbolt is released by either knob. The dead bolt is activated by a key on the outside (Fig. 19-14).

HARDWARE

Figure 19-15 illustrates a sample of inside and outside door trim offered by one manufacturer. Figure 19-16 shows five

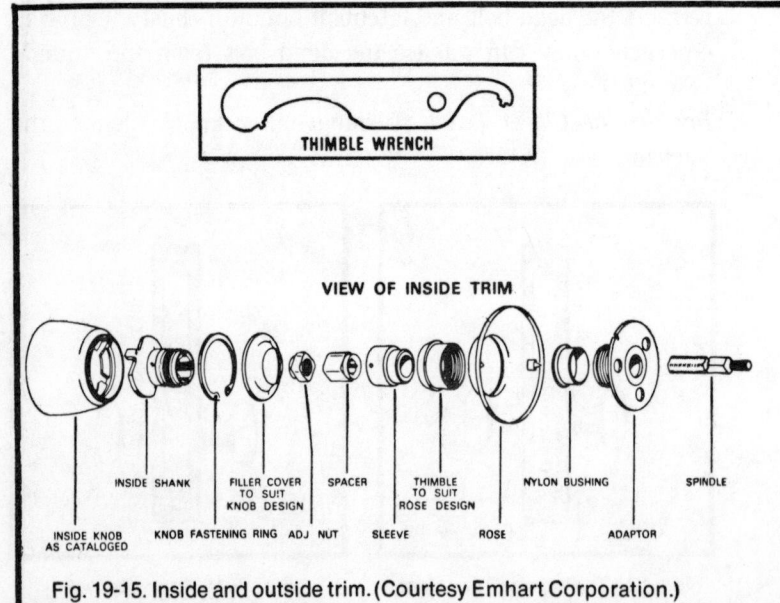

THIMBLE WRENCH

VIEW OF INSIDE TRIM

INSIDE SHANK | FILLER COVER TO SUIT KNOB DESIGN | SPACER | THIMBLE TO SUIT ROSE DESIGN | NYLON BUSHING | SPINDLE

INSIDE KNOB AS CATALOGED | KNOB FASTENING RING | ADJ NUT | SLEEVE | ROSE | ADAPTOR

Fig. 19-15. Inside and outside trim. (Courtesy Emhart Corporation.)

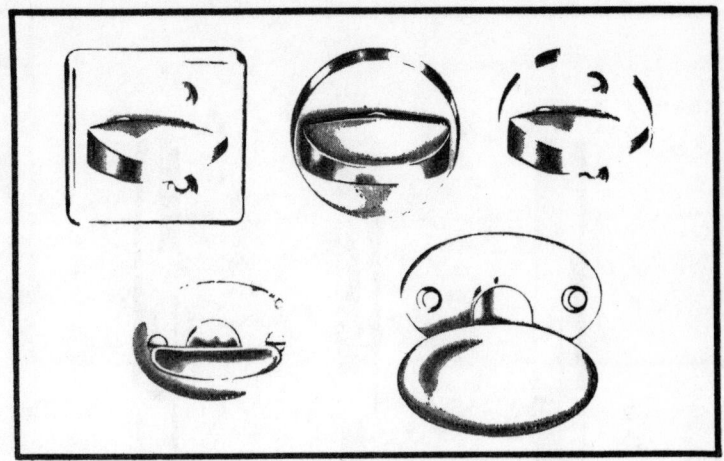

Fig. 19-16. Turnpieces by Corbin.

turnpiece (thumb knob or turn knob) options from the hundreds that are available. Many of these options are purely cosmetic: others are mechanical parts that must be replaced with the exact equivalent. Prime examples of this are knob and lever spindles (Fig. 19-17). You would do well to stock up on a variety of these parts.

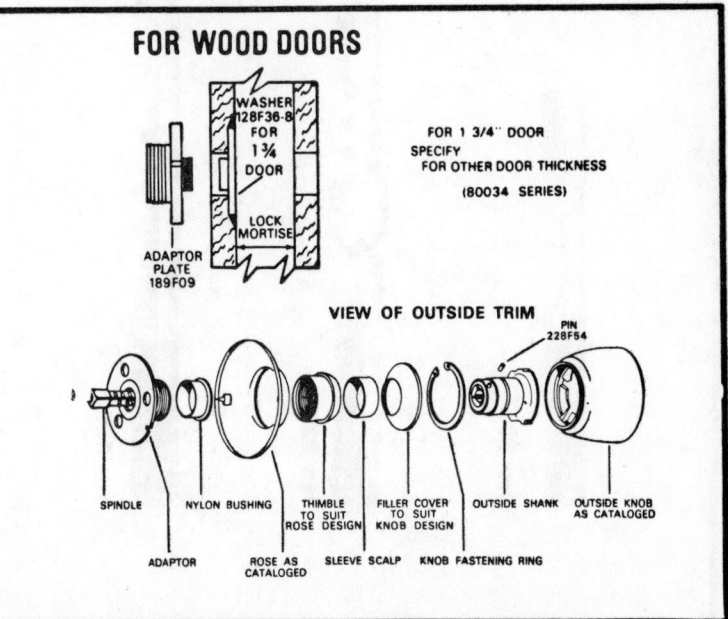

FOR WOOD DOORS

WASHER 128F36-8 FOR 1 ¾ DOOR

LOCK MORTISE

ADAPTOR PLATE 189F09

FOR 1 3/4" DOOR
SPECIFY
FOR OTHER DOOR THICKNESS
(80034 SERIES)

VIEW OF OUTSIDE TRIM

PIN 228F54

SPINDLE   NYLON BUSHING   THIMBLE TO SUIT ROSE DESIGN   FILLER COVER TO SUIT KNOB DESIGN   OUTSIDE SHANK   OUTSIDE KNOB AS CATALOGED

ADAPTOR   ROSE AS CATALOGED   SLEEVE SCALP   KNOB FASTENING RING

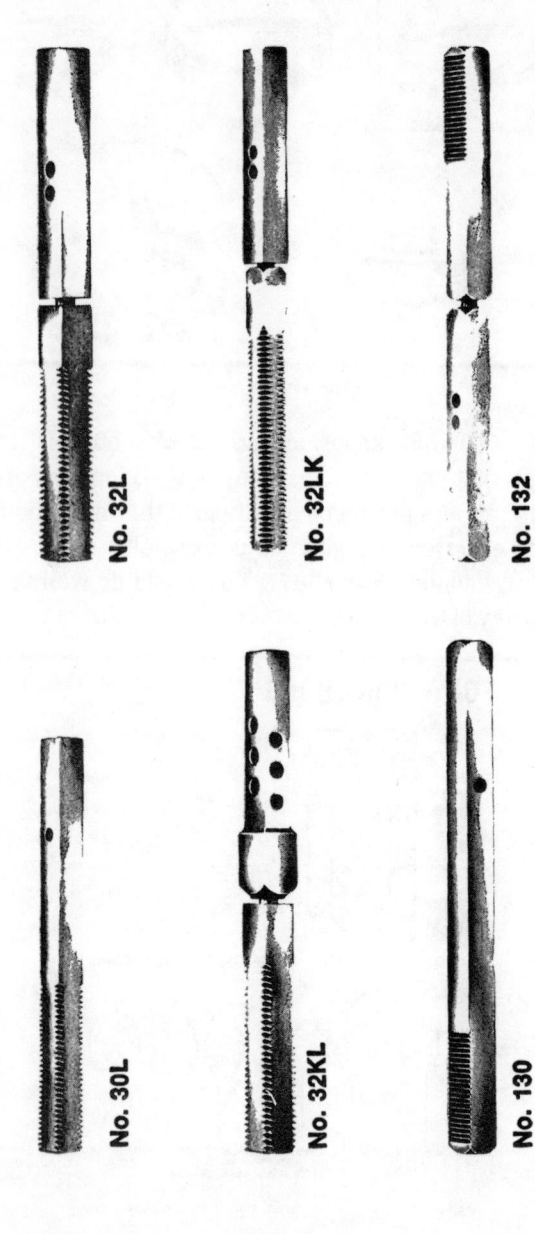

No. 32L

No. 32LK

No. 132

No. 30L

No. 32KL

No. 130

298

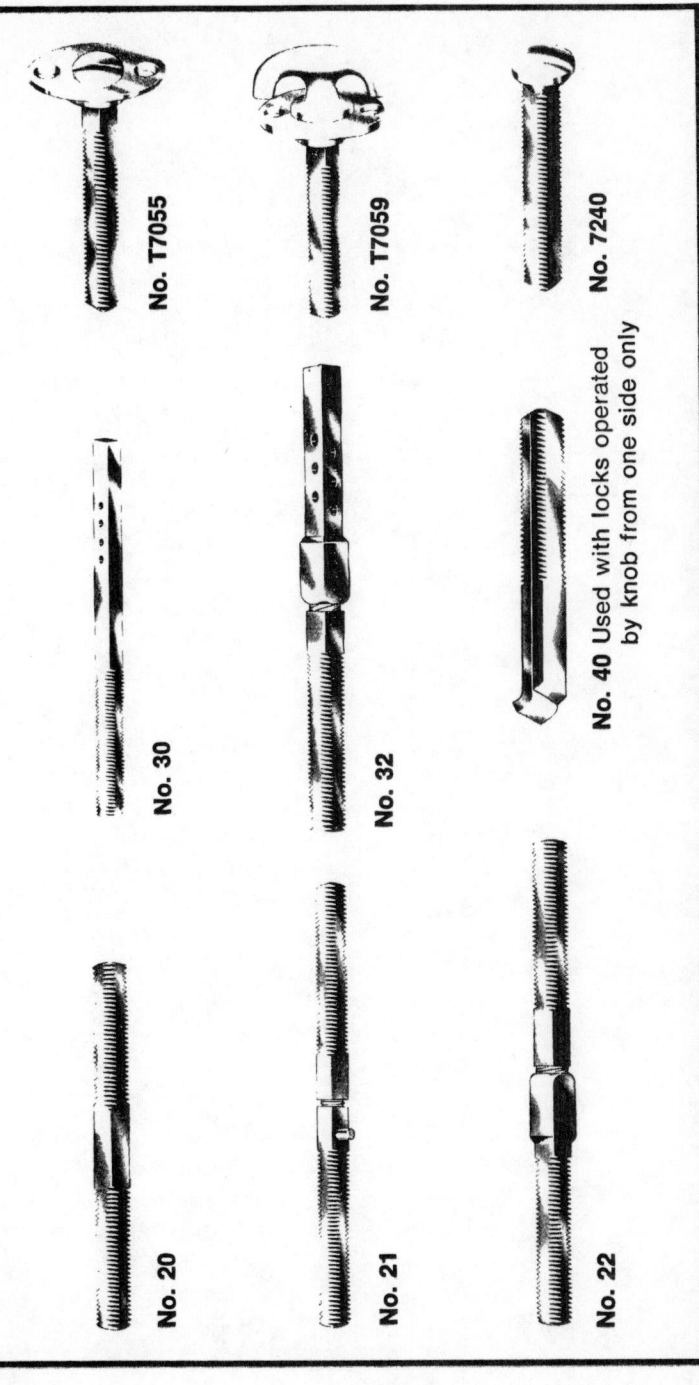

No. T7055

No. T7059

No. 7240

No. 30

No. 32

No. 40 Used with locks operated by knob from one side only

No. 20

No. 21

No. 22

Fig. 19-17. Knob and lever spindles by Corbin.

299

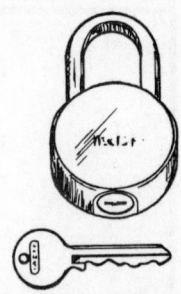

Chapter 20
Office Locks

Office locks represent an important and lucrative market and one that is easy to overlook. An important part of the security of every business depends upon office locks. A profusion of manufacturers are in this field, making thousands of different locks, each with more or less unique features. But almost all of these locks fall into three groups: rotary-cam locks, sliding-bolt locks, and plunger locks.

ROTARY-CAM LOCKS

Simplest of office equipment locks, most rotary-cam locks employ disc tumblers,although you will occasionally run across one with pin tumblers. The cam is oval and rotates with the cylinder. Those with a 90° throw must be locked for key removal; 180° throw locks can be unkeyed in the locked or unlocked position (Fig. 20-1).

The lock body may be threaded (as shown in the illustration), or it may be secured by a retaining clip. The cam may be integral with the lock, or it can be secured by a screw and washer. A removable cam is an advantage since it allows the same lock body to be used in a variety of applications. Locks that turn 90° have a single tumbler chamber; locks that turn 180° employ a double chamber and may have a stopping

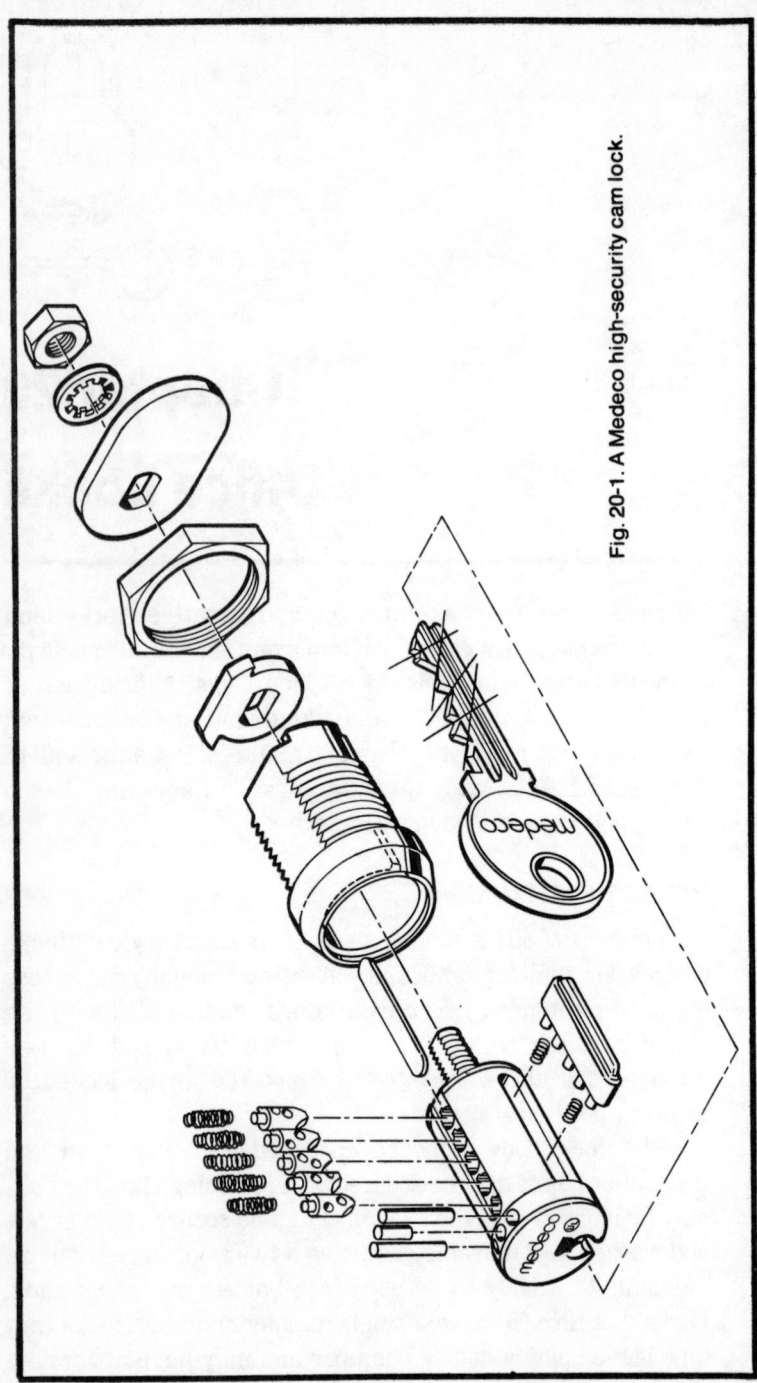

Fig. 20-1. A Medeco high-security cam lock.

pin, notch, or blocking member at the rear that limits rotation. In some cases it is possible to change the rotation in the field. The lock may then be used in other applications.

SLIDING-BOLT LOCK

Most of these locks are disc tumblers, although pin tumblers are not unknown. There is no cam: The bolt is grooved to accept a projection on the back of the plug (Fig. 20-2). The projection engages the groove and converts the rotary motion of the plug into reciprocating (to and fro) motion. The bolt is generally heavier and more durable than those of other office locks. This arrangement allows the key to rotate 180°: the key can be withdrawn in either the locked or unlocked position.

The most serious sliding-bolt lock is known as the T-bolt because of its shape (Fig. 20-3). These locks almost always

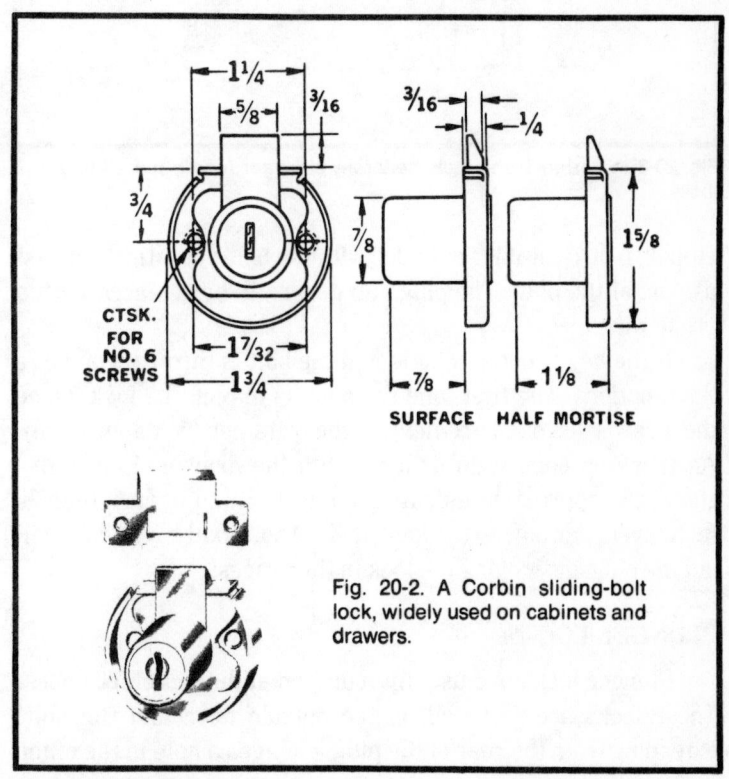

Fig. 20-2. A Corbin sliding-bolt lock, widely used on cabinets and drawers.

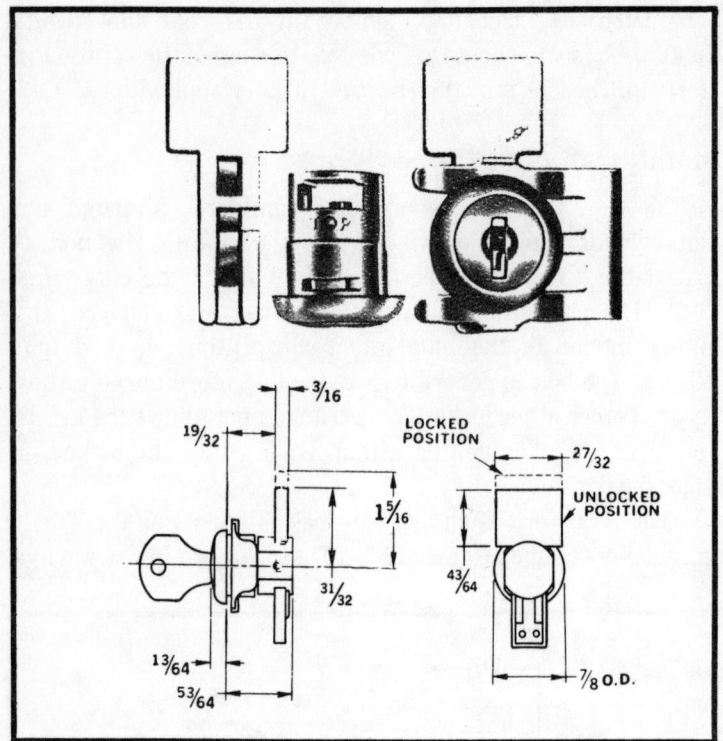

Fig. 20-3. A Corbin T-bolt lock, generally stronger than other sliding-bolt locks.

employ a disc-tumbler cylinder with the bolt-actuating pin cast as part of the plug. The plug can generally be released with a probe wire.

If the key is not available and the bolt is thrown, you have three options. The first, and the best, is to pick the lock. Once the drawer is open, removing the retainer is no problem. Another approach is to drill through the drawer. Figure 20-4 shows the approximate drilling point. Drill no deeper than ⅞ in. to avoid damage to the lock itself. And, finally, you can drill out the plug, destroying the lock in the process.

PLUNGER LOCKS

Plunger locks are usually found on sliding cabinet doors. These locks are mounted on the outside door and the bolt, extending from the rear of the plug, engages a hole in the other

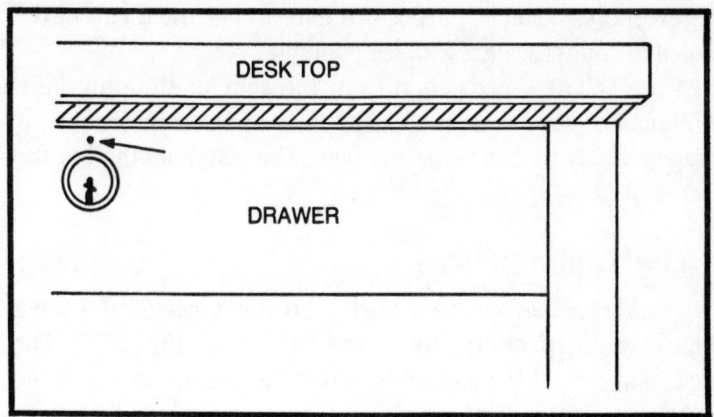

Fig. 20-4. Occasionally it is necessary to drill a desk lock for access to the retaining clip. A $^1/_{16}$ in. hole, $^7/_8$ in. deep or less, is adequate.

door (Fig. 20-5). The doors are locked against each other so neither can be moved. A spring returns the plug to the open position when the key is turned. Although disc-tumbler types once were popular, pin-tumblers are almost universal today. Locksmiths new to the trade may never have seen a disc-plunger lock.

The plug is usually secured by a screw located at the side of the cylinder and set in a notch in the lock (Fig. 20-5). The

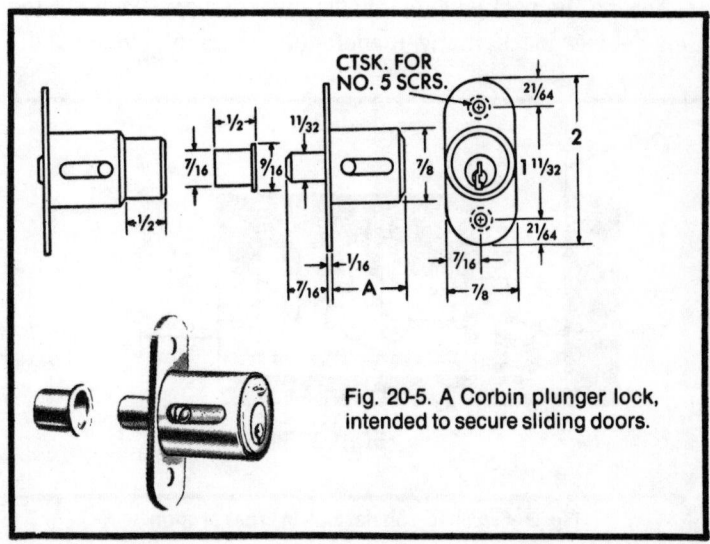

Fig. 20-5. A Corbin plunger lock, intended to secure sliding doors.

screw serves a double purpose: It determines the throw of the bolt and aligns the plug with the cylinder.

The bolt is a projection from the rear of the plug unit. Without the key you can pick the lock or force the doors far enough apart to disengage the bolt. The latter method is the faster of the two.

FILING-CABINET LOCKS

Filing-cabinet locks are fairly well standardized and use a bolt extending from the top of the lock body (Fig. 20-6). The bolt manipulates the control bars that lock each drawer of the cabinet.

The bolt is spring-driven; unless the plug is turned, the bolt extends out of the locking body. In other words, the drawers are automatically locked.

The bolt may be round or rectangular in cross section. The lower face of the bolt—the part buried in the cylinder—is recessed for the spring. The plug cam works in a notch on the side of the bolt.

"Open" keyways are generally used, as opposed to "blind," or "masked," keyways. The keyway runs from the face through the plug body. You can open the lock with relative ease. Use the tool shown in Fig. 20-7 to work the bolt directly.

During the past decade, new standards have been set for filing-cabinet locks; many manufacturers have revised their

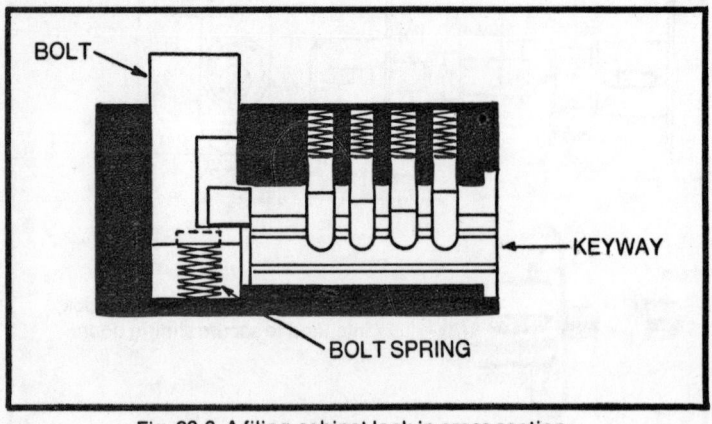

Fig. 20-6. A filing-cabinet lock in cross section.

Fig. 20-7. A jimmy for open-keyway locks. The tool is made of spring steel $1/8$ to $3/16$ in. wide and slips through the keyway for direct manipulation of the bolt.

dies to do away with the open keyway and the breach in security it represents. This means that today many filing-cabinet locks have bolts that are isolated from the keyway. A few manufacturers use their original dies but add a fifth pin. This pin is longer than the others and has no upper chamber. It cannot be lifted and so blocks access to the bolt. Another option has been to block the keyway end with a horizontal pin. The jimmy no longer works, and we must turn to the more traditional methods of opening a keyless lock. There are three choices: Pick the lock, drill the plug, or force the bolt.

Picking is the most desirable of these alternatives, but requires patience and skill. Drilling is fast and means the loss of the lock, but it is preferable to forcing the bolt. If the customer accepts the risk, you can pry one drawer open far enough to manipulate the bolt with a letter opener.

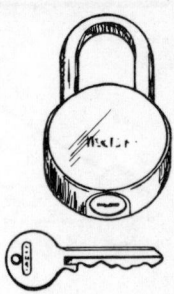

Chapter 21

Automotive Locks

While the automobile has been with us since the start of the century, the lock was slow to be adopted. However, by the late 1920's nearly every auto had an ignition lock, and closed cars had door locks as well. Current models may be secured with half a dozen locks.

SIDEBAR LOCKS

While automobiles may be fitted with pin- or disc-tumbler cylinders, the sidebar lock is most typical and the only one that merits special attention. The other two have been discussed earlier.

Figure 21-1 illustrates this lock in simplified form. Notice that the wafers have V-shaped notches in their sides. Inserting the correct keys aligns the notches, and the spring-loaded sidebar moves out of the cylinder and into the plug. Once the sidebar passes the shear line, the plug is free to rotate.

KEY FITTING

Key fitting for disc- and pin-tumbler locks has been described in earlier chapters. The techniques used for automotive variations of these locks are no different.

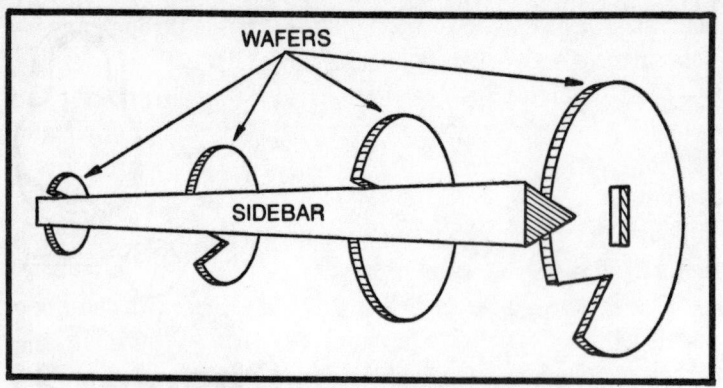

Fig. 21-1. A simplified drawing of a sidebar lock.

Key fitting sidebar locks requires that the lock be disassembled. Remove the lock from the vehicle and, with a small screwdriver or other sharp tool, release the staking that holds the sidebar and associated spring in place.

Remove the tumblers and their springs. Insert the appropriate blank into the keyway and replace the tumbler at the far end of the plug. Working from the sidebar chamber, scribe a mark on the blank where the blank touches the tumbler. This is the guide mark for the cut. File the mark lightly and reinsert the blank; the file mark should be at or under the tumbler. File slowly until the V-notch on the side of the tumbler appears to align with the sidebar. Hold the sidebar and spring temporarily in place with your fingers and try the key. If all is right, the sidebar will extend into the notch, freeing the plug. The key should turn.

Repeat this operation for each of the remaining tumblers.

DOOR LOCKS

Modern door locks are disassembled from inside the door. Figure 21-2 shows a typical example. Remove the arm rest (shown as No. 7 in the drawing) and inside handles. Most handles are secured by a spring clip that is accessible with the tools shown in Fig. 21-3. Note the positions of the handles on their spindles for assembly.

Remove the Phillips screws holding the door trim (6) to the door body (2). Gently pry the trim away from the body and set it aside. The door lock mechanism is visible through access holes in the inner door-body panel.

Most problems involve the lock mechanism and not the cylinder. Check for broken springs, missing retainer clips, and rust on working parts. The cylinder (20) is secured by a U-shaped retainer clip. Cylinders can be disassembled, although the modern practice is to replace the cylinder as a unit, rather than to attempt repairs.

The striker can be moved to align the door with the body panels and to compensate for door sag. Loosen the mounting screws a quarter turn or less and, using a soft mallet, tap the striker into position (Fig.21-4). Shims are available from the dealer to move the striker laterally, out from the door post.

While this is out of the province of locksmithing, as such, it is useful to know that the door can be adjusted at the hinge on all automobiles except Vega (Fig. 21-5). Loosen the pillar-side bolts slightly and manipulate the door.

IGNITION LOCKS

Early production ignition locks are secured to the dash by bezel nuts or retainer clips and feature a "poke hole" on the lock face or in the back of the cylinder that gives access to the retaining ring. In a few cases the lock is held by screws from the underside. Late-model locks, the kind that lock the steering wheel as well as the ignition, are not something that the average locksmith should become involved with. These locks can be serviced, but special tools and special knowledge are required. The upper portion of the steering column must be dismantled and, in some cases, the steering column must be dropped from the dash hanger. A mistake in assembly—even so much as using the wrong screw—can defeat the collapsible steering column. So, if you want to get into this aspect of the trade, by all means spend a few weeks at dealer schools and invest in the proper tools.

GLOVE COMPARTMENT LOCKS

Modern glove-compartment locks are typically secured by a retainer accessible from the front of the lock. Figure 21-6

Key to Fig. 21-2

(1) Door hinge (upper)
(2) Door body
(3) Damper
(4) Door hinge (lower)
(5) Door weatherstrip
(6) Door trim
(7) Arm rest
(8) Door glass
(9) Main roller guide (front)
(10) Channel
(11) Regulator assembly
(12) Regulator handle
(13) Upper stopper (lower)
(14) Sub-roller guide
(15) Upper stopper
(16) Main roller guide (rear)
(17) Door inside handle
(18) Door inside lock
(19) Door outside handle
(20) Cylinder lock
(21) Door lock
(22) Door striker
(23) Shim

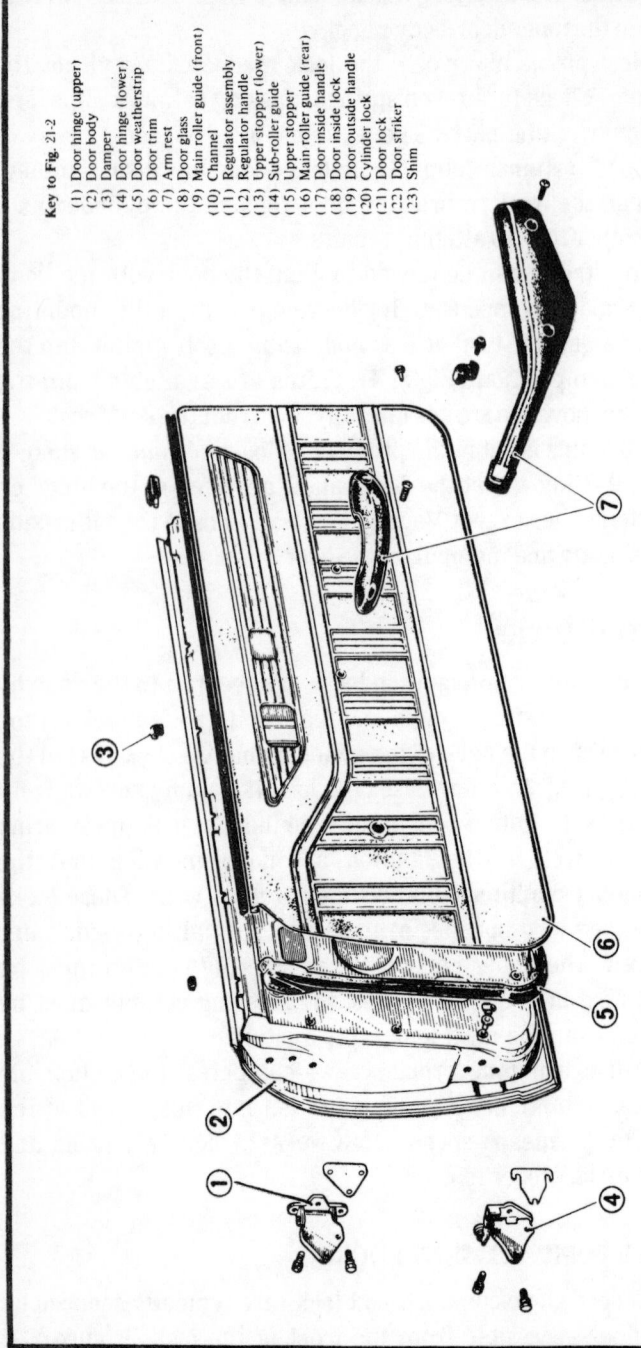

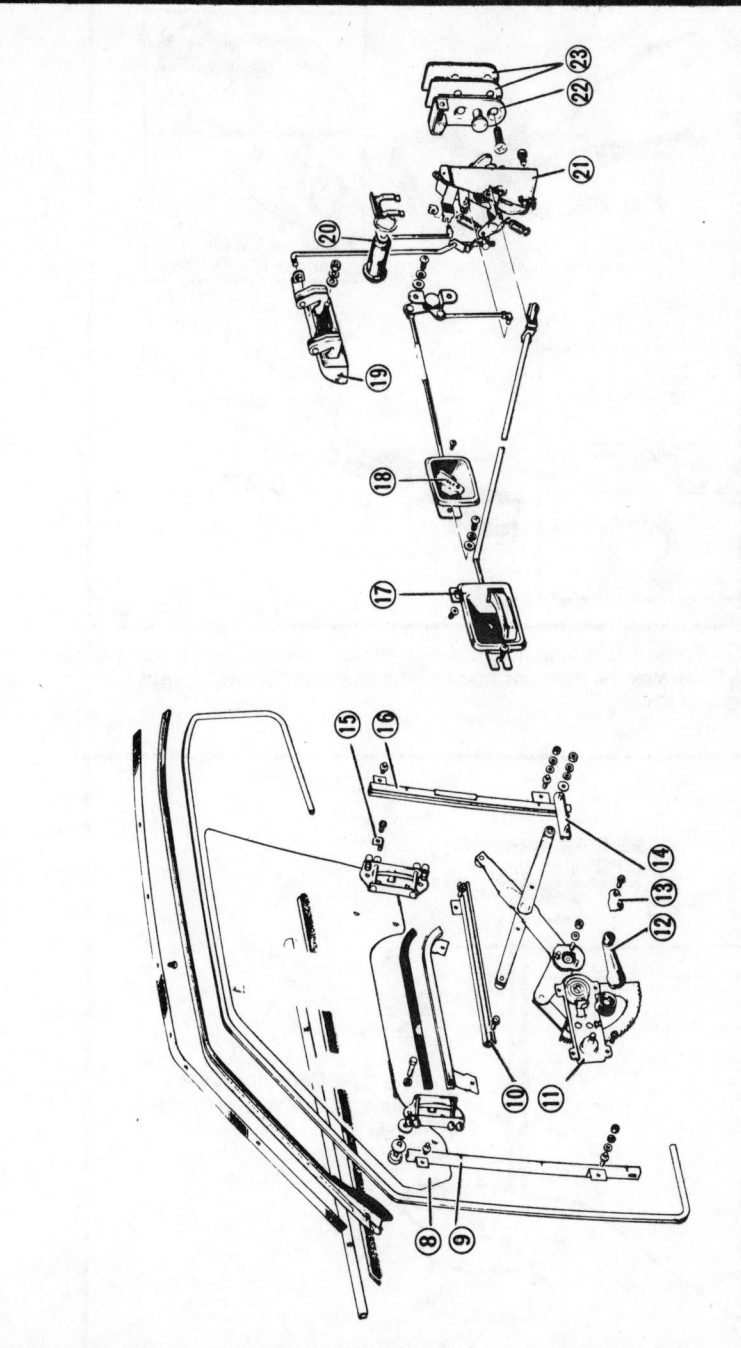

Fig. 21-2. A typical front-door assembly. (Courtesy Chrysler Corp.)

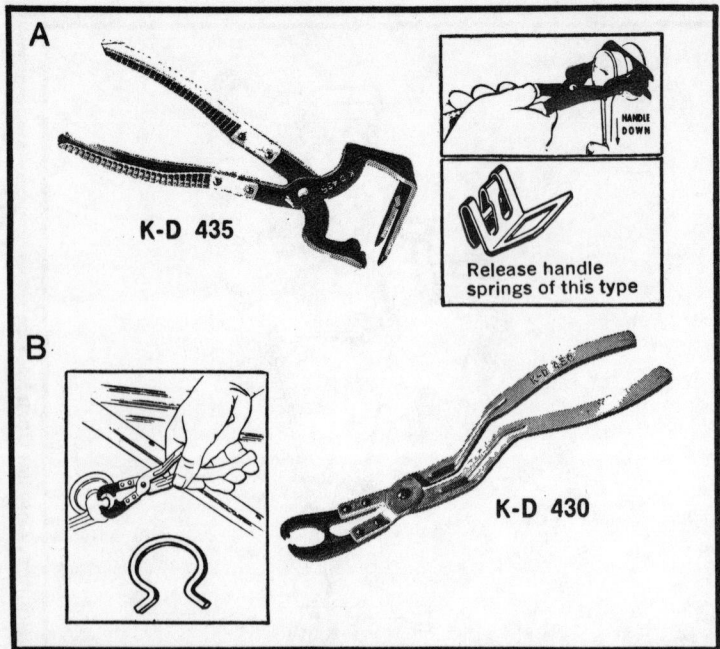

Fig. 21-3. These K-D door-handle tools are available from auto parts stores. View A illustrates the tool for Chrysler retainers; view B, the tool for Ford and General Motors.

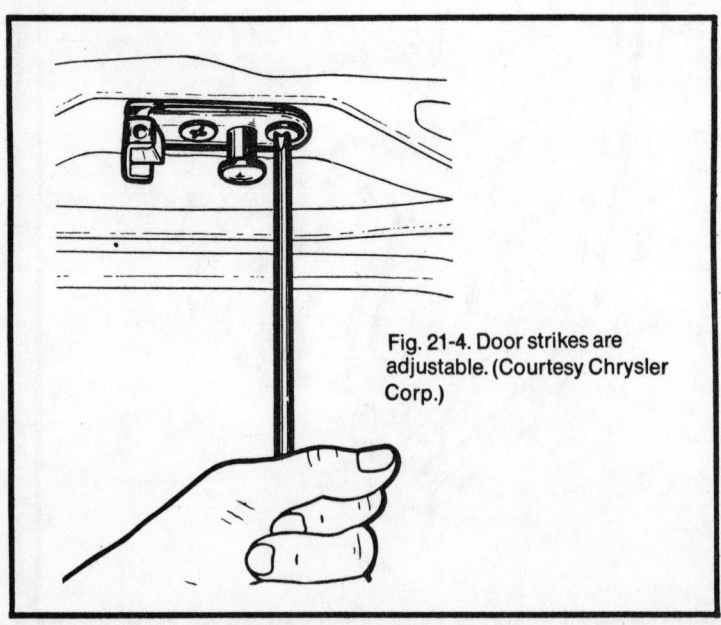

Fig. 21-4. Door strikes are adjustable. (Courtesy Chrysler Corp.)

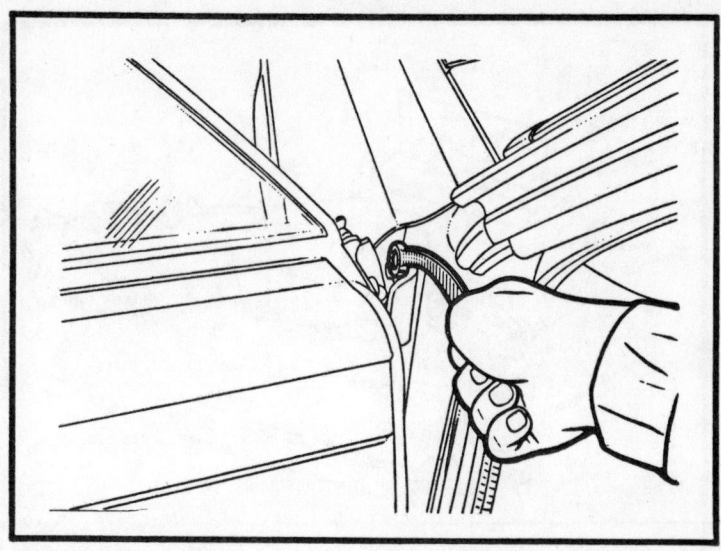

Fig. 21-5. The door adjustment is at the pillar side of the hinges for most vehicles. (Courtesy Chrysler Corp.)

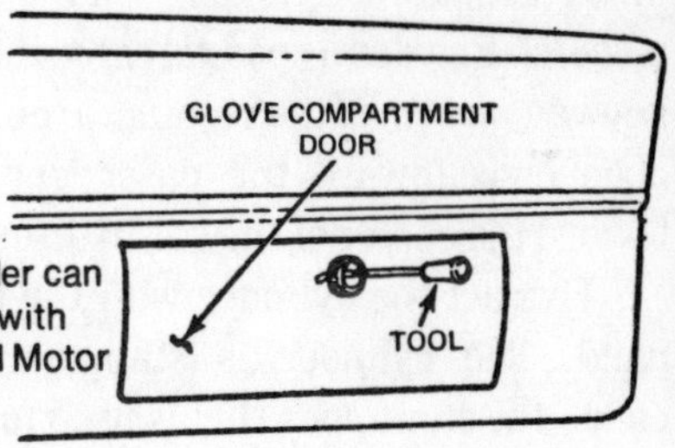

Fig. 21-6. The glove-box cylinder can be removed from the outside with the tool shown (Courtesy Ford Motor Co.)

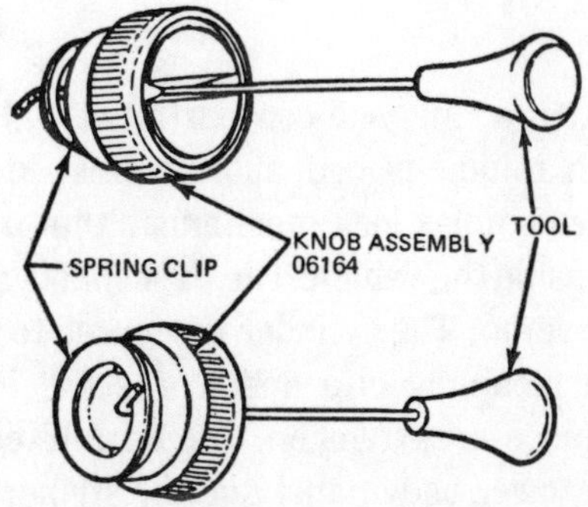

GLOVE COMPARTMENT DOOR

TOOL

SPRING CLIP

KNOB ASSEMBLY 06164

TOOL

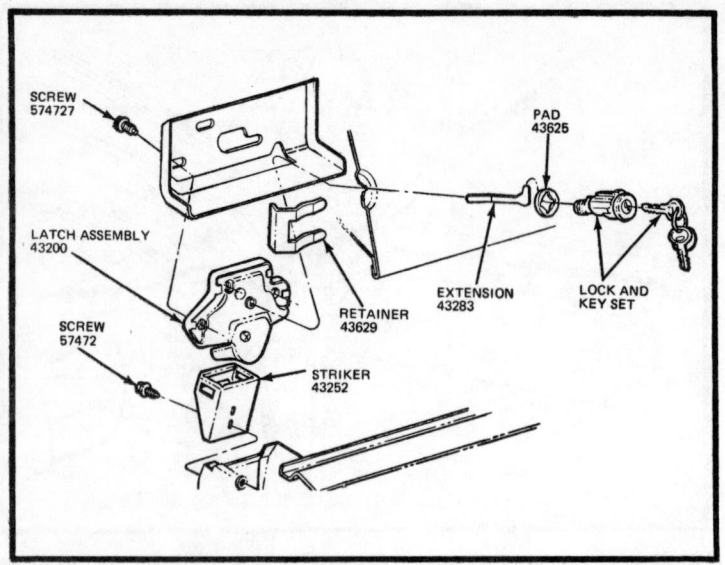

Fig. 21-7. The trunk lock assembly used on Mustang and Cougar autos. (Courtesy Ford Motor Co.)

illustrates how this is done. Make a ¼ in. hook in the end of a piece of Bowden (hood-latch) cable. Insert the hook into the keyway and work the retainer free. Most are installed with the open ends toward the passenger door; nudge the retainer toward the center of the car to disengage it.

Extract the cylinder with the key and release the bolt by hand. The cylinder is stamped with a numerical keying code—the same code that is used for the trunk or, in the case of station wagons, the tailgate.

TRUNK LOCKS

Figure 21-7 illustrates a luggage-compartment latch, typical of those found in middle-priced automobiles. More expensive cars use a more complex lock mechanism that may be solenoid-tripped from inside the vehicle.The basic parts and their relationships are the same: The cylinder is secured to the trunk lid by means of a retaining clip and works the lock assembly through a tailpiece or extension. (Some few cars mount the cylinder on the lower body panel and the striker on the lid.) The striker is located on the aft body panel.

Most lock assemblies are adjusted only at the striker plate. The plate is mounted on elongated bolt holes and can be moved up and down, left and right. Other types have an additional adjustment at the latch. The idea is to position these parts so the trunk lid closes without interference from the striker and closes tightly enough to make a waterproof seal between its lower edge and the weatherstripping.

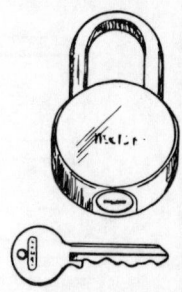

Chapter 22

Lockpicking

The purpose of this chapter is to introduce you to the tools and techniques of lockpicking. Lockpicking is the ultimate test of a locksmith: It tries his knowledge as well as his skill. To many customers, lockpicking is the one ability that distinguishes a true locksmith from someone who merely tinkers with locks.

LOCKPICKING TOOLS

Basic lockpicks are illustrated in Fig. 22-1 and include the following:

- Half-round feeler
- Round feeler
- Rake
- Half-rake
- Diamond
- Double-diamond
- Circular
- Reader
- Extractor
- Mailbox
- Flat lever

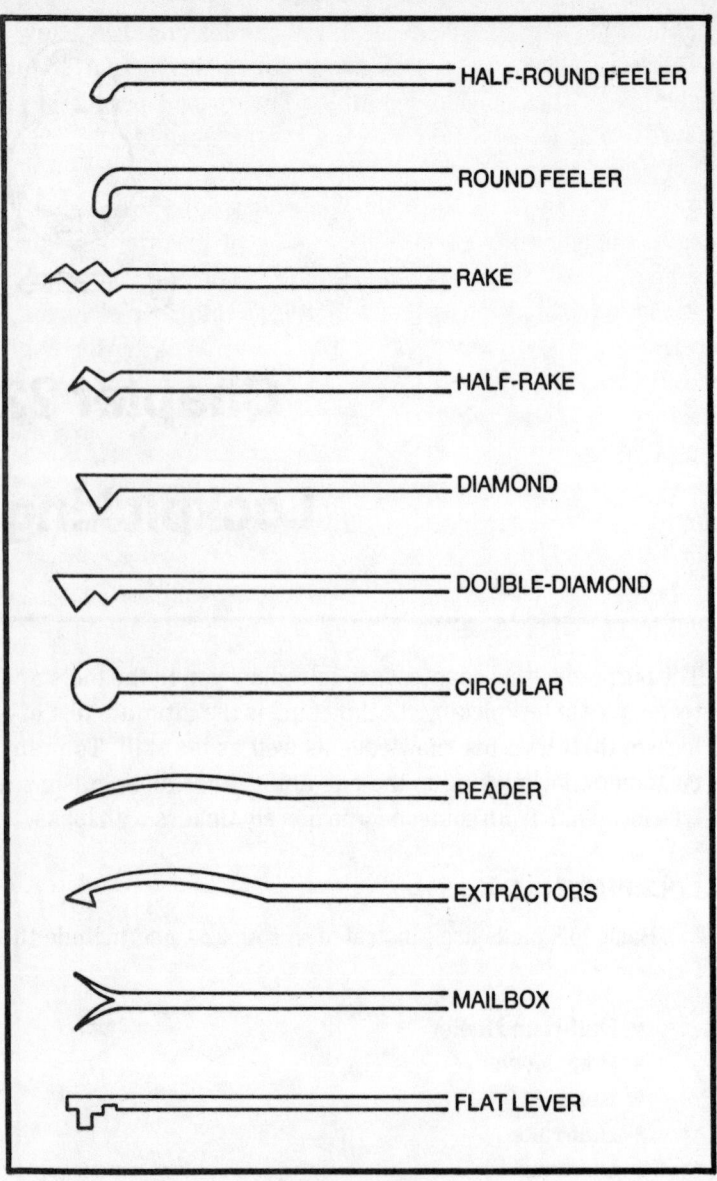

Fig. 22-1. Basic lockpicks.

You can make your own lockpicks from flat cold-rolled steel, 0.020 to 0.025 in. thick. The strip should be about 6 in. long and at least ¹₃ in. wide. One end may be fitted with a handle, or both ends may carry working surfaces.

Picks for warded locks come in various designs also. A few of them are shown in Fig. 22-2. It's usually a simple matter to make one of these picks. Sometimes a standard precut key shaved to pass the keyhole wards is all that is needed.

Tension tools for use with various picks are made of hardened spring steel. Normally they are from 3 to 5½ in. in length, excluding the parts that actually fit into the keyway opening.

A letter box tension tool is made from a thin strip of spring steel about 4 in. long and has a ⅛ to ¼ in. probe at its working end. It is best to have several with different probe lengths to insure that you always have the correct one on hand.

Picks and tension tools for double-sided disc-tumbler locks are unique in their design (Fig. 22-3). They should be purchased from your locksmith supply house; it's cheaper to buy them than to make them.

Lever lockpicks come in two varieties—those for true lever locks and those that double as key picks. The true lever lockpick is made from 0.095 in. or thinner spring steel and has a tension tool that is the same thickness (Fig. 22-4). The second variety is cut from spring steel, or else made from standard key blanks.

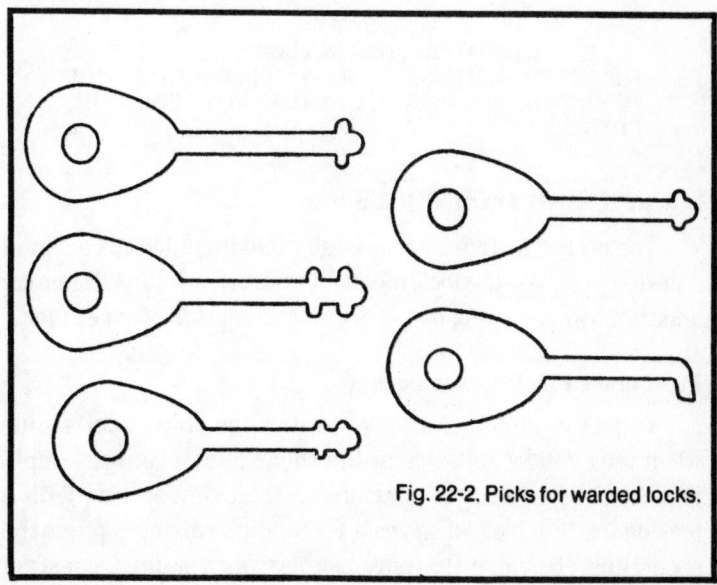

Fig. 22-2. Picks for warded locks.

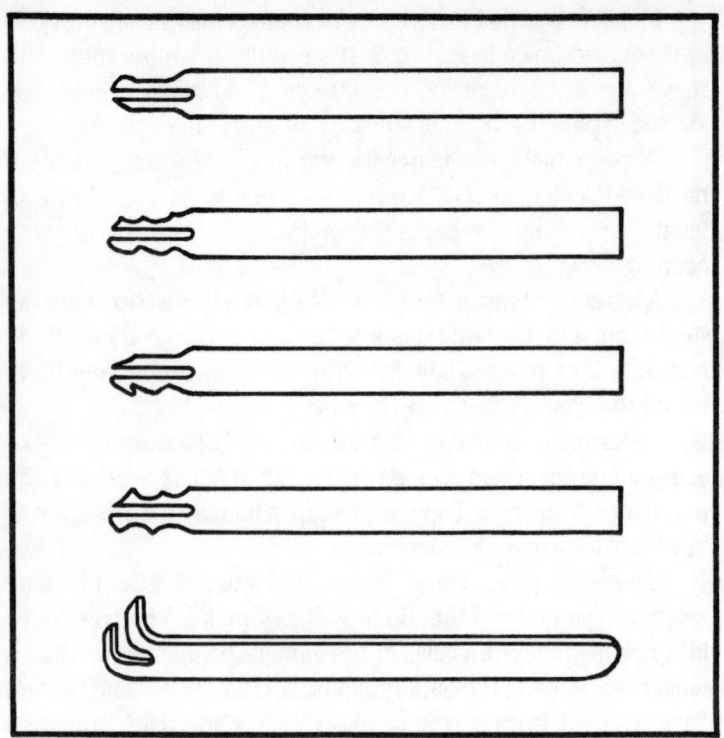

Fig. 22-3. Picks for double-sided disc-tumbler locks.

LOCKPICKING TECHNIQUES

The prerequisite for successfully picking a lock is a sound knowledge of how the lock mechanism works. Without such an understanding, picking a lock becomes a matter of sheer luck.

Picking the Pin-Tumbler Lock

To pick a pin-tumbler lock you must be able to determine when the cylinder pins are at the shear line. You must apply just the right amount of pressure on the cylinder core with a tension tool. Figure 22-5 shows a feeler pick raising a pin to the shear line. To raise the pins in this way requires constant

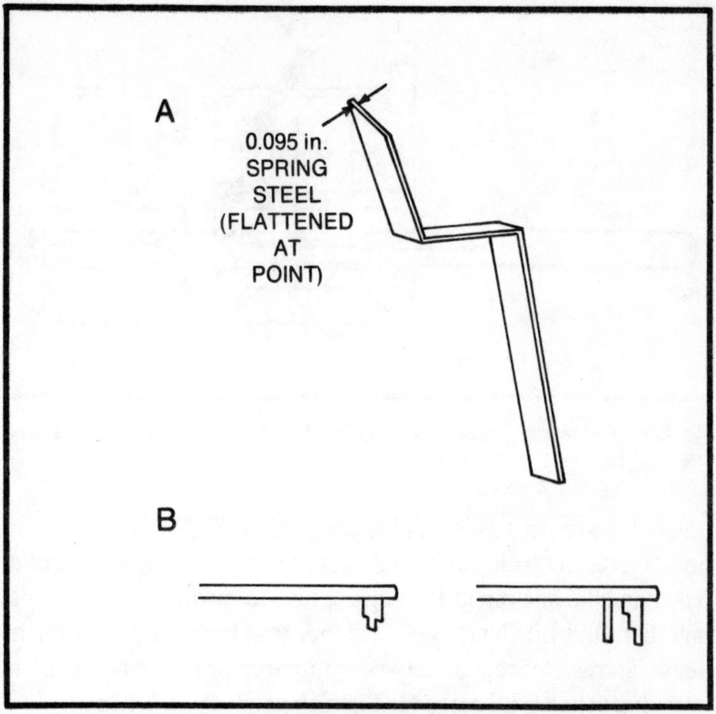

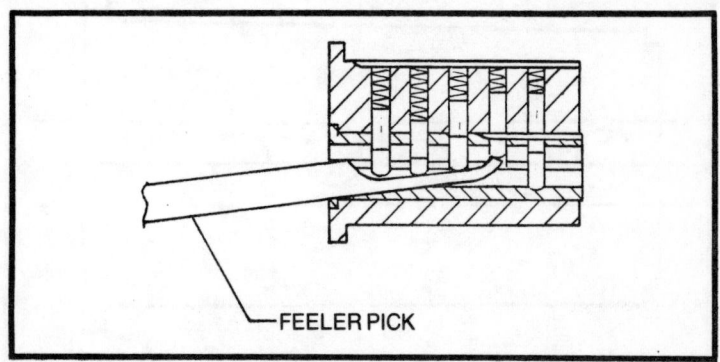

Fig. 22-4. A tension tool (A) and picks for warded locks (B).

practice. Too much or too little pressure on the pick will cause the pin to stick at the wrong place (Fig. 22-6).

Factory precision in drilling the cylinder pin holes is never perfect. An enlarged view of the top of a plug would show that

Fig. 22-5. A feeler pick raising a pin to the shear line. (Courtesy Desert Publications.)

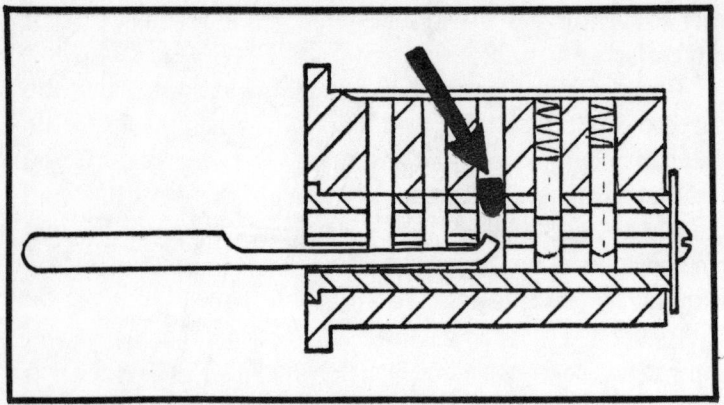

Fig. 22-6. Too much pressure with a pick will raise a pin above the shear line.

the holes are not really in alignment (Fig. 22-7). This is a help in opening the lock. Since the holes are not perfectly aligned, you can hold one pin at the shear line while working on others. another built-in advantage is the fact that locks always have at least some clearance between moving parts. And a little working space is all a good locksmith needs to pick a lock.

In raising the various pins, it is preferable to start with those of the greatest length. By using this method, you progress from the smallest amount of pick movement up to the greatest. As each pin reaches the shear line, the core moves

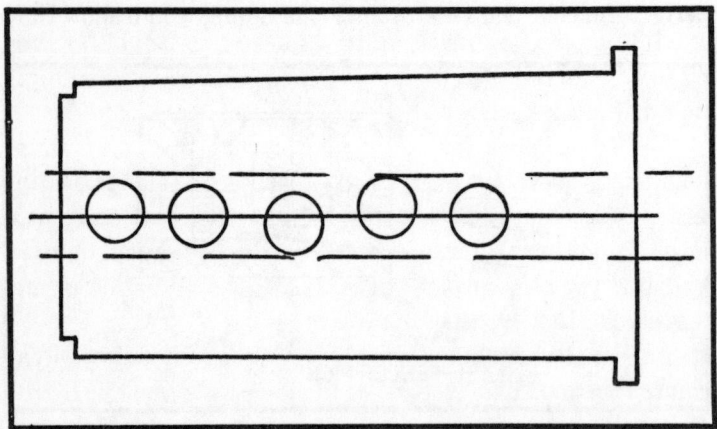

Fig. 22-7. Cylinder pin holes are never in perfect alignment.

ever so slightly; you will be able to feel this movement when it is transferred to your tension tool.

Using a feeler pick, move one pin at a time. Since the purpose of the pick is to move the individual pins up to the shear line, not above them, you must take your time. If you feel that all but one of the pins have reached the shear line, and the last one just won't go, you have probably pushed the last pin above the shear line. Release the tension and start again, with slightly less tension than before.

Remember that you must vary the amount of pressure you apply to the pick, depending upon the amount of resistance you receive from each pin. Also, different locks require different amounts of pressure. What works on one lock will not necessarily work on the next. Even two locks of the same make and style may not require the same amount of pressure.

It is sometimes possible to use picks (especially the rakes) to "bounce" the pins to the shear line. To do this, insert the pick fully then withdraw it quickly while holding light tension on the plug. Simply forcing the pick rapidly in and out of the cylinder won't do the job: Such a technique merely bounces the pins *above* the shear line.

Picking Warded Locks

Warded locks are very easy to pick. Sometimes you can even use a pair of wires: one for throwing the bolt, the other for adjusting the lock mechanism to the proper height for the bolt to be moved. Here, as with any other kind of lockpicking, it is a matter of practice on a variety of locks.

Warded padlocks have only one or two obstacles, thus simplifying your choice of pick. Some warded locks (such as the Master) which have keyways with corrugated cross sections, can be opened in no time at all by having several precut blanks available. The regular picks are usually just a fraction of an inch too wide to fit into the keyway.

Smaller locks with correspondingly smaller keyways require that you either get a second set of tools and cut them down to size, or use several blanks to obtain the proper picks for these locks.

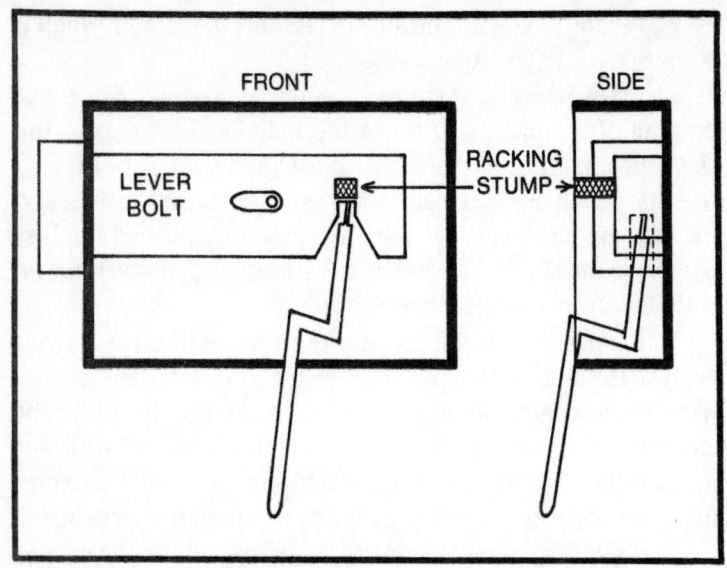

Fig. 22-8. A tension tool exerting pressure on lever tumblers.

Picking Lever-Tumbler Locks

Figure 22-8 shows a lock with the frontplate removed, indicating the position of the tension tool in relation to the bolt. Holding the levers with the tension tool (as shown in the side view) enables you to manipulate the levers with the pick.

In learning to pick the lever lock, it is best to start with a lock with only one lever in place. At first, work on the lock with the faceplate removed so you can get an idea of how much pressure to apply to the bolt, and how much movement is required of the pick to move the lever into position for the bolt to move through the lever gating.

As you progress to locks with several levers, keep the faceplate on. When you encounter a problem, remove the faceplate so you can see what you're doing wrong. Always remember to insert the tension tool first. Push it to the lowest point within the keyway so the pick will have maximum working space.

As with the pin-tumbler lock, one tumbler tends to take up most of the tension; work on this one first. As you move this tumbler slowly upward to the proper position for the bolt to pass through the gate, you will feel a slight slackening in tension from the bolt as it attempts to force its way into the gating. This tension will be transmitted to you through the tension tool. Stop at this point and do the same to the next lever with the greatest amount of tension. Continue until all the levers reach this point. Then by shifting the tension tool against the bolt, the bolt will pass through the gate, opening the lock.

There must be at least some pressure on the tension tool at all times. A lack of pressure will cause any levers in position to drop back into their original locations.

Picking Disc-Tumbler Locks

Standard disc-tumbler locks, those using a single-bitted key, take the same picks as pin-tumbler locks. Also like the pin-tumbler lock, a disc-tumbler lock can be picked by bouncing the tumblers to the shear line. Usually a rake is the best tool for this job, but other picks can be used too.

In order to develop your proficiency, you should also try opening the disc-tumbler lock with the feeler pick, working on each tumbler individually. By doing this, you learn which picks you are most comfortable with and which ones you will prefer for different types of locks.

When working on double-bitted disc-tumber locks, the bounce method is best. Insert the tension tool into the keyway and apply a slight pressure to the core as the pick is pulled out of the lock. These locks can also be opened with a standard pick set and tension tools—but it can take from 10 minutes to half an hour.

CONSIDERATIONS IN LOCKPICKING

The following are points to remember when picking any lock:

- It is to your advantage to use the narrowest pick available in order to give yourself maximum working space.
- Hold the pick as you would a pen or pencil. Don't use wrist action; your fingers work much better in manipulating the pick in the lock.

- Steady your hand with your little finger against the door. When working on a key-in-the-knob cylinder, steady your hand against the edge of the knob.
- Your pick should enter the keyway above the tension tool without moving any of the tumblers. If it doesn't the tension tool is either too high in the lock, or the keyway grooves are such that the tension tool must go at the very top of the keyway.
- Worn cylinders and loose plugs frequently open easier and quicker with the bounce method—normally within three to five tries.

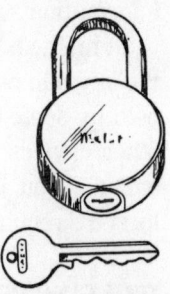

Chapter 23
Emergency Entry
Procedures (EEP)

Sometimes locksmiths gain entry into locked premises by picking open the lock. But a good locksmith can also gain entry through special techniques called Emergency Entry Procedures (EEP). In some instances, a lock resists picking. It is then that the EEP come into play.

Emergency Entry Procedures are also called special techniques, emergency measures, and quick entry. Much of the information and illustrations in this chapter come from Desert Publications' book *Lock Out*.

As a professional locksmith, you should have the proper equipment to make emergency entries as quickly and as neatly as possible. Some entry tools are manufactured; you should purchase a set of these for your use. Some tools you can fashion yourself. Experience will show you which tools and techniques work best for you.

Using Emergency Entry Procedures can account for a substantial percentage of a locksmith's income. Thus, it is wise to be knowledgeable in a variety of entry techniques. It may not seem like good locksmithing practice to force a lock open, but economics, time, and other requirements may make it necessary.

Several preliminaries should be undertaken when preparing to make an entry using EEP. Try to obtain as much information as possible. With this information, you can decide what methods will work. Ask questions about the type of lock, the key numbers, nearby open windows, a possible extra key held by someone else, the condition of the lock, alarms, and other devices attached to the door, etc. All this information will aid you in preparing for the entry. Fortunately, most lockouts require only that you pick the lock or make a new key.

DRILLING PIN-TUMBLER LOCKS

Drill a lock only as a last resort. There are two drilling methods used with pin-tumbler locks. The first method is the drilling of the cylinder plug. This has the advantage of saving the cylinder itself. The inner core can be replaced. This method destroys the lower set of pins just below the shear line. allowing the plug to be turned. Follow these procedures for this method:

1. Drill the plug below the shear line.
2. Insert a key blank and a wire through the drill hole to keep the upper pins above the shear line and the destroyed pins below it.
3. Turn the cylinder core and open the lock.
4. Once the door is opened. dismantle the lock: Remove and replace the core and fit new pins to the lock. If you used the plug follower when removing the core. you will only have to fit new lower pins to the lock to match a key.

The second method involves drilling just above the shear line into the upper pins.

1. Insert a key blank into the lock first to force all the pins to the upper parts of their chambers.
2. Drill about ⅛ in. above the shear line or shoulder of the plug and directly above the top of the keyway (Fig. 23-1). You can use a drilling jig to ensure drilling at the proper point on the lockface. Use a ⅛ in. or $^3/_{32}$ in. drill.
3. Poke a thin wire or needle into the hole and withdraw the key to just inside the keyway. You can then use the

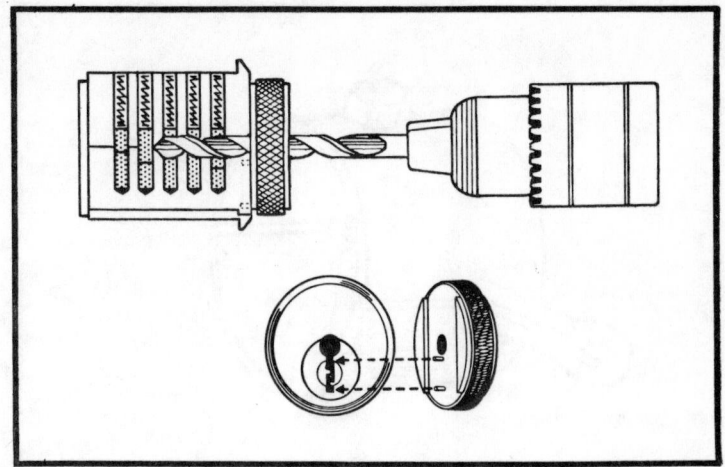

Fig. 23-1. Drilling above the shear line of a pin-tumbler lock. (Courtesy Desert Publications.)

tip of the key to turn the core. Removing the key part way allows the bottom pins to drop below the shear line, while the wire or needle keeps the upper pins above the shear line.

4. Once the door is open, dismantle the lock, remove the cylinder, and replace it with a new one.

CYLINDER REMOVAL

Sometimes the cylinder must be removed in order to open a lock.

This means shearing off the screws that hold the cylinder in place. Several different rim cylinder removal tools are available. The one in Fig. 23-2 can be made in your own shop. It's called a cylinder removal clamp.

To make one, follow these steps:

1. Use a piece of steel tubing about ⅛ in. larger in diameter than the cylinder. The tubing should be no more than 2½ in. long.
2. Cut into the tubing as shown in Fig. 23-2 about 1 ³/₁₆ in.
3. Cut a center hole at the end of both cuts ¼ to ⅓ in. in diameter and insert a steel rod about 10 in. long. The rod becomes your handle for turning the cylinder.

Fig. 23-2. A cylinder removal clamp. (Courtesy Desert Publications.)

4. Drill two holes in the tubing so you can insert a bolt perpendicular to the handle. Insure that the bolt will be ½ in. in front of the handle. The bolt threads must extend on both sides of the tubing.

To remove the cylinder, follow this procedure:

1. Set the removal clamp over the edge of the cylinder after first removing the cylinder rim collar. Tighten down the tension bolt. This provides the gripping pressure to the cylinder.
2. Twist the cylinder clamp by the handles and force the cylinder to rotate, shearing off the retaining screw ends.
3. Remove the cylinder from the lock. When removed, reach in and open the door by reversing the bolt.

You can also use the standard Stillson wrench to remove the cylinder.

Cylinders can also be pulled out of the lock unit. But this method ruins the cylinder threads, requiring you to replace the entire unit. The tool used is called a nutcracker. The sharp pincer points are pushed in behind the front of the cylinder face. Clamp down and pull the cylinder out. Many times an unskilled locksmith can ruin the entire lock and a portion of the door by not knowig how to use this tool properly.

Some cylinders have to be drilled. Follow this procedure:

1. Drill two $^3/_{16}$ in. diameter holes about $^7/_8$ to $1^1/_8$ in. apart on the face of the cylinder.
2. Drive two heavy bolts into these holes so that at least $1^1/_2$ in. is sticking out.
3. Place a pry bar or heavy screwdriver between the bolts and use it as you would a wrench to force the cylinder to turn, shearing off the long screws.

WINDOW ENTRANCES

Window entrances are relatively easy. The old butter knife trick is usually successful. Since most window latches are located between the upper and lower windows, sliding a knife up between the windows allows you to open the latch.

Should the area be too narrow for a knife, shim, or other device, drill a $^1/_{16}$ in. hole at an angle through the wood molding to the base of the catch. Insert a stiff wire and push back the latch.

OFFICE LOCKS

Most office equipment can be opened using a few basic methods and a handful of tools.

Filing Cabinets

Though most filing cabinets have the lock in essentially the same position, the locking bar arrangement for the various drawers will vary. These variations have to be considered in working on the locks.

One method is to work directly on the lock itself·

1. Slide a thin strip of spring steel $^1/_8$ in. thick into the keyway and pull the bolt downward. This will unlock the lock (Fig. 23-3).
2. If the lock has a piece of metal or pin blocking access to the locking bolt, use a piece of stiff wire with one end turned 90°. Insert the wire between the drawer and the

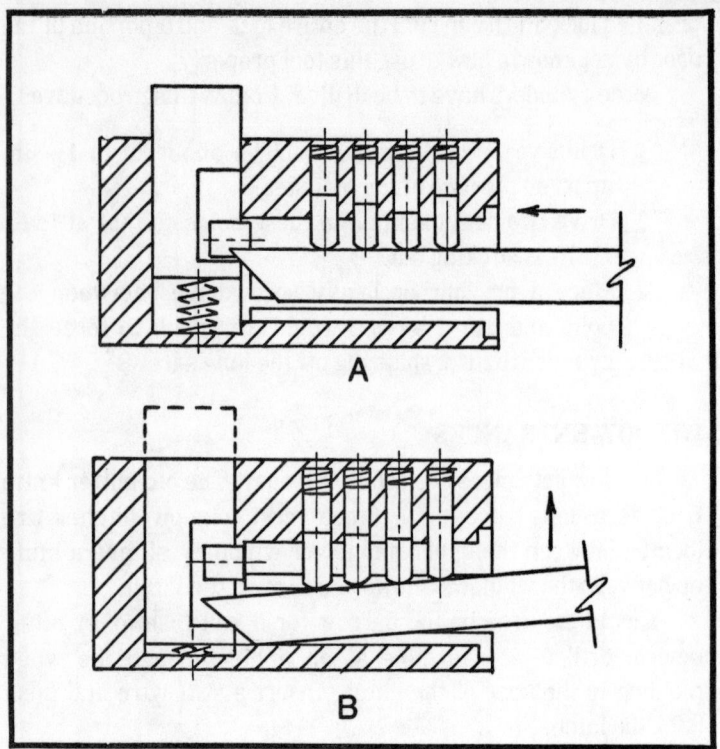

Fig. 23-3. Sometimes a thin strip of spring steel inserted into the keyway of a filing cabinet lock (A) can be used to pull the bolt down (B). (Coutesy Desert Publications.)

        cabinet face and force the bolt down with the wire. This will allow you to open the drawers.

3. If there is not enough room for you to work with the wire, use a piece of thin steel or a small screwdriver to pry back the drawer from the cabinet face to allow you to see and work with the wire.

You can also open the drawers individually if necessary:

1. Use a thin piece of spring steel or a wedge to spread the drawer slightly away from the edge.
2. As your opening tool use another strip of steel about 18 in. long, ½ to 1 in. wide, and 0.020 in. thick.
3. Insert the opening tool between the drawer catch and the bolt mechanism (Fig. 23-4).

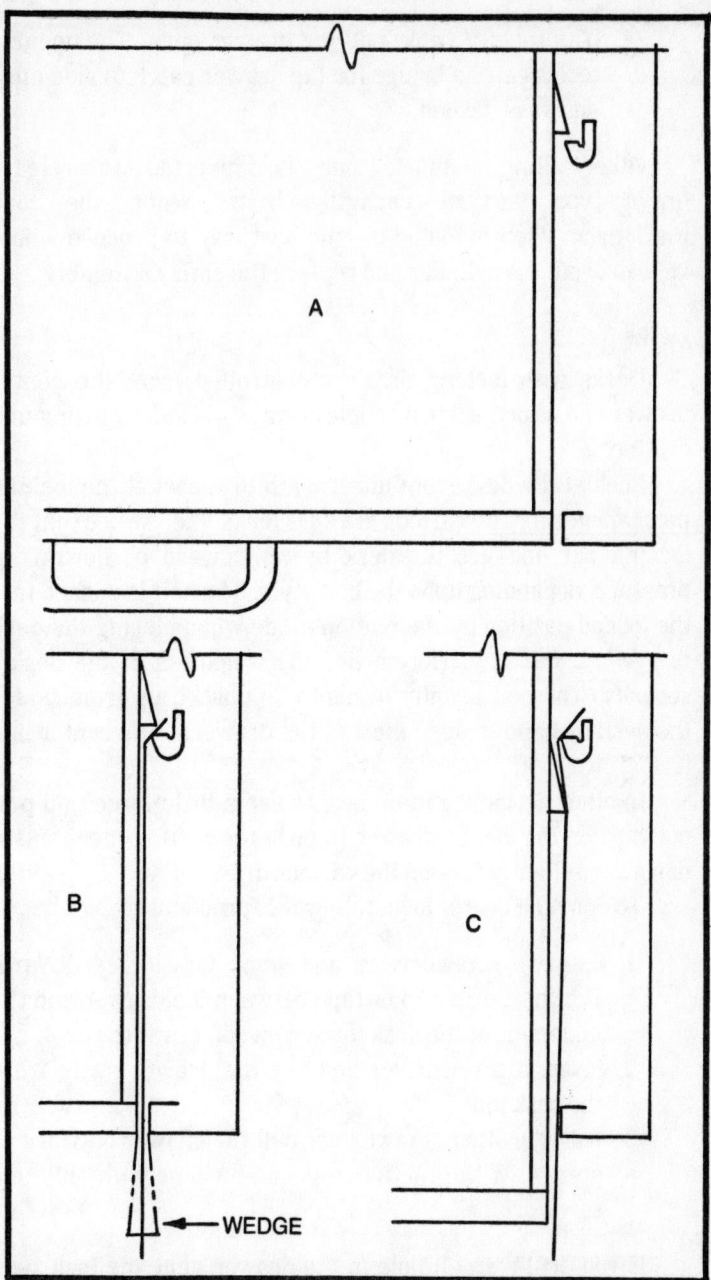

Fig. 23-4. Sometimes the drawer of a filing cabinet (A) can be opened by wedging the drawer away from the cabinet frame (B) and inserting a strip of steel between the drawer catch and the bolt mechanism (C). (Courtesy Desert Publications.)

4. With a hefty yank, pull the drawer open. The opening tool creates a bridge for the drawer catch to ride upon and pass the bolt.

Other filing cabinets can be inverted to release gravity-type vertical engaging bolts. When the lock mechanism itself is fouled up, the best way to proceed would be to drill out the cylinder and replace the entire assembly.

Desks

Desks with locking drawers controlled from the center drawer can be opened in a couple of ways—besides picking and drilling.

Look at the desk from underneath to see what the locking mechanisms for the various drawers looks like. Notice that the locking bar engages the desk by an upward or downward pressure, depending upon the bolt style. The bolt is pushed into the locked position by the motion made when closing the desk drawer all the way. Herein lies the weakness in the desk's security. The bolt usually needs to be pushed up from under the desk by hand to open most of the drawers. The center one has its own lock.

In other desks you may have to use a little force and pull outward on the center drawer to push the bolting mechanism downward slightly to open the various drawers.

To open the center lock, follow this procedure:

1. Use two screwdrivers and some tape or cardboard. Put the cardboard or tape between the drawer and the underside of the desk top so you don't mar the desk.
2. Insert a screwdriver and pry the drawer away from the desk top.
3. With the other screwdriver pull the drawer outward to open it. With practice, this can be done with only one screwdriver.

If you drill a small hole in the drawer near the lock, you can insert a stiff piece of wire, such as a paper clip, to push down the plug retainer ring. In doing so, you pull the plug free of the lock, causing the bolt to drop down into the open position.

OPENING DOORS

Possibly the simplest way to open most doors is with a pry bar and a linoleum knife:

1. Insert the linoleum knife between the door and the jamb with the point tipped upward.
2. Insert the pry bar as shown in Fig. 23-5. Exerting a downward motion on the pry bar spreads the door slightly and allows the locking safety latch to be disengaged.
3. When this is done, bring the linoleum knife forward, pushing the latchbolt into the locking assembly, and opening the door. If there is no safety catch, the knife alone can be used to move the bolt inward. It's also possible to use a standard shove knife or even a kitchen knife.

Sometimes the deadlatch plunger is in the lock but there isn't room to insert a pry bar. What do you do? You use wooden

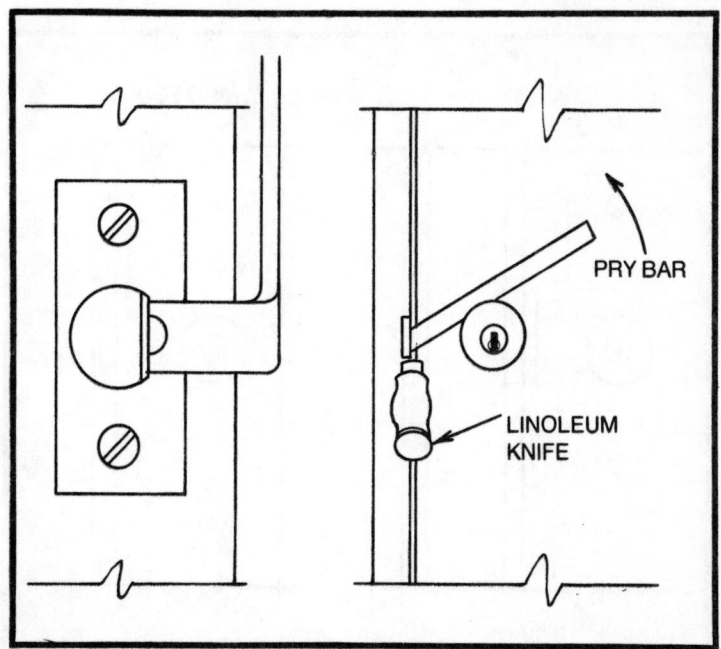

Fig. 23-5. Opening a door using a pry bar and linoleum knife. (Courtesy Desert Publications.)

wedges. Insert one of them on each side of the bolt—about 4 to 6 in. from the bolt assembly. Spread the door away from the jamb. Then use a linoleum knife to work the bolt back.

Some doors and frames have such close clearances that you cannot insert a wooden wedge or pry bar. In an instance such as this, you can use a stainless steel door shim. Merely force it into the very narrow crevice between the door and the frame and work back the bolt.

Many times a door lock can be opened with a Z-wire. This tool is made from a wire at least 0.062 in. thick and 10 to 12 in. long (Fig. 23-6). The Z-wire is inserted between the door and the jamb. When the short end is all the way in, it is rotated toward you at the top. As you do this, the opposite end will rotate between the door and the jamb. It will contact the bolt and retract it. If the bolt should bind, exert pressure on the knob to force the door in the direction required.

Sometimes you may be required to open locked chain latches. You can, of course, force the door and break the chain, but there are better ways.

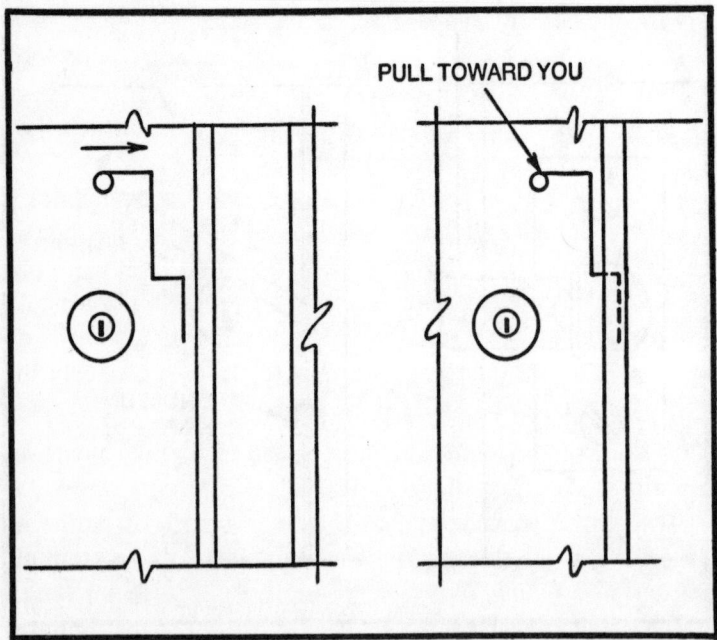

Fig. 23-6. Opening a door with a Z-wire. (Courtesy Desert Publications.)

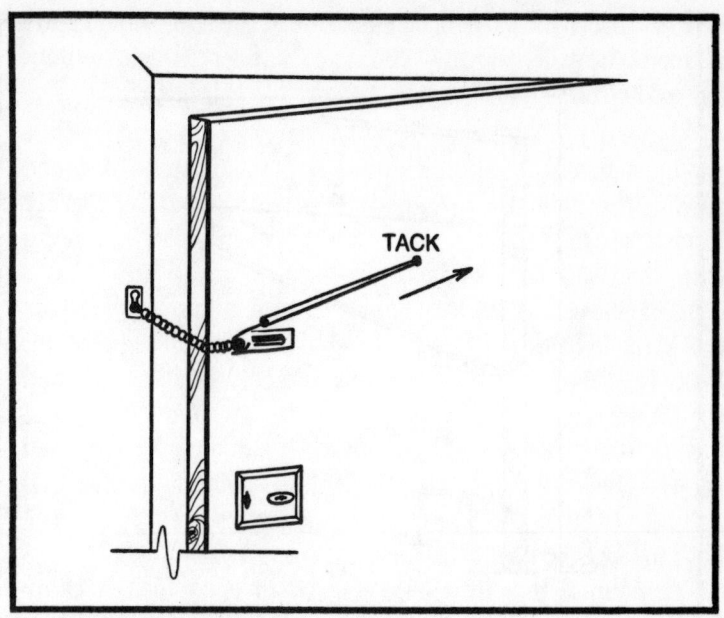

TACK

Fig. 23-7. The rubberband technique is usually the easiest way to unlock most chain locks.

The rubberband technique works most of time:

1. Reach inside and stick and tack in the door behind the chain assembly (Fig. 23-7).
2. Attach one end of a rubberband to the tack; attach the other end to the end of the chain.
3. Close the door. The rubberband will pull the chain back. If it doesn't pull the chain off the slide. shake the door a little.
4. If the door's surface will not receive a tack. use a bent coathanger to stretch the rubberband (Fig. 23-8). Make sure the coathanger is long enough and bent properly so you can close the door as far as possible.

If the door and jamb are even and there is enough space, a thin wire can be inserted to move the chain back.

Sometimes you can open a door very easily if it has a transom. You can use two long pieces of string and a strip of rubber inner tubing. The tubing should be 8 to 10 in. long. Attach a string to each end of the tubing so you can manipulate the tubing from the open transom. Lower the tubing and wrap

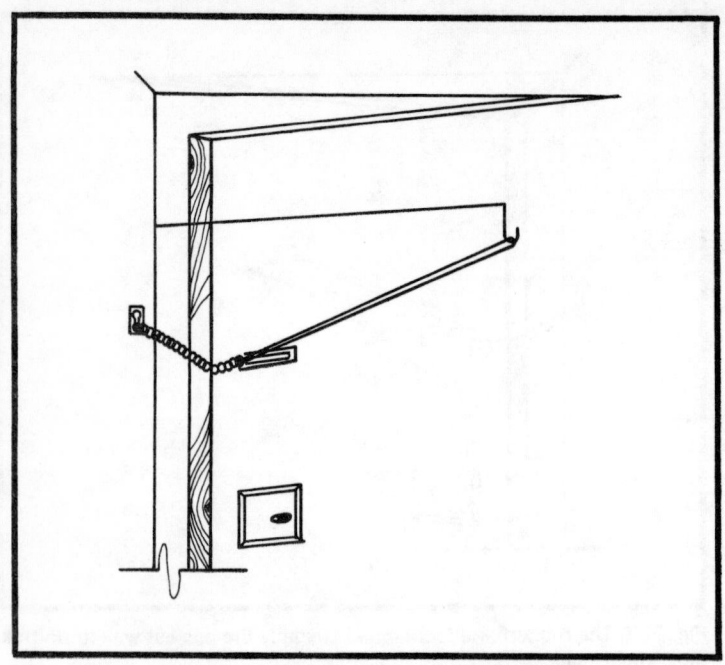

Fig. 23-8. A variation on the rubberband technique.

it around the knob. Pull up firmly on both strings to maintain tension and turn the knob. This method can be used with either a regular door knob or an auxiliary latch unit.

OPENING AUTOMOBILE LOCKS

Automibile EEP can be used on front and rear ventilation windows, door windows, doors, and trunks. Automotile locks vary considerably in their design and reliability. Despite this, you can use several methods to gain access. Most of them are faster and easier than picking the lock. In some automobile locks, picking must be ruled out entirely because of the sidebar cylinder which makes picking a very time-consuming, tedious job.

Keyless entry into an automobile is most commonly done at the windows, the most vulnerable part of a car's security. The door release pushbottom lever is often the easiest area to attack, especially on the newer car models that have a single pane of glass. The procedure is fairly simple:

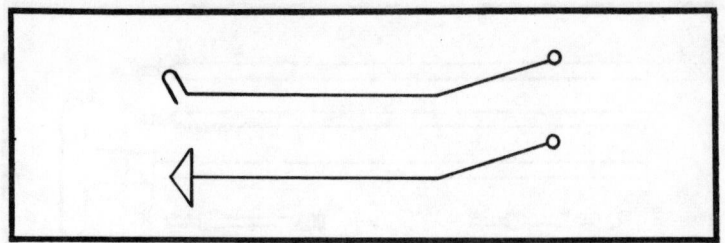

Fig. 23-9. Bent coathangers used to lift automobile pushbutton door locks.

1. Use a straightened coathanger with a loop or triangle in its end as shown in Fig. 23-9.
2. If the window is rolled up tight. force a paint scraper between the edge of the window and the insulation (Fig. 23-10).
3. Insert the coathanger in the crevice made by the paint scraper and pull the pushbutton lever up with the loop or triangle. Of course,.a coathanger isn't the only tool that could work here. Any instrument that could slip through the crevice and lift the lever could do the job.

Two .22 caliber gun cleaning sets can help you lift either the pushbutton lever or the door handle (Fig. 23-11). The

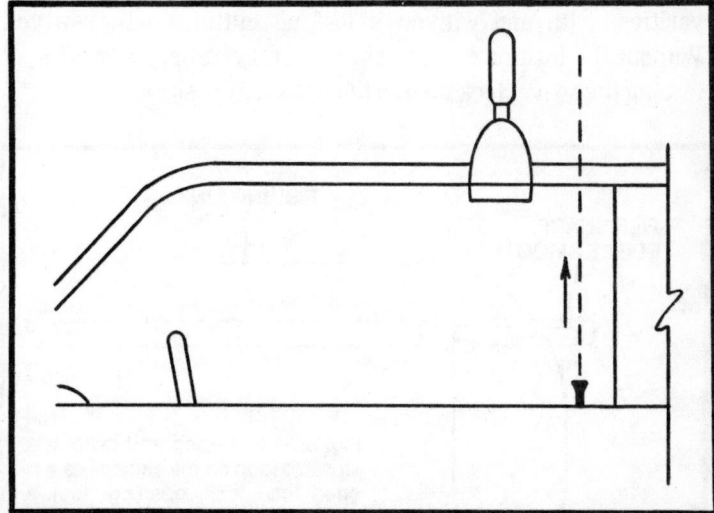

Fig. 23-10. Sometimes a paint scraper can be used to make room for a coathanger that can pull open a car pushbutton lock. (Courtesy Desert Publicatins.)

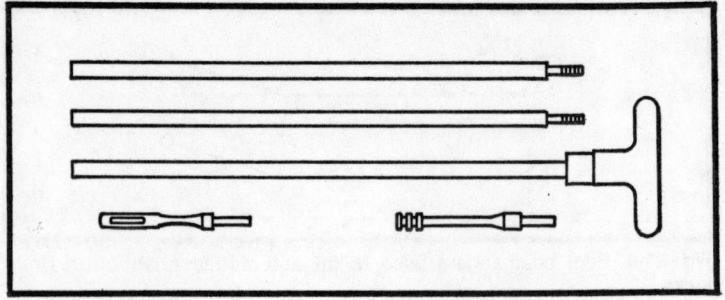

Fig. 23-11. A .22 caliber gun cleaning set. (Courtesy Desert Publications.)

cleaning tools come in sections so you can make a variety of rod lengths.

1. Use the two sets to make a rod long enough to reach across the inside of the car.
2. Use the slotted cleaning attachment and some fishing line to make a loop in the end of the rod (Fig. 23-12).
3. Work on the window or door opposite the side you are on. Raise the lever or the handle with the loop by pulling on the line and lifting the rod up.

The rear ventilation, or wing, window comes in two varities: with and without a locking button on the swivel. Without the locking button, gaining entry is only a matter of forcing the swivel lock up into the unlocked position.

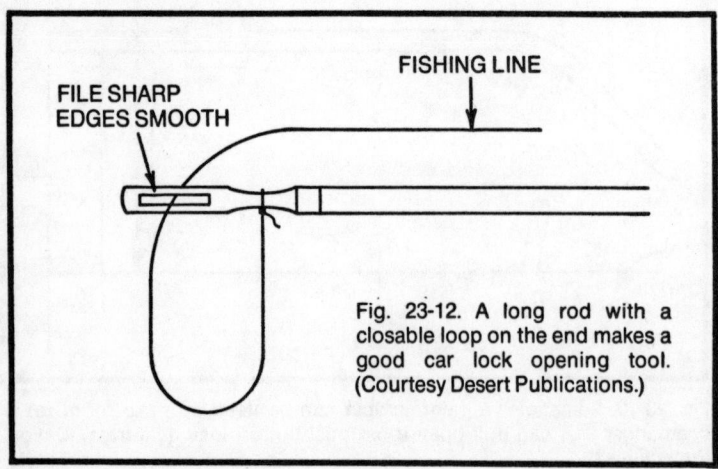

FISHING LINE

FILE SHARP
EDGES SMOOTH

Fig. 23-12. A long rod with a closable loop on the end makes a good car lock opening tool. (Courtesy Desert Publications.)

To open rear ventilation windows without locking buttons on the swivel, follow this procedure:

1. Insert a putty knife or paint scraper between the window and the frame, bending the knife slightly to allow the unlocking tool to enter. In this case, the unlocking tool could be a thin piece or wire.
2. Loop the wire around the swivel latch and pull upward. The wire should be bent slightly on the end to ensure a better upward motion without it slipping off the lever.

Most cars now have locking pushbuttons in addition to the swivel levers. To unlock these mechanisms you need standard automobile opening tools. These tools are inserted on different sides of the lock (Fig. 23-13). One is used to depress the button

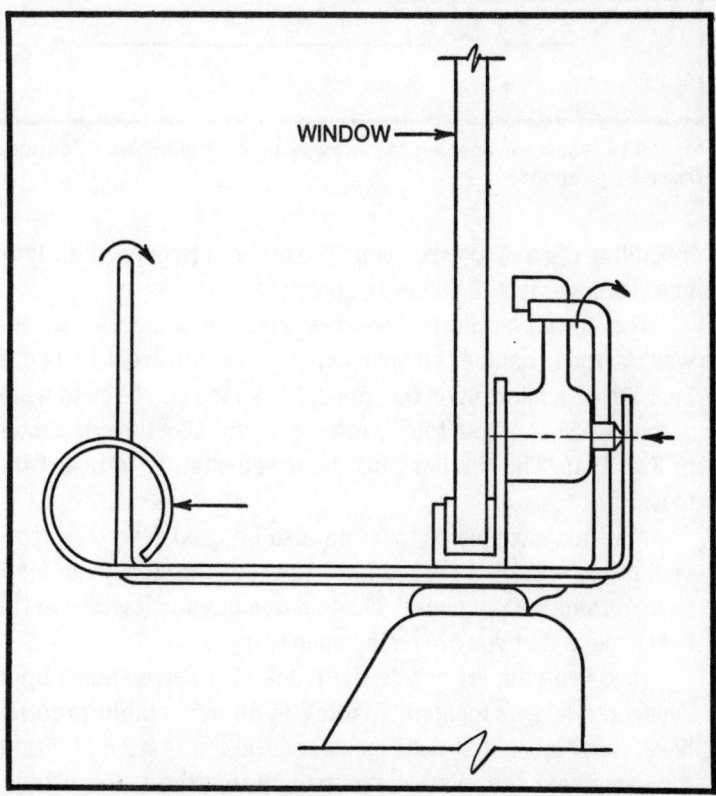

WINDOW

Fig. 23-13. Standard automotive opening tools are used to open rear ventilation windows with locking buttons. (Courtesy Desert Publications.)

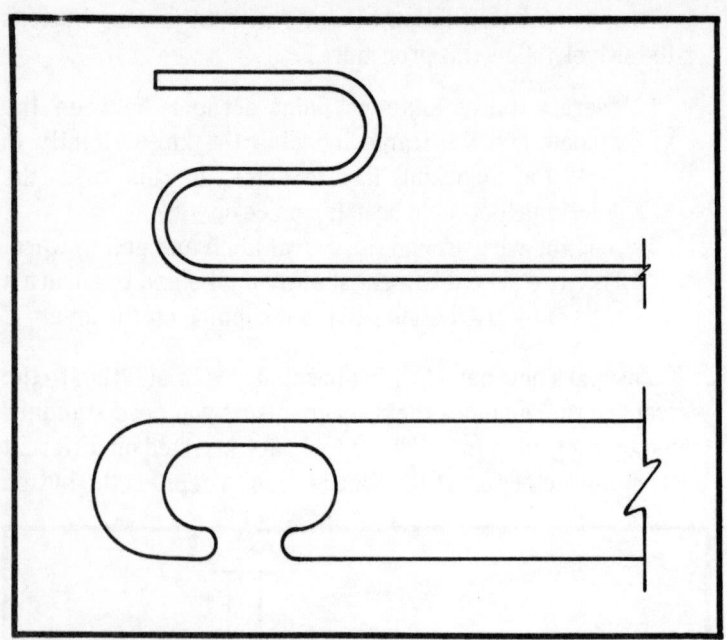

Fig. 23-14. Tools for opening the front ventilation windows. (Courtesy Desert Publications.)

by pulling the tool towards you. The other is twisted slightly to push the lever into the unlocked position.

The front ventilation window gives you access to the window roller handle the door handle. Two different tools can be used, depending upon the amount of space you have to work with (Fig. 23-14). Both tools work something like the one shown in Fig. 23-15. The window must be pryed slightly to insert the tool.

A Lenco car opener tool can also be used from the front ventilation window. You can get the tool through your local locksmithing supply house. The tool can pay for itself with the first lockout that you are called upon to open.

How you gain entry into the trunk of a car depends upon where the lock is located. Drilling is an acceptable practice here. Remember that door locks and the ignition usually take the same key; the glove compartment and the trunk usually take the same key also. Sometimes you can open the trunk lock by inserting the door key and pulling it out as you turn it.

Picking the trunk lock is simple. But it can be time consuming if you have not worked with very many automobile locks.

You can knock the lock out by forcing it inward with a hammer and a heavy screwdriver. But this ruins the lock and, sometimes, associated parts on the inside.

Drilling can have its drawbacks also. You can drill out the plug by shearing off the pins. But this means replacement of the lock unit.

Lately manufacturers have improved the security of trunk locks by installing a steel plate in front of the catch to prevent drilling. However, it is possible to drill about 2 in. to the side of the catch bolt. This enables you to use a bent piece of wire to force the catch backwards to open the trunk.

In some trunk cylinders there is a retainer. An L-shaped wire inserted into the keyway will catch the retainer, and you can pull it down and pull the plug out of the lock (Fig. 23-16). After that a small screwdriver or an awl can be used to push back the catch mechanism.

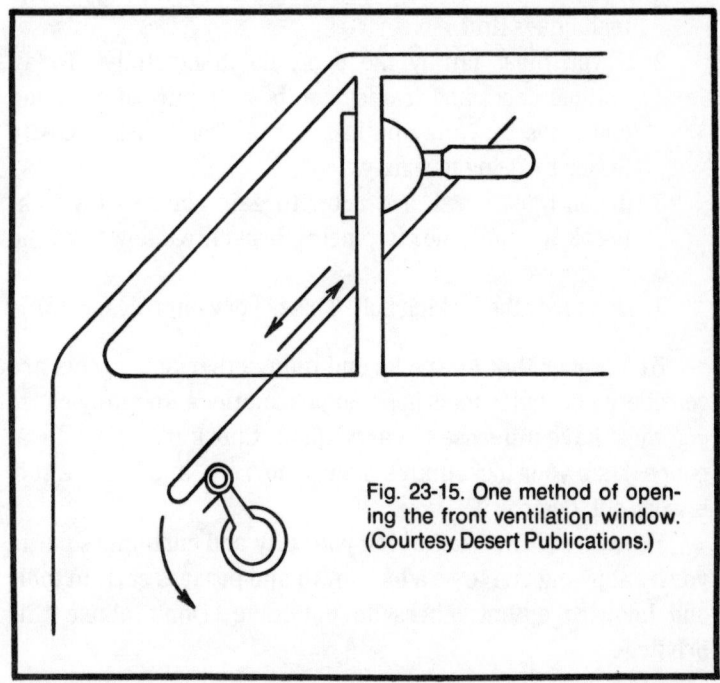

Fig. 23-15. One method of opening the front ventilation window. (Courtesy Desert Publications.)

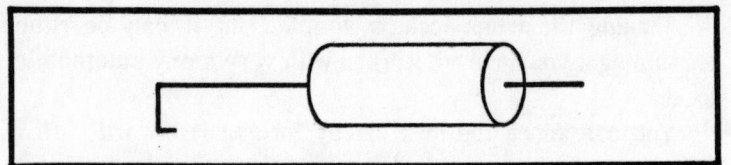

Fig. 23-16. An L-shaped wire used to open trunk locks.

The easiest of all automobile locks to open is the glove compartment lock. Insert an L-shaped wire into the keyway and pull down the retainer. Actually, this wire works on two distinct types of locks. In one, you remove the plug; in the other, the wire engages the glove compartment catch mechanism and opens it with a downward movement.

FORCED ENTRY

Forced entry in place of proper professional techniques is never recommended, except in emergencies or when authorization is given by the owner. There are some simple rules about forced entry you should consider:

1. Attempt forced entry only as a last resort. Try other techniques first.
2. If you must jimmy the door, do it carefully. Today antique doors and frames can be only part of a valued collection that may be in a home. Don't risk a costly lesson by being too hasty.
3. If you must break a window to gain access to a lock, break a small one. Replacing broken windows can be expensive.
4. Don't saw the locking bolt. This is very unprofessional.

Remember that lockpicks and many other entry tools are considered burgular tools in some jurisdictions. In many cases you must have a license to carry them. Check with your local police, keep your locksmith's license current, and when a tool is worn out, destroy it!

Be aware of the trust which your city and customers put in you by allowing you to be a locksmith and possess certain tools and knowledge that others do not have. Don't abuse this privilege.

Chapter 24

Combination Locks

Like other locks, the combination lock is available in a variety of types and styles. In this chapter, we will consider a number of representative types. Some of these locks are more secure than others, but each has its role. For example, it would be inappropriate to purchase a high-quality combination padlock for a woodshed; by the same token, it would be foolish to protect family heirlooms with a cheap lock. And it should also be pointed out that even the best lock is only as good as the total security of the system. A thief always looks for the easiest entry point.

PARTS OF THE COMBINATION LOCK

Though combination locks have some internal differences, all operate on the same principle; that is, rotation of the combination dial rotates the internal wheel pack. The pack consists of three and, sometimes, four wheels. Each wheel is "programmed" to align its gate with the bolt-release mechanism after so many degrees of rotation. Programming may be determined at the factory, or it may be subject to change in the field. As the wheels are rotated in order (normally three turns, then reverse direction for two turns, and then reverse direction again for one turn), the gates are aligned by stops, one for each wheel and one on the wheel-pack mounting plate. When all the gates are aligned, the bolt is free to release. Padlocks are built so the shackle disengages automatically, or manually by pulling down on the lock body. The captive side of the shackle, the part known as the heel, has a stop so the shackle will not come free of the lock body.

One weakness of low-priced locks is their ease of manipulation. With practice you can discriminate between the clicks as the wheels rotate. Once you learn to do this, manipulation of the lock is child's play. Fortunately, manufacturers of better locks make false gates in the wheels; these gates make manipulation much more difficult. True, a real expert will not be thwarted. But experts in this arcane art are few and far between. It takes training and intensive practice to distinguish between three or more false gates on each wheel and the true gate.

MANIPULATION

Manipulation of combination locks is a high skill—almost an art. A deep understanding of combination-lock mechanisms and many, many hours of practice are needed to master this skill. But the rewards—both in terms of the business this skill will bring to your shop and the personal satisfaction gained—are great.

No book can teach you to manipulate a lock, anymore than a book can teach you to swim, box, or do anything else that is essentially an exercise in manual dexterity. But a book can teach the rudiments, the skeleton, as it were, that you can flesh out by practice and personal instruction from an expert.

At the risk of redundancy, allow me to repeat that manipulation is a matter of touch and hearing. It also involves an understanding of how the lock works. Electronic amplification devices are available to assist the locksmith. These devices are useful, especially as a training aid.

Padlocks usually do not have false gates and so are ideal for beginners (Fig. 24-1). Pull out the shackle as you rotate the wheels. This tends to give better definition to the clicks. Work only for one number at a time. Stop when the bolt hesitates and

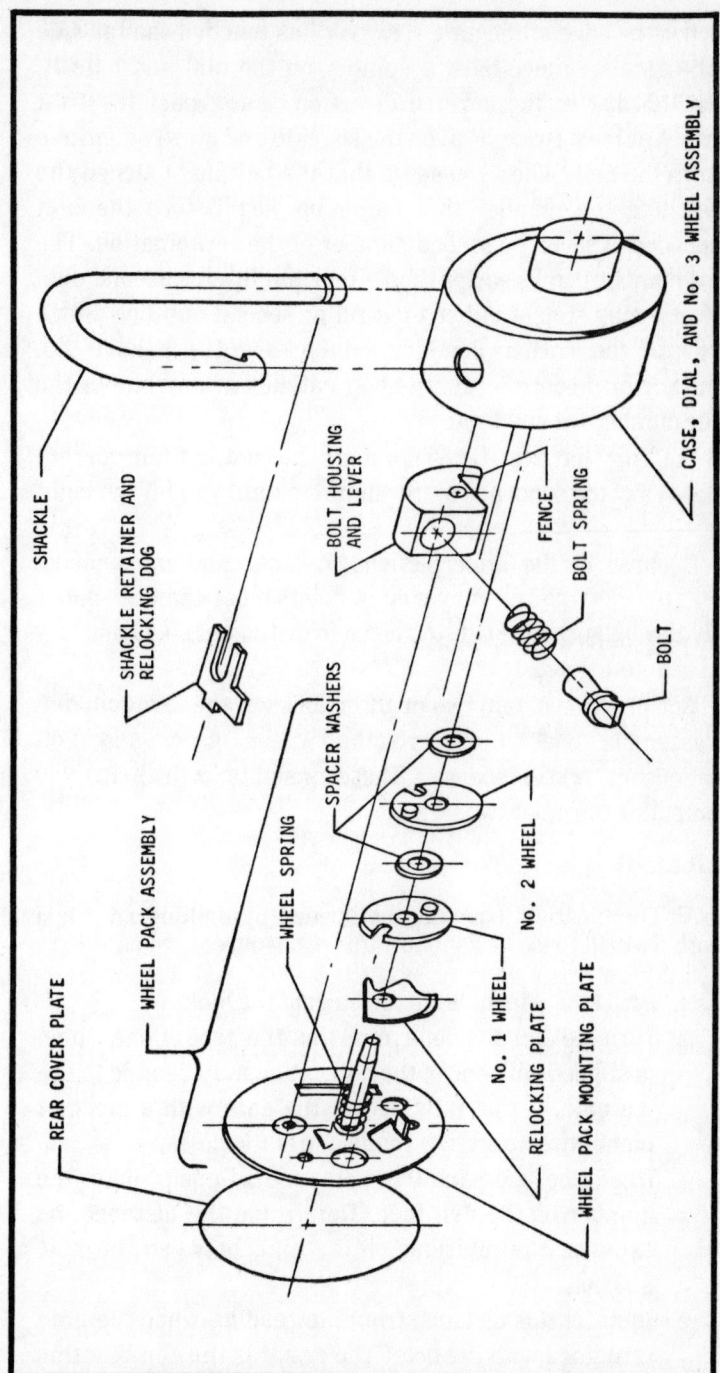

SHACKLE

SHACKLE RETAINER AND RELOCKING DOG

BOLT HOUSING AND LEVER

FENCE

BOLT SPRING

BOLT

CASE, DIAL, AND No. 3 WHEEL ASSEMBLY

REAR COVER PLATE

WHEEL PACK ASSEMBLY

WHEEL SPRING

SPACER WASHERS

No. 2 WHEEL

No. 1 WHEEL

RELOCKING PLATE

WHEEL PACK MOUNTING PLATE

Fig. 24-1. A typical combination padlock in exploded view. (Courtesy Desert Publications.)

touches the edge of the gate. The bolt has touched the far side of the gate, so move back a number on the dial and note it. Turn the dial in the reverse direction, going past the first number at least twice. Pull on the shackle and slowly continue to turn the dial. When you sense that the bolt has touched the gate, note the number that came up just before the bolt responded. This is the second number of the combination. The third number comes easier than either the first or second one. The dial may stop at either the first or second number. Since many of the earlier combination locks did not have the accuracy of modern locks, the bolt catches at any one of the three numbers at any time.

You may have the three numbers, but not in their correct order. Vary the sequence of the numbers until you hit the right one.

Because of the imperfections in older and inexpensive modern locks, the bolt may stop at points other than the gates. Only through practice can you learn to distinguish between true and phantom gates.

For practice obtain two or three locks of the same model. Disassemble one to observe the wheel, gate, and bolt relationship and response. These insights will help you manipulate the other two locks.

DRILLING

All combination locks can be opened by drilling as a last resort. To drill these locks follow this procedure:

1. Drill two ⅛ in. holes in the back of the lock.
2. Turn the dial and determine that the gate of one wheel is aligned with one of the holes. You may be able to see the gate. If that fails, locate the gate with a piece of piano wire inserted through one of the holes.
3. When the gate and the hole are aligned, note the number on the dial face. Determine the distance, as expressed in divisions on the dial, between the hole and bolt.
4. Subtract this distance from the reading when the gate is aligned with the hole. The result is the combination number for that wheel.

5. Reverse dial rotation and find the number for the second wheel. Do the same for the third.

CHANGING COMBINATIONS

Many locks are designed for combination changes in the field. I will discuss three of them.

Sargent and Greenleaf

S & G padlock combinations are changed by key.
1. Turn the numbers of the original combination to the change-key mark located 10 digits to the left of the zero mark on the dial face.
2. Raise the knob on the back of the lock to reveal the keyway. Insert the key and turn 90°.
3. Repeat step 1. but use the new combination and the change-key mark as zero.
4. Remove the change key and test the new combination.

Simplex

The Simplex is a unique combination lock, employing a vertical row of pushbuttons rather than the more usual dial (Fig. 24-2).

1. Turn the control knob left to activate the buttons.
2. Release the knob and push the existing combination (Fig. 24-3).
3. Push down the combination change slide on the back of the lock (Fig. 24-4).
4. Turn the control knob left to clear the existing combination (Fig. 24-5).
5. Push the button for the new combination—firmly and in sequence (Fig. 24-6).
6. Turn the control arm right to set the new sequence (Fig. 24-7).

Dialoc

Partial disassembly is required to change the combination. Refer the parts numbers in the following steps to Fig. 24-8.

Fig. 24-2. The Simplex pushbutton combination lock. (Courtesy Simplex Security Systems, Inc.)

To disassemble:

1. Remove the Dialoc from the door by withdrawing the two mounting screws 487-1, holding the inside plate 821 against the door.
2. Place the outside plate 820 face down on a smooth work surface so that housing 414 is facing up.

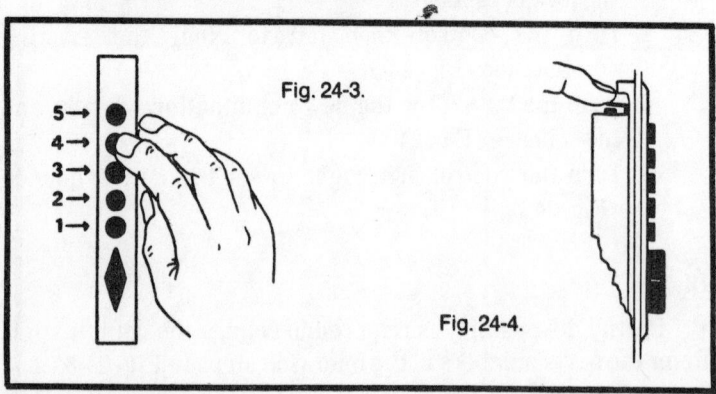

Fig. 24-3.

5→
4→
3→
2→
1→

Fig. 24-4.

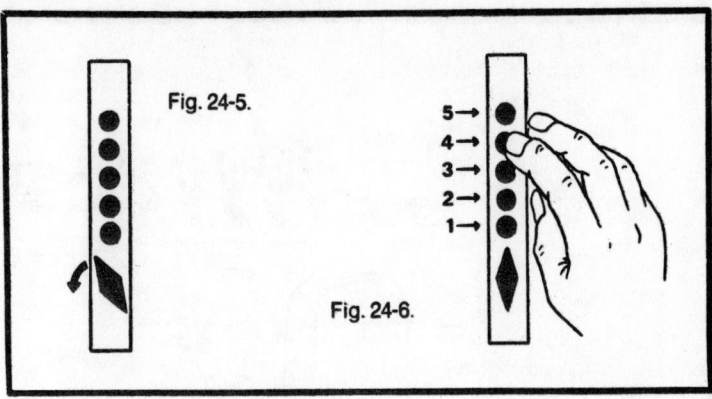

Fig. 24-5.

5→
4→
3→
2→
1→

Fig. 24-6.

3. Remove the three nuts 484-5 holding the housing cover 745.

4. After removing the nuts. gently lift the housing cover up and away from the housing. Take care that control-bar spring 744 is not dislodged and lost in the process. Lay it where it won't get lost.

5. Now. remove the secondary arm 402.

6. Grasp the end of the control bar 726 and lift out the four ratchet assemblies. Observe how they are arranged before dismantling.

7. Now you can rearrange these four ratchet assemblies to give you a new combination. (Should you want to use numbers that are not in the present combination. your distributor or the Dialoc Corporation can readily supply these.) Remove the ratchet assemblies 469 from the control bar and change your sequence as you desire. Put the ratchet assemblies back on the control

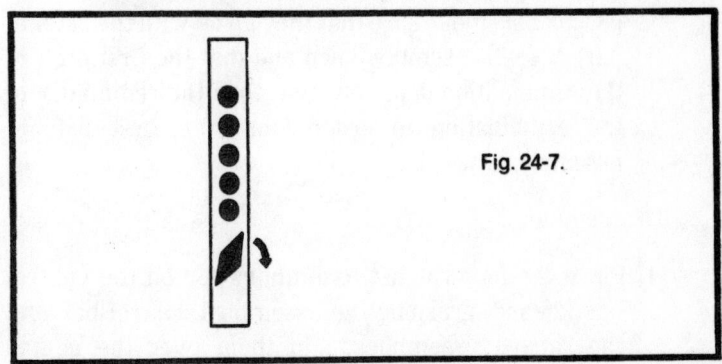

Fig. 24-7.

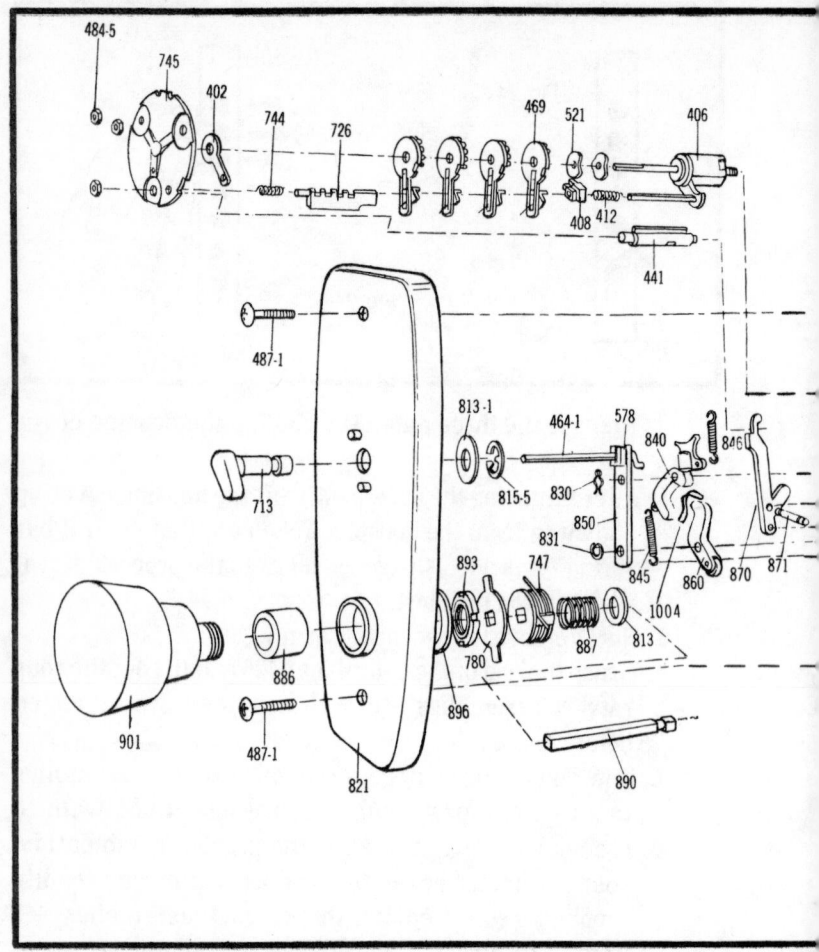

bar. When replacing the ratchet assemblies on the control bar, make sure that they go on with the ratchet part of each assembly down and that the first digit of the combination is put on first, then the remainder of the combination in order from the first ratchet assembly outward.

To assemble:

1. Place the four ratchet assemblies 469 on the control bar 726 and, grasping the assembled control bar and the ratchet assemblies, slip them over the center

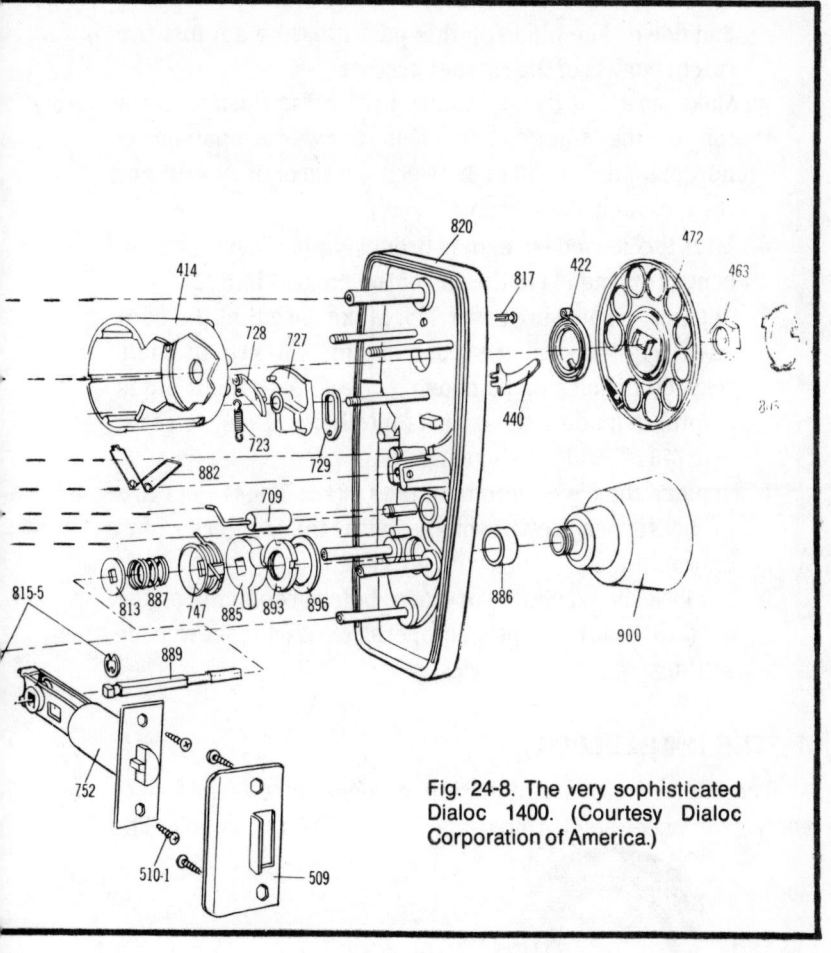

Fig. 24-8. The very sophisticated Dialoc 1400. (Courtesy Dialoc Corporation of America.)

spindle of the primary arm 406. The ratchet assemblies can be guided over the lower mounting pin protruding from the outside plate 820.

2. Care should be taken that the control bar 726 has passed through the elongated hole in the bottom of the housing 414.

3. Replace the secondary arm 402 on the primary arm 406. Make sure that the finger 408 and the finger follower spring 412 are in position.

4. If the cam pawl trip 441 has been dislodged during the combination change, it can now be replaced by

inserting it into the housing with the small protruding end down. The blade on this part must be against the ratchet pawls of the ratchet assemblies.

5. Make sure that the secondary arm 402 is flush with the top of the square bar. This allows a spacing of approximately 0.020 in. between the secondary arm and the top ratchet assembly.

6. After the secondary arm is properly placed, return the control spring 744 to the end of the control bar 726.

7. Replace the housing cover 745. Make sure that the cam pawl trip 441 and the control bar 726 are in their respective holes in the housing cover. A straight pin is helpful in guiding these parts into holes. Be sure reset arm 870 is inside the housing 414.

8. Replace the cover-retaining nuts 484-5. These nuts are to be turned down snugly with the fingers, plus approximately $\frac{1}{2}$ turn with a wrench.

9. The lock should now be operated before replacing on the door to insure its proper operation on the new code settings.

MASTER 1500 PADLOCK

The Master 1500 combination padlock is covered with heavy-gage sheet steel, rolled and pressed at the edges. The

Fig. 24-9. The Master 1525 masterkeyed combination lock. (Courtesy Master Lock Company.)

net result of this somewhat unorthodox construction is a tight-fitting lock and one that is not pried open easily. Disassembly is recommended only as a training exercise. It is not practical to open this lock for repair.

Three wheels are employed, each with a factory-determined number. If you receive the lock with the shackle open you can determine the combination by looking through the shackle hole. Note the dial reading as you align each wheel. If you read the numbers properly and if the wheel gates are in alignment, you should only have to add 11 to the dial readings to get the true combination. If you are slightly off, compensate by adding 10 or 12 to the original readings.

The Master 1500 series may be masterkeyed (Fig. 24-9). This lock is popular in schools. While the combination can be obtained by manipulation, picking the keyway is faster. Once the lock is open, determine the combination by the method described above.

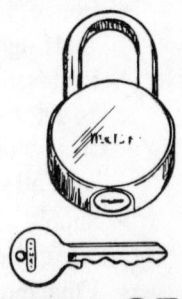

Chapter 25

The Business
of Locksmithing

Locksmithing is a profession involving skill, competence, and public trust. It is a career that can be highly rewarding. This chapter will give you the background necessary to use the basics of what you have already learned and apply it in a business.

The *Occupational Outlook Handbook* indicates that in 1972 there were approximately 8000 full-time locksmiths in the U.S. Many of these 8000 individuals operated their own locksmithing business, but others were in a shop which employed one to three persons. The situation is very much the same today. This means that in this country, the demand for competent locksmiths is overwhelming.

Fifty to seventy-five million keys are lost and must be replaced each year. Every household, automobile, and business has two to five locks. In a medium-sized city of 75,000 to 100,000 persons, there are easily over 250,000 locks. This means there are literally billions of locks and keys in this country—plenty of business for a good locksmith. Statistics show that locksmiths make good money too.

CONSIDERATIONS

If you're thinking about launching a career in locksmithing, there are several things you should consider:

- Police clearance—Many jurisdictions require licensing and fingerprinting.
- Professional qualifications—Some cities insist on tests to determine professional competence.
- Basic tools and equipment—You should ensure that you have all you need to tackle any locksmithing job.
- Supply stock—It's possible to work out special arrangements with some local supply houses.
- Financing—Setting up shop involves an initial cash outlay.
- Bookkeeping—Records must be kept of all transactions.
- Vehicles—Necessary for out-of-shop calls.
- Advertising—Sometimes it take more than just locksmithing competence to bring in the business. Good advertising usually includes signs, business cards, window displays, display racks, display materials, etc.
- Literature—Locksmith supply house catalogs, manufacturers' literature, reference books, business forms, etc.
- Contacts—It pays to know other locksmiths, factory representatives, etc. They can help you from time to time with locksmithing problems.

SELECTION OF THE BUSINESS SITE

Selecting the business site is a major decision, a decision that will affect the business throughout its infancy.

In determining the location, consider the types of businesses you will be near and what affect they will have upon your customers. Shopping malls are ideal for locksmithing shops because of the walkin business that is generated by people who need one or two keys duplicated. Should you undertake home servicing, you would want a location that enables you to get to your customers quickly.

The type of building must also be considered. It would be unwise to obtain a building that is in need of massive repairs. The building should also be large enough for your needs.

Thought should be given to the building's potential for future expansion.

Demographic and economic factors of the surrounding area should be considered too. Make a study of population density, population growth, projected economic growth, the tax structure, etc. Examine any potential competition with other locksmiths.

SHOP LAYOUT

Window displays are a good way to attract customers. The display itself should be simple and effective, neat and clean. Window displays cannot be cluttered; you must be discriminating and orderly in developing them. The display should tell the viewer something, something interesting and appealing about your products or services.

The front counter area should be carefully organized: this is where sales are made. Insure that the key machines are near the counter—visible from the street. Spread out your keys on a key board; such a display stays in your customers' mind longer than a pile of keys on a table. Code books usually are kept near the key machines.

The workbench usually should be out of the customers' sight, preferably in the back room. If it is kept out front, it should be clean at all times.

Lighting of the work and customer service areas is most important. It should be adequate—not overlighted. The lights should be of the nonglare variety. The work area can have additional small lights placed several feet above the bench.

MERCHANDISING THROUGH ADVERTISING

Advertising takes many forms, from radio, television, and handbills to putting your name on each key duplicated. There are no hard and fast rules. Good advertising is advertising that gets results.

The selection of the proper advertising can be important for a beginning locksmith. Consider the population; do they listen to the radio, watch television, read the paper more than in other areas, or look for gimmicks? What are the costs of the various advertising media? You can approach the advertising staffs of the local media to determine these answers.

Once an advertising scheme is carried out, evaluate the results. Did the advertising increase business? Was the advertising worth the investment?

Ultimately, the locksmith himself is the most effective advertising. Responsible locksmithing, a neat appearance, and a little courtesy go a long way. In spite of all the sales talk and advertising hoopla, the customer is only interested in one thing—quality service.

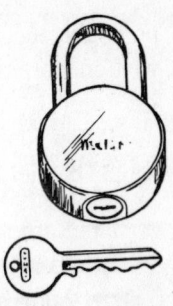

Chapter 26
The Locksmith
and the Law

The locksmith, because of the uniqueness of his profession, must have a better understanding of various laws than most persons. In this regard, when setting up a locksmithing business it is always prudent to consult with your attorney regarding all laws that concern you as a locksmith. In many jurisdictions there are laws covering licensing, control of locksmithing tools, and registration of code books. Some local laws regulate the conduct of certain locksmithing business practices, such as duplicating masterkeys, making bank deposit box keys, opening automobiles, etc.

When first entering into the business, aside from consulting with your lawyer and possibly the police, you should also contact your area locksmithing association, if one exists. Information can be obtained from the national locksmithing organizations too—through their newsletters and publications. These are excellent sources of information about your legal responsibilities.

Your legal responsibilities demand that you be very careful about the jobs you accept. You must be certain that you are not breaking the law by complying with a customer's wishes. Authorization statements from supervisors, such as from a bank or post office, for duplicating a key should always be doublechecked with the main offices. Verification of such written statements is a must. They should also be filed with the job order.

When jobs of this type arise, the following information should be included in your files:

- Name of person who brought the job in
- Identification (Social Security card or driver's license)
- Address of person who brought the job in
- Business telephone (call to doublecheck)
- Name of firm
- Business address
- Type of service performed
- Type of payment (if a check, it should be a business check, not a personal check)
- Automobile make, model, serial number, license number, state it is registered in

You may also have some forms printed up for the individual to fill out. A single form can be used to cover all the situations you may run into. The form should be kept along with the work order.

In some states or cities, laws require that you take an examination before your locksmithing peers to show that you are a competent locksmith meeting professional standards. Besides this, police checks of your reputation, qualifications, background, and previous employers may be made to ascertain that you are of good moral character.

Various laws have been enacted to protect the public from unscrupulous individuals posing as locksmiths and to protect the locksmithing trade itself from such individuals. The following is extracted from the Los Angeles Code, Ordinance No. 83,128, as an example of a local law regulating the locksmithing profession:

SEC 27.11 LOCKSMITH—REGULATIONS APPLICABLE TO
A. Definitions

"Locksmith" shall mean any person whose trade or occupation, in whole or in part, is the making or fashioning of keys for locks, or similar devices, or who constructs, reconstructs or repairs or adjusts locks, or who opens or closes locks for others by mechanical means other than with the regular keys furnished for the purpose by the manufacturers of the locks.

B. Trade of Locksmith—Permit Required

No person shall engage in the business of locksmith, or practice or follow the trade or occupation of locksmith without a permit therefore from the Board of Police Commissioners.

C. Permit—Application

Such permit shall be issued only upon the verified application of the individual seeking the permit. The application shall be upon a form prescribed by the Board and shall set forth the proposed location of the applicant's place of business, the names and addresses of five character references and such other things as the Board may require to determine the character, honesty, and trustworthiness of the applicant. Specimen fingerprints of the applicant shall be furnished with the application.

D. Permits—Fees—Expiration

Each application for a permit shall be accompanied by a fee of $10 and each application for the annual renewal thereof, by a fee of $5. Each permit, unless sooner revoked, shall expire on December 31st, following the date of issuance. Each permit shall bear a serial number.

E. Permits—Issuance and Denial

The Board shall cause an investigation to be made upon each application and if the Board finds that the applicant's reputation for honesty is good, that he has not used his skill or knowledge as a locksmith to commit or aid in the commission of burglaries, larcenies, thefts, or other crimes, that he intends honestly and fairly to practice the trade of locksmith in a lawful manner, and that he has not been convicted of a felony, then the permit shall issue. Otherwise it shall be denied.

F. Permittee—Must Keep Record

Each permittee must keep a book, which shall be open to inspection by any police officer at all times, in which the following must be entered:

1. The name and address of every person for whom a key is made by code or number.

2. The name and address of every person for whom a locked automobile. building. structure. house. or store. whether vacant or occupied. is opened. or a key fitted thereto.

G. Keys to Be Stamped

It shall be unlawful for any locksmith to fail to stamp the serial number of his permit upon any key made. repaired. sold. or given away by him.

H. Signs to be Displayed

Every locksmith shall display in a conspicuous manner in the place where he is carrying on such business. trade. or occupation. a sign of a style. size. and color to be prescribed by said Board. reading. "Licensed Locksmith." together with the official permit number.

I. Permits—Revocation

The Board. upon proceedings had as prescribed in Section 22.02. may revoke or suspend any permit isued hereunder upon any of the following grounds:

1. Misrepresentation in obtaining such permit.
2. Violation of any provision of this section.
3. That the permittee has committed or aided in the commission of or in the preparation for the commission of any crime by the use of his skill or knowledge as a locksmith or by using or letting the use of his tools. equipment. facilities. or supplies.

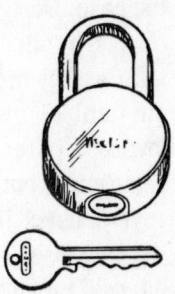

Appendix A

Selecting a Padlock

Though initially developed by the Master Lock Company with their products in mind, this appendix is a real gold mine of information concerning what anyone should look for when considering the purchase of any padlock.

What is the real key to security when you pick out a padlock? The fact is, locks vary tremendously in the amount of protection they provide. Important differences may be hard for you to spot. Keep the following in mind when you're shopping:

Wrought-Steel Padlocks—You Get What You Pay For

Wrought-steel or "shell" padlocks are the lowest priced locks you can buy. A great buy if you understand they're intended mainly for nuisance protection—keeping the kids out of your tool box; locking power tools against tampering; restricting access to a mailbox, storage room, cage, etc. Costing little, they can prevent thoughtless misadventures or injury from hazardous household and industrial items.

Warded Padlocks—Good Strong Middle Ground

Laminated padlocks are far sturdier than shell types—much more resistant if anyone tries to hammer open the case. Design permits a wider range of key variations than offered by shell locks.

Costing more than shell types—but less than those with pin-tumbler mechanisms—warded padlocks give dependable low-cost pilfer protection. Use them where property is of significant but limited value—securing oil tank caps, well covers, beach lockers, duffel bags, barn doors, etc.

Because of the relatively large clearances between internal moving parts, warded padlocks are frequently chosen for applications where sand, water, ice, or other contaminants are a problem.

WHERE CORROSION EATS AT PADLOCKS BRASS HAS BETTER STAYING POWER

Hard-wrought brass padlocks are designed to withstand severe corrosion problems encountered along seacoasts, aboard boats, at refineries, and in areas of high humidity and atmospheric pollution. For theft deterrence the shackle typically will be chrome-plated hardened steel. But for extreme conditions, even the shackle can be brass. Expect the price of maximum-security brass padlocks to run substantially higher than for steel.

The solid-brass padlock is a lower cost medium-security alternative for the average user. Again, look for pin-tumbler locking and case-hardened shackle to protect valued property. Priced substantially less than heavy-duty laminated brass padlocks, solid-brass locks are naturals for many coastal uses, boats, outdoor lockers, gates, and similar applications.

Combination Padlocks—the Key Is in Your Head

The chief reason for this kind of padlock is keyless convenience—particularly with children where lost keys may be a problem. Protection features to look for include reinforced double-wall construction—tough stainless steel outer case over a sturdy wrought-steel innercase—and hardened steel shackle for added resistance against cutting and sawing. Combination padlocks are classed as medium security, with strength equivalent to a quality warded padlock.

No Stronger Than the Weakest Link

People should consider more than just the strength of the padlock they choose. The finest lock affords little protection if burglars find it hung from an undersized or unhardened hasp that can be cut with ease. To avoid a weak link in security, get a hasp that matches the quality of the lock. Look for adequate size, pinless hinge, concealed screws, case-hardened staple (the metal loop the lock passes through), and steel ribbing for added strength.

An evolutionary step beyond the hasp is the hasplock. Here instead of using a separate padlock, the lock and hasp are one, permanently joined so the lock can't be misplaced or stolen. Hasplocks give built-in convenience akin to the dead bolt door lock while offering much easier installation. This accounts for their popularity with builders, contractors, farmers, and do-it-yourselfers. Their uses range from garage doors and sheds, to boat houses, truck warehouses, and a host of other places.

Another Weak Link in Security May Be the Wrong Choice of Chain or Cable

When it comes to movable property, the use of a strong padlock with a properly matched chain or steel cable is the way to go. They enable owners of bicycles, boats, motorcycles, and other easily stolen goods to obtain maximum protection by securing them to a post, tree, dock, or other immovable object. Common chains available in hardware stores should be passed over in favor of chains specifically designed for locking applications. Be sure it's case hardened for high resistance to cutters, saws, and files. Individual links should be welded, not just twisted, to resist being pried apart. Multistranded security cable is available for equivalent protection, with the added benefit of light weight. In either case, the thicker the chain or cable, the greater the protection. Examine cable closely: Some manufacturers add a thick coating of vinyl to make a small steel strand look bigger.

For greatest protection, position the lock and cable (or chain) as high above the ground as possible. This makes it difficult for thieves to gain extra leverage by bracing one leg of a bolt cutter against the ground.

When in Doubt About a Lock...

The best place to turn with questions of security is your local police department. Most law enforcement agencies have experts to help you and may have displays of security locks. When in doubt and unable to get expert advice. the rules to know are:

1. Buy the best padlock protection you can afford (strength and cost generally coincide).
2. When possible avoid leaving locked items in out-of-the-way places where thieves have time to work on the lock unseen.
3. If locked property has strong appeal to the light-fingered. back up your locks with some insurance.

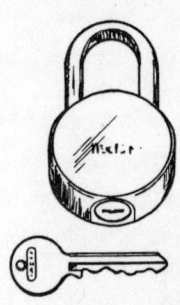

Appendix B

Locksmithing Supplies

Allen and Company
10 Smithfield St.
Pittsburgh. PA 15222

Blaydes Lock Company
1815 Bryant St.. N.E.
Washington. D.C. 20018

Capital Lock and Hardware
3730 Georgia Ave.. N.W.
Washington. D.C. 20010

Commonwealth Lock Company
1853 Massachusetts Ave.
Cambridge. MA 02140

H. Hoffman Company
6245 W. Western Ave.
Chicago. IL 60650

Kenco Supply Company
Box 4151 Benson Station
Omaha. NB 68104

M.D. Kramer
535 Liberty Ave.
Brooklyn. NY 11207

Kipf Lock Company
Box 987
Columbus. OH 43216

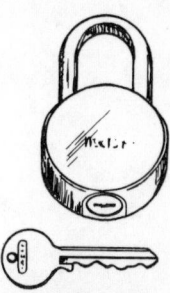

Appendix C
Plug Follower and Holder
Diameters for Popular Locks

Manufacturer	Diameter (in inches)			
	0.395	0.495	0.500	0.550
Acrolock		X		
Corbin		X		
Corbin oversize				X
Corbin small pin	X			
Eagle		X		
Eagle small pin	X			
ILCO		X		
ILCO 4019 rim cylinder			X	
ILCO "peanut" cylinder	X			
Keil			X	
Kwikset		X		
Lockwood		X		
National (Rockford)		X		
National (Ozone Park)		X		
Russwin oversize				X
Sargent		X		
Segal		X		
Taylor		X		
Weslock (except KNK)		X		
Yale		X		
Yale small pin	X			

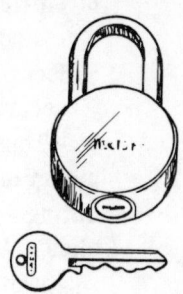

Glossary

Ace lock—A lock in which the tumbler pins are arranged in a circle.

antipick latch—A spring latch fitted with a parallel bar that prevents the latch from responding to external pressure from a tool that is not a proper key.

armored front—A plate covering the bolts or set screws holding a cylinder to its lock. These bolts are normally accessible when the door is ajar.

backplate (rim cylinder)—A small plate applied to the inside of a door through which the cylinder connecting screws and bar are passed.

backset (of a lock)—The horizontal distance from the face of a lock to the center line of its cylinder, keyhole, or knob hub. On locks with beveled fronts, this distance is measured from the center of the face. On locks with rabbeted fronts, it is measured from the upper step at the center of the lockface.

barrel key—A key with a round stem that is hollow at its end. The hollow fits over a pin in a lock keyway and helps keep the key aligned. Also known as a hollow post key or pipe key.

bicentric cylinder—A special lock having two pin-tumbler cylinders. The correct key in either cylinder opens the lock. This permits masterkeying without reducing the security of the locking mechanism.

bit—That part of a key that is cut to fit a lock.

bit key lock—A lock operated by a key having a wing bit.

bitting—A cut, or series of cuts, in a bit.

blank—A key that has not been cut or shaped to fit a specific locking mechanism.

bow—That portion of a key that is held between the fingers when the key is being used.

burglarproof—A term used to describe locks or doors designed to be impregnable against a thief without explosives or an unlimited amount of time.

burglar resistant—A term used to describe locks or doors capable of resisting attack by prowlers and thieves for a limited time.

cam—That portion of a lock that turns when its key is turned or when the bolt knob is turned. Unlike a tang or tail, a cam has no play and operates around a central axis.

case (of a lock)—The box containing the lock-operating mechanism.

case ward—A ward or obstruction integral to the case of a warded lock.

change key—A key that operates only one lock in a series, as distinguished from a masterkey which operates all the locks in a series.

clean opening—A locksmith's term for opening a lock without obvious force, i.e., a skilled entry.

closet spindle—A knob spindle having a turn knob made fast to one end and with provision for attaching a knob to the other end. For use on the inside of closet doors.

combination lock—A lock that does not use a key but requires that certain parts be placed in a specific juxtaposition by use of a turning knob on the face of the lock. Generally the parts are coded with digits and the correct digit series sets the internal parts in the right order so the lock will open.

connecting bar—A bar attached to the cylinder rear of a plug lock to operate the locking bar mechanism.

core—The plug of a cylinder lock.

corrugated key—A key with pressed longitudinal corrugations in its shank to correspond to an irregularly shaped keyway.

cuts—The indentations made in a key blank in order to fit the lock.

cylinder—A lock housing containing a tumbler mechanism and a keyway.

cylinder lock—A lock having mechanisms operated by a cylinder.

cylinder ring—A collar or washer used under the head of a cylinder.

cylinder screw—The set screw in the front of a cylinder lock which prevents the cylinder from being turned after installation.

dead bolt—A lock bolt having no spring action, usually rectangular and actuated by a key or turn knob.

deadlatch—A lock with a beveled latchbolt that is automatically or manually locked against end pressure when projected.

dead lock—A lock with a dead bolt only.

depth key—A special key that enables a locksmith to cut blanks made for a lock according to a code.

derivative code—A special numerical code that relates a lock's tumbler arrangement to the depth of the key cuts necessary to operate the lock.

disc tumbler—A circular or oval-shaped disc with a rectangular hole and one or more side projections. A number of these are used in a disc-tumbler lock.

door check—A device consisting of a heavy spring and arm coupled to an air or oil cylinder that prevents the spring from closing the door too rapidly. Also called a door closer.

double-bitted key—A key with cuts on both sides of its blade.

drill pin—A round pin projecting from the inside of a lock case to receive a barrel key. Also called a barrel post.

drivers—The upper set of spring-activated pins in a pin-tumbler cylinder lock.

dummy cylinder—A cylinder containing one operating mechanism.

emergency key—A key capable of opening any lock in a building even though a door may be locked from the inside.

escutcheon—A plate, either protective or ornamental, containing openings for all of the controlling members of a lock such as knob, handle, cylinder, keyhole, etc.

fence—Another name for the post in a lever-tumbler lock.

finished key—A key that has been cut to fit a lock.

following tool—A tool used to hold the pin tumblers and springs in place while a pin-tumbler cylinder is being assembled, disassembled, inserted, or removed.

gate—The opening in lever tumblers that allows them to pass the post, or fence. Also called gating.

graphite—A lubricant especially useful for the finer parts of a lock.

heel (of a padlock shackle)—The end of a padlock shackle that is not removable from the case.

jamb—The inside vertical face of a doorway or window frame.

keeper—Another term for strike.

key blank—An uncut key. Once cut, a key blank becomes a finished key.

key caliper—Any small caliper capable of measuring the height of a cut on a key.

key extractor—A device used to remove broken key pieces that have broken off in a lock keyway.

key plate—A small plate or escutcheon having a keyhole only.

keyway—The opening in a lock mechanism into which the key is inserted.

latch—A door fastening device, usually with a sliding or spring bolt.

latchbolt—A beveled spring bolt, usually operated by knob, lever handle, or thumb piece.

layout board—A board with a number of parallel grooves used to hold pin-tumbler parts in order when the locksmith is working on the lock.

lever handle—A horizontal handle for operating the latchbolt of a lock.

lockface—A plate that shows in the edge of a door after lock installation.

locking dog (of a padlock)—The part that engages the shackle and holds it in the locked position.

masterkey—A key capable of operating several different locks, each lock being operated by its own change key.

mortise—An opening made in a door to receive a lock or other hardware. Also the act of making such an opening.

mushroom pins—A mushroom-shaped driver.

night latch—An auxiliary lock with a spring latchbolt and functioning independently of, and providing additional security to, the regular lock on the door.

panic bolt—A type of lock fitted with a long bar placed horizontally across the inside of a door. Pressure on the bar releases the lock. Used in theaters, schools, and other public buildings. Also called panic bar.

passkey—A masterkey or skeleton key.

pick—Any tool or device (other than a key), either from a locksmith supply house or homemade, that is used to open a lock.

pin tumblers—Small sliding pins in a lock cylinder that work against coil springs and prevent the cylinder plug from rotating until the correct key is inserted.

plug (of a lock cylinder)—The cylindrical mechanism in a lock cylinder that houses the keyway.

plug retainer—A device that retains the plug in a cylinder lock. Also called a retaining ring.

post (of a key)—The cylindrical portion of a bit key to which the bit is attached.

reversible lock—A lock in which the latchbolt can be turned over so as to adapt it to doors of either hand.

opening in or out. This does not apply to beveled front locks and certain cylinder mortise locks.

security—In the locksmithing trade, the ability of a lock to withstand attempts at unlawful entry.

shackle—The curved portion of a padlock that passes through the hasp.

shackle spring—The spring that projects the shackle from a padlock case when unlocked.

shank (of a key)—The part of a bit key between the bow and the bit.

shear line—The space between the shell and the plug of a lock cylinder.

skeleton key—A warded lock key cut especially thin to bypass the wards in several warded locks so the locks can be opened.

spacing—The distance between a lock's keyhole and its spindle hole.

stop pin (of a padlock)—A pin that retains the shackle in a padlock case when unlocked.

strike—The part of a locking arrangement that receives the bolt, latch, or fastener. Usually recessed in the door frame. Sometimes called a keeper.

tension wrench—A device used to maintain constant pressure on a lock while its tumblers are being manipulated by a pick. It assists in picking the lock and moving the plug in the lock after it has been picked.

thimble—A tool used to hold the plug of a pin-tumbler lock while it is being worked on.

Index

Index

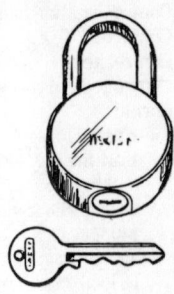

Index